KNOWLEDGE STRUCTURES FOR COMMUNICATIONS IN HUMAN-COMPUTER SYSTEMS

KNOWLEDGE STRUCTURES FOR COMMUNICATIONS IN HUMAN-COMPUTER SYSTEMS

GENERAL AUTOMATA-BASED

Eldo C. Koenig

IEEE computer society
60TH anniversary

BICENTENNIAL 1807 WILEY 2007 BICENTENNIAL

WILEY-INTERSCIENCE
A John Wiley & Sons, Inc., Publication

Published by John Wiley & Sons, Inc., Hoboken, New Jersey.
Published simultaneously in Canada.

Library of Congress Cataloging-in-Publication Data is available.

ISBN-13: 978-0-471-99813-6
ISBN-10: 0-471-99813-3

10 9 8 7 6 5 4 3 2 1

To Gloria

CONTENTS

PREFACE

The book presents knowledge structures for communications in human-computer systems (HCS) based on general automata. The general automaton was considered basic in disciplining the natural language, in that knowledge to be communicated is about automata and histories of automata. The work of others on finite automata does not include the detail required for finite general automata. To accomplish completeness and unification for the broad concept of a general automaton, the analysis applies the algebra of sets and functions and follows with the application of combinatorial mathematics for graph theory to define a general graph model for knowledge representation. Since interacting automata must be interfaced in space and time, the graph model is fundamental to the analysis of systems of automata and is supported by more than 50 theorems and proofs. The material presented makes reference to 34 publications by the author or jointly by the author and graduate students.

The resulting model provides knowledge representations for software engineering. Of the many features required of a method to achieve the desired communication in HCS, six are identified and illustrations presented for achieving them by the general automata method (GAM):

1. Extracting and storing the knowledge of sentences
2. Knowledge association
3. Deductive processes
4. Inferences
5. Feedback
6. Sequencing of knowledge

After the analysis for each feature is presented, the result is illustrated with practical applications. Algorithms and programs are described in achieving some of the features, and additional algorithms and further research are indicated.

The material is presented in the style and form that is suitable for either an academic or an industrial setting, and for either self-study or group learning. It has been taught to advanced undergraduate and graduate students. For those with interest and background in applied computer science and software engineering, the book describes algorithms and programs, and suggests others.

Emphasis for this group would be placed on Chapters 1, 4, 6, 7, and 8. Some knowledge of algebraic language and systems programming, algebra of sets and functions, and combinatorial mathematics for graph theory would be helpful. For those with interest in research and advanced development, the book supplies guiding principles and suggests additional research, as in e-translation and human-computer interactions. Emphasis for this group would be placed on Chapters 2, 3, 5, and the Appendices. Some knowledge of algebra of sets and functions, combinatorial mathematics for graph theory, recursive theory, and logical foundations is desirable.

The author acknowledges the various support efforts of a number of groups at the University of Wisconsin–Madison, especially the Computer Sciences Department, the Electrical and Computer Engineering Department, and the School of Education in making available the time and the facilities for establishing many of the results presented. The author also acknowledges and thanks the many groups and individuals throughout the world for their expressed interest and approval of the work as it progressed over the years. Also acknowledged and appreciated are the efforts of the author's wife, Gloria, in putting the material of this book in its final form and style for publication.

Department of Computer Sciences
University of Wisconsin–Madison
(Retired status)

ELDO C. KOENIG, PH.D.

CHAPTER 1

Introduction

1.1 CONSIDERATIONS FOR ESTABLISHING KNOWLEDGE STRUCTURES FOR COMPUTERS

Thousands of languages have been developed by as many isolated societies over thousands of years (Wuethrich 2000). The object of any society in developing a language is to communicate perceived knowledge about its environment. Environments are considered to contain living things (automata) that perform individually and in groups (interactive automata) and that make changes in the environments through their responses (histories of automata). That is, perceived knowledge is about automata and histories of automata, and languages are established to communicate the knowledge.

A model for communication in human-computer systems involves more than a syntactical analysis for extracting the meanings of sentences as is demonstrated by the following two pairs of sentences:

"John struck the table with a hammer."
"John struck the table with a glass top."
"The deer came out of the wood."
"The worm came out of the wood."

A model should also accommodate the duplication of the following example performances:

1. An observer sees Mrs. Bee shop at Supermarket Dee one day, the next day sees Mrs. Cee shop at the same supermarket, and then says,

   "Mrs. Bee and Mrs. Cee shop at Supermarket Dee."

2. An observer one day reads that during the Renaissance period, the people of Venice built the Library of St. Mark, and years later reads that

Knowledge Structures for Communications in Human-Computer Systems: General Automata-Based, by Eldo C. Koenig

during the Renaissance period, the Romans built the Church of St. Peter, and then says,

> "During the Renaissance Period, the people of Venice built the Library of St. Mark, and the Romans built the Church of St. Peter."

3. A person at one time hears the sentence

   > "John ate a nut."

   and at a later time says to a person of little experience,

   > "John ate a nut. He put it into his mouth. He chewed the nut with his teeth and swallowed it."

4. An observer sees and almost simultaneously hears the oral sound of an animal one day, hears and almost simultaneously smells (but does not see) that same animal at some later time, and still later, only smells that animal, and then draws a picture of that animal in answer to the question

   > "What is that?"

5. A person hears the statement

   > "Copper is a metal."

   and then responds with the statement

   > "Copper is malleable."

   without having previously received that responding statement.
6. A person hears the statement

   > "Mary put the nut into her mouth."

   and then responds with the statement

   > "Mary probably chewed the nut."

1.2 KNOWLEDGE ABOUT AUTOMATA AS A SUBSET OF WORLD KNOWLEDGE

The above discussion presents some important considerations for establishing a model for knowledge structures for computers in application to highly interactive human-computer systems. The desire is to replace a human with a computer in an interactive system and have it perform in the same manner. To achieve a high degree of success, one must discover the order and organization of the universe that a human wishes to describe in the natural language. A force opposed to order is the desire for quick use of a computer.

Any degree of order suggests a mathematical model for knowledge structures for computers. One advantage of a mathematical model involving functions is that knowledge to be stored may be restricted to that which meets the requirements for single-valueness of the functions. To a casual observer, it is reasonable to expect that a computer operating on these principles may appear superior in intelligence compared with a human who is not so rigidly disciplined.

A mathematical model for knowledge structures in the form of a graph is very desirable. One can make use of many existing graph properties by properly interpreting them in physical terms. There are also many existing algorithms associated with graphs for calculating properties.

The technological advances in hardware are encouraging highly interactive human-computer systems that involve a large amount of memory and computing. Parsing methods may be giving way to template matching methods or to a combination of the two. Parsing methods favor single processors while template matching methods favor parallel processors. It is economically practical to have many microprocessors operating in parallel (Koenig 1994b). A million processors in an array is not considered outrageous (Cray 1978).

The subset of world knowledge that has been chosen for consideration in establishing a model for knowledge structures for computers is knowledge about automata. It represents order and organization of the universe. To the author's knowledge, no other model for knowledge structures uses the order of automata to describe knowledge, and no other model employs the mathematical treatment presented here.

Briefly, an automaton is defined as anything that can move or act of itself. For the past several centuries, society has witnessed various individuals and groups in their efforts to study and produce automata (Chapuis and Droz 1958, Sabliere 1966, Von Neumann 1961). In the 17th century, lower animals were viewed as natural automata by Decartes, and mechanical devices reflecting the technology of the times were viewed as artificial automata. Leibnitz regarded clockwork as the model for all automata, and, as a result, automata was not considered to be affected by or to affect the outside world. Beginning the 19th century, both natural and artificial automata were studied in relation to energy. An automaton at that time was considered a "heat engine burning some combustible fuel instead of the glycogen of the human muscles" (Wiener 1948). Babbage, early in the 19th century, had the first idea of a computing automaton.

The 20th century has witnessed a broad concept of automata viewed in total in its interaction with its environment. The concept played an important role in the natural sciences in a qualitative sense at the beginning of the century. In the 1920s, Bush (1929) established the first computing automata of the analog class operating on the continuous principle, and, in the 1930s, Turing

(1936) did pioneering work on the digital class built around the all-or-none concept and operating at discrete moments of time. Prior to the decade of the 1960s, the concern was generally for computing automata as it performed independent of other automata. Since then, interest grew rapidly in interacting automata operating in real time. Systems of non-computing artificial automata operating in the manufacturing processes, called automation, have been made to interact with systems of computing automata. Also prevalent today are interactive computing automata and interactive human-computer systems. It is in highly interactive human-computer systems where models for knowledge structures for computers become particularly important. Here, knowledge about automata will encompass knowledge about general automata, general systems of interactive automata, and about histories of general automata (Koenig 1972a, 1997b).

1.2.1 General Automata

Previous analysis for finite automata did not include the detail required for finite general automata. Normally, two equations are used to describe finite automata, and the state-transition diagram is used to study the operations. Twelve equations are required to describe finite general automata, general systems of interactive automata, and histories of general automata. This set of equations is established and discussed in Chapters 2, 3, and 5. The analysis takes into account component responses, distinguishable receptors and effectors, nonhomogeneous environments, and responses interpreted as stimuli for the five senses. Three graphs are used to study the operations of a general automaton and a general system of interactive automata. They are

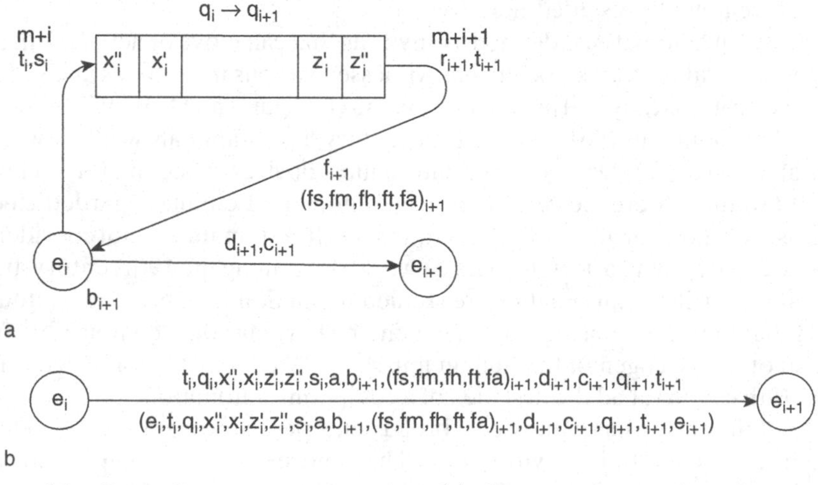

Figure 1.1 General automaton operation: (a) diagram of operation of a general automaton; (b) environment graph containing the information of (a) and the corresponding graph tuple.

1. Processor graph with points representing states (extended state-transition diagram)
2. Environment graph with points representing environments (stimulus locations)
3. Time graph with points representing times

The operation of a general automaton is diagrammed to operate as shown in Figure 1.1a and is described as follows: Given a discrete moment $m+i$, at time t_i, a general automaton a is accessing one and only one environment e_i (stimulus location), receiving one and only one stimulus s_i, through one and only one auxiliary receptor x_i'' and only one principal receptor x_i', and is in one and only one state q_i. Multiple accesses of a same environment during a period of consideration can be made only by the subject automaton. (For a system of automata, multiple accesses of a same environment can be made by different component automata of the system.) Furthermore, during the next moment $m+(i+1)$, at the time t_{i+1}, the automaton a produces one and only one response r_{i+1} through one and only one principal effector z_i' and one and only one auxiliary effector z_i'' based on the aforementioned conditions. The response r_{i+1} consists of a transformation response f_{i+1}, a spatial change response d_{i+1}, and a time change response c_{i+1}. The transformation response consists of responses interpreted as stimuli for sight fs_{i+1}, smell fm_{i+1}, hearing fh_{i+1}, touch ft_{i+1}, taste fa_{i+1}, and is recorded as a response b_{i+1} at the environment e_i accessed at the first moment $m+i$. At moment $m+(i+1)$, the automaton is in state q_{i+1} and at environment e_{i+1}.

If the automaton is removed from the diagram of Figure 1.1a and if the information it contains is included in the label for the connecting line of two points e_i and e_{i+1}, Figure 1.1b results. This is the basic automaton graph of two points and an adjacent line, which is the environment graph for storing knowledge. The corresponding graph tuple is also shown in Figure 1.1b. The time graph is the same except the points represent times, and the times t_i and t_{i+1} of the line label are replaced with the environments e_i and e_{i+1}. For the processor graph, the points represent states, and the number of terms of the line label is reduced.

The sequence of terms of the line label of each type of graph is not arbitrary but is determined by a graph function that is discussed in Chapters 2, 3, and 5. The requirement for single-valueness of the function must always be met. There are features like this resulting from the rigorous treatment of the problem of knowledge structures based on general automata that should give a computer following these principles much greater power than a human for displaying intelligence.

Feedback can be readily represented by the graph model for knowledge structures. Since the spatial change response component d_{i+1} can be zero, an automaton can be told of its immediate past performance; i.e., it can interpret its own response as a stimulus from the current environment before receiving a stimulus from a new environment. This immediate feedback is represented in an environment graph by loops. A detailed analysis of the graph model for feedback is presented in Section 3.4.

This model for knowledge structures for computers, based on general automata, accommodates procedures for

1. extracting and storing the meanings of sentences
2. associating knowledge
3. establishing conclusions and inferences

These are discussed in the following sections. It will also be shown how the model for knowledge structures accommodates the duplication of the example performances listed in Section 1.1.

1.2.2 Extracting and Storing the Meanings of Sentences

The meanings of sentences can be stored in the environment, time, and processor graphs. A graph of two points and an adjacent directed line is normally required to store the meaning of a sentence. An environment graph of two points and an adjacent directed line and a corresponding graph tuple are shown in Figure 1.1b. Not all of the automaton terms of a graph tuple will have word values when the meaning of a single sentence is stored. Each of the following sentences is an example of a class of sentence whose meaning can be extracted and stored as a knowledge structure based on general automata. The seven classes of sentences are discussed in Section 4.1.

1. "John struck the table."
2. "Today, John drove the car carefully three blocks from the house to the store in five minutes."
3. "John became angry."
4. "The table has a glass top."
5. "Today, Jim saw the dent in the table that John made yesterday."
6. "Today, Jim read that John struck the table with a hammer yesterday."
7. "Today, Jim wrote that John struck the table with a hammer yesterday."

Consider, for example, the knowledge structure for the first sentence:

1. "John (a) struck (f_{i+1}) the table (e_i)."

Behind each content word (knowledge element) there is written an automaton term in parentheses that will take on that word as a value in the knowledge structure. The meaning of the sentence is stored in the 16-tuple for the environment graph of Figure 1.1b in the following manner:

$$(e_i = \text{table}, -, -, -, -, -, -, -, a = \text{John},$$
$$-, f_{i+1} = (fs = \text{struck}, -, fh = \text{struck}, -, -), -, -, -, -, -)$$

John is the observed performing automaton *a*. **Struck** is the transformation response f_{i+1} of a, and since it is observable through the receptors for sight and hearing, it takes the first and third position within the parentheses of the five elements. **Table** is the environment e_i. The symbol, -, indicates an unknown value for each of the remaining elements of the 16-tuple. The time and processor graphs can be determined in a similar manner to complete the knowledge structure for the sentence.

The knowledge of this first sentence relates to procedural-type knowledge. Procedural knowledge is defined as the knowledge of *knowing how*. Algorithms for extracting and storing the meanings of sentences of this first class are discussed in Section 4.2.

Consider for a second example the knowledge structure for the fourth sentence:

4. "The table (E'_i) has a glass top (e_i)."

Its meaning can be stored in the 16-tuple for the environment graph in the following manner:

$$(e_i = (\text{glass top}, -, -, \text{table}), -, -, -, -, -, -, -, -, -, (-, -, -, -, -), -, -, -, -, -)$$

The four elements contained in the parentheses (glass top, -, -, table), identify the environment e_i. This four-element parentheses is discussed in Section 4.1 and has the general form, $e_i = (e_i, f_i(e_i), g_i(e_i), E'_i)$. e_i is a part of, or element of E'_i. $f_i(e_i)$ pertains to a functional use of the environment; e.g., $f_i(e_i)$=reading, if the table with the glass top is used as a reading table. $g_i(e_i)$ describes the physical appearance of the environment; e.g., $g_i(e_i)$=round, if the table with the glass top is round. As before, the symbol, -, indicates an unknown value for each of the remaining elements of the tuple.

The knowledge of this fourth class of sentence is commonly classified as declarative. Declarative knowledge is defined as the knowledge of *knowing that* and is acquired suddenly.

The above discussion pertained to knowledge structures for sentences describing single automata. There are also knowledge structures for sentences describing systems of interactive automata. The automata interact by sharing environments, where a recorded response of one automaton is later interpreted as a stimulus by another automaton of the system. Sentences describing interactive automata often contain words of *give*, *receive* and *sell*, *buy*. The knowledge structures for these sentences are discussed in Chapter 6.

An example sentence containing *give* will be analyzed here, and its meaning will be extracted and stored as a knowledge structure. The example sentence is

"John gives Mary a book."

Two knowledge structures sharing a single point are required to store the meaning of the sentence. The shared point represents an environment. A first structure consisting of two points and an adjacent line, is required for **John**, the giver of the book, and a second, also consisting of two points and an adjacent line, is required by **Mary**, the receiver of the book. The two graph tuples for these two knowledge structures are shown in Figure 1.2. The two corresponding knowledge structures are joined at the common environment point $E'_{k+1} = E'_{i+1}$. The resulting structure is also shown in Figure 1.2. The part of the knowledge structure with the solid lines pertains to **John** and the part with the broken lines pertains to **Mary**. A shared environment is required if a transfer of the book from **John** to **Mary** is to take place. When the path of **Mary** is followed through the structure, an equivalent sentence is generated.

"Mary receives a book from John."

Tuple 1

$(e_{i+1}=(-,-,-,E'_{i+1}),-,-,-,-,-,-,-,a=John,b_{i+2}=book,$

$f_{i+2}=(fs=gives,-,-,-,-),-,-,-,-,-)$

Tuple 2

$(e_{k+1}=(book,-,-,E'_{k+1}=E'_{i+1}),-,-,-,-,-,-,-,a=Mary,b_{k+2}=(-book),$

$f_{k+2}=(fs=receives,-,-,-,-),-,-,-,-,-)$

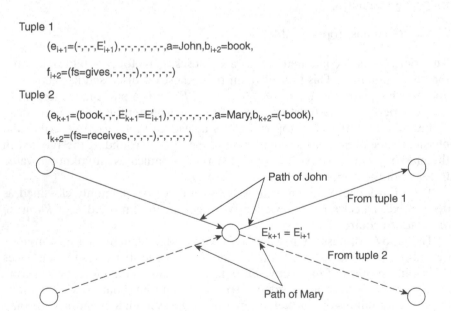

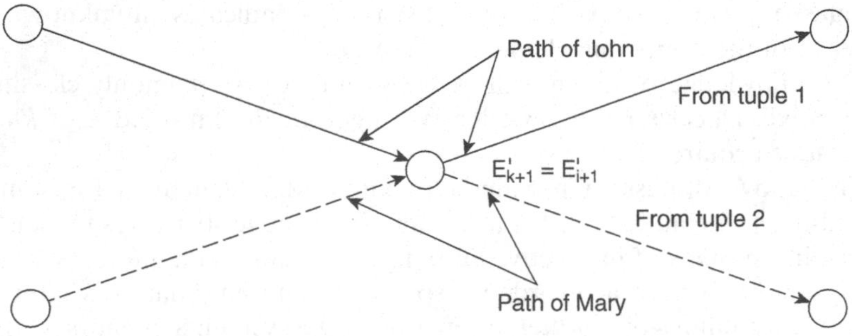

Figure 1.2 Environment graph representation for the knowledge structure of the sentence: "John gives a book to Mary."

The knowledge contained in a sequence of sentences describing a sequence of observations of performing automata, or histories of automata, is represented by a sequence, point, line, point, line, ---, point in a graph representing that knowledge. Such a sequence is called an effective operation path. A similar type of sequence for a knowledge structure may be determined and expressed as a sequence of sentences as in storytelling or as in the conveyance of plans for achieving a goal. Derivations of disciplines describing effective operation paths in general automaton graphs and graphs of general systems of interactive automata are presented in Appendices A and B.

1.2.3 Associating Knowledge

The association of current knowledge and previous knowledge is accomplished by combining separate knowledge structures; i.e., by establishing connected graphs. This eliminates a duplication of structures and facilitates operating on the structures. An algorithm for performing associations is discussed in Section 4.3.

Now return to the first pair of example sentences of Section 1.1 and see how their meanings are distinguishable when association of knowledge is employed. The pair of sentences is

"John struck the table with a hammer."
"John struck the table with a glass top."

The sentences are syntactically identical but they have different meanings. Assume that some time before the pair is received by an automaton for interpretation, the following sentence was received, and its meaning was stored as a knowledge structure:

"The table (E_i') has a glass top (e_i)."

$$(e_i = (\text{glass top}, \text{-}, \text{-}, \text{table}), \text{-}, \text{-}, \text{-}, \text{-}, \text{-}, \text{-}, \text{-}, \text{-}, \text{-}, (\text{-}, \text{-}, \text{-}, \text{-}, \text{-}), \text{-}, \text{-}, \text{-}, \text{-}, \text{-})$$

This knowledge structure, now in the memory of the automaton, is an aid in the interpretation of the pair of sentences. The knowledge structure for the sentences become

"John (a) struck (f_{i+1}) the table (e_i) with a hammer (z_i'')."

$$(e_i = \text{table}, \text{-}, \text{-}, \text{-}, \text{-}, \text{-}, z_i'' = \text{hammer}, \text{-}, a = \text{John},$$
$$\text{-}, f_{i+1} = (\text{struck}, \text{-}, \text{struck}, \text{-}, \text{-}), \text{-}, \text{-}, \text{-}, \text{-}, \text{-})$$

"John (a) struck (f_{i+1}) the table (E_i') with a glass top (e_i)."

$$(e_i = (\text{glass top}, \text{-}, \text{-}, \text{table}) = \text{-}, \text{-}, \text{-}, \text{-}, \text{-}, \text{-}, \text{-}, a = \text{John}$$
$$\text{-}, f_{i+1} = (\text{struck}, \text{-}, \text{struck}, \text{-}, \text{-}), \text{-}, \text{-}, \text{-}, \text{-}, \text{-})$$

Thus, the true meanings of the pair of sentences are stored as knowledge structures and are readily distinguishable. That is, when this pair of sentences is received by the automaton, it is clear from previously stored knowledge that the **glass top** is a part of the **table** and the **hammer** is an auxiliary effector of **John**.

There is also a need to refer to past knowledge in establishing the meanings of the second pair of example sentences of Section 1.1. The sentences are

"The deer came out of the wood."
"The worm came out of the wood."

For these sentences, **wood** has different meanings. Assume that some time before the pair is received by an automaton for interpretation, the following sentences were received, and their meanings were stored as knowledge structures:

"A deer is greater than three feet in height."

$$(\text{-}, \text{-}, \text{-}, \text{-}, \text{-}, \text{-}, \text{-}, \text{-}, a = (\text{deer}, \text{-}, >3\text{ ft. high}, \text{-}), \text{-}, (\text{-}, \text{-}, \text{-}, \text{-}, \text{-}), \text{-}, \text{-}, \text{-}, \text{-}, \text{-})$$

"A worm is less than one foot long."

$$(\text{-}, \text{-}, \text{-}, \text{-}, \text{-}, \text{-}, \text{-}, \text{-}, a = (\text{worm}, \text{-}, <1\text{ ft. long}, \text{-}), \text{-}, (\text{-}, \text{-}, \text{-}, \text{-}, \text{-}), \text{-}, \text{-}, \text{-}, \text{-}, \text{-})$$

"A wood is greater than 100 feet square."

$$(e_i = (\text{wood}, \text{-}, >100\text{ ft. sq.}, \text{-}), \text{-}, \text{-}, \text{-}, \text{-}, \text{-}, \text{-}, \text{-}, \text{-}, \text{-}, (\text{-}, \text{-}, \text{-}, \text{-}, \text{-}), \text{-}, \text{-}, \text{-}, \text{-}, \text{-})$$

"A wood is less than three feet in diameter."

$$(e_i = (\text{wood}, \text{-}, >3\text{ ft. diam.}, \text{-}), \text{-}, \text{-}, \text{-}, \text{-}, \text{-}, \text{-}, \text{-}, \text{-}, \text{-}, (\text{-}, \text{-}, \text{-}, \text{-}, \text{-}), \text{-}, \text{-}, \text{-}, \text{-}, \text{-})$$

For the first two sentences, a has a general form that is similar to the general form of e_i for the last two sentences. The environment was previously defined as $e_i=(e_i, f_i\ (e_i), g_i\ (e_i), E_i')$. The automaton is defined as $a=(a, f\ (a), g\ (a), A')$.

These knowledge structures, now in memory, aid in the interpretation of the pair of sentences. The sentences and their knowledge structures are

"The deer (a) came out (f_{i+1}) of the wood (e_i)."

(e_i =(wood, -, >100 ft. sq., -), -, -, -, -, -, -, -, a=(deep, -, >3 ft. high, -),
-, f_{i+1} =(fs=came out, -, -, -, -), -, -, -, -, -)

"The worm (a) came out (f_{i+1}) of the wood (e_i)."

(e_i =(wood, -, <3 ft. diam., -)-, -, -, -, -, -, -, a=(worm, -, <1 ft. long, -),
-, f_{i+1} =(fs=came out, -, -, -, -), -, -, -, -, -)

The proper environment (**wood**) was selected in each case from past knowledge of the size of the involved automata, **deer** and **worm**. You, the reader of Section 1.1, very likely had already selected the proper environments for the automata from your past knowledge.

It will now be shown how the model for knowledge structures based on general automata accommodates the duplication of the example performances described in Section 1.1. The first four require the procedures for associating knowledge and will be discussed here. The last two will be discussed in the next section. The environment graph and time graph will be used as knowledge structures, and knowledge associations will take place on these structures.

Recall the first performance.

1. An observer sees Mrs. Bee shop at Supermarket Dee one day, the next day sees Mrs. Cee shop at the same supermarket, and then says,

   "Mrs. Bee and Mrs. Cee shop at Supermarket Dee."

The observer first establishes a knowledge structure, in the form of an environment graph, of **Mrs. Bee** shopping at **Supermarket Dee**, and the graph tuple is

(e_i =Supermarket Dee, -, -, -, -, -, -, -, a=Mrs. Bee,
-, f_{i+1} =(shops, -, -, -, -), -, -, -, -, -)

The next day the observer establishes a second knowledge structure of **Mrs. Cee** shopping at the same **Supermarket Dee**. The graph tuple is

(e_i =Supermarket Dee, -, -, -, -, -, -, -, a=Mrs. Cee,
-, f_{i+1} =(shops, -, -, -, -), -, -, -, -, -)

Supermarket Dee is an environment for both the automata, **Mrs. Bee** and **Mrs. Cee**, so that the two separate graphs have a common point, e_i=Supermarket Dee. The common point gives a single connected environment graph, and the knowledge association process is complete. Operations on the connected structure lead immediately to the sentence

"Mrs. Bee and Mrs. Cee shop at Supermarket Dee."

This type of knowledge association can appropriately be called *space association*.

Consider the second performance of Section 1.1.

2. An observer one day reads that during the Renaissance period, the people of Venice built the Library of St. Mark, and years later reads that during the Renaissance period, the Romans built the Church of St. Peter, and then says,

   "During the Renaissance period, the people of Venice built the Library of St. Mark, and the Romans built the Church of St. Peter."

Knowledge association takes place in a manner similar to that of the first performance except the graphs as knowledge structures are time graphs instead of environment graphs. The observer first establishes a knowledge structure, in the form of a time graph, of the **people of Venice** building the **Library of St. Mark during the Renaissance period**, and the tuple is

(t_i = Renaissance Period, e_i = Venice, -, -, -, -, -, -, a = people of Venice, b_{i+1} = Library of St. Mark, f_{i+1} = (build, -, -, -, -), -, -, -, -, -)

Years later, the observer establishes a second Knowledge Structure of the **Romans** building the **Church of St. Peter during the Renaissance Period**, and the corresponding graph tuple is

(t_i = Renaissance period, -, -, -, -, -, -, -, a = Romans, b_{i+1} = Church of St. Peter, f_{i+1} = (build, -, -, -, -), -, -, -, -, -)

The **Renaissance period** represents time for both the building of the **Library of St. Mark** and the **Church of St. Peter**, so that the two separate time graphs have a common point, t_i = Renaissance period. The common point gives a single connected time graph, and the knowledge association process is complete. Operations on the connected structure leads immediately to the sentence

"During the Renaissance period, the people of Venice built the Library of St. Mark, and the Romans built the Church of St. Peter."

This type of knowledge association can appropriately be called *time association*.

Consider now how the model for knowledge structures accommodates the duplication of the example Performance 3 of Section 1.1. Performance 3 is

3. A person, at one time, hears the sentence

> "John ate a nut."

and at a later time says to a person of little experience,

> "John ate a nut. He put it into his mouth. He chewed the nut with his teeth and swallowed it."

The person who elaborated on eating may be said to have previously observed or read about a number of people eating a variety of foods at different times and to have stored the detailed knowledge about eating in the form of graph tuples. Much of the knowledge contained in the graph tuples can be described by the following sentences:

1. "*a* ate e_i."
2. "*a* put e_i into *a*'s mouth."
3. "*a* chewed the e_i with *a*'s teeth."
4. "*a* swallowed e_i."

a is a variable and is any one of the people who was observed eating food. e_i is also a variable and is the corresponding food that was observed being eaten.

Figure 1.3 shows the four graph tuples for the above sentences combined into a single environment graph, and the knowledge association process is complete. The numbers on the graph correspond to the numbers identifying the above sentences. The graph is described in greater detail in Section 4.3.

When the sentence, "John ate a nut," is received, the variable *a* is given the word value of **John**, and e_i is given the word value of **nut**. The person is then able to operate on the stored knowledge structure and to respond with the details

> "John ate a nut. He put the nut into his mouth. He chewed the nut with his teeth and swallowed it."

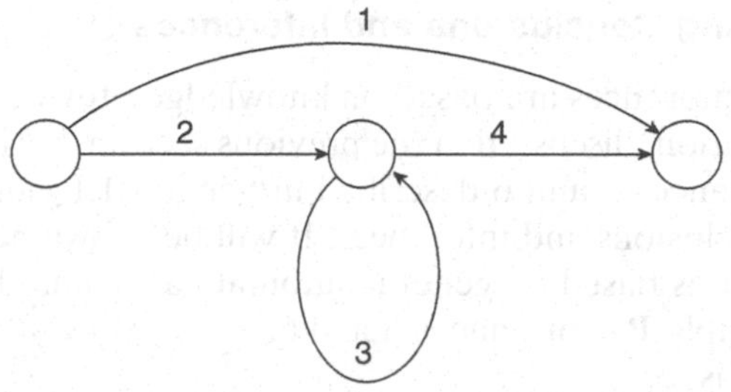

Figure 1.3 Four graph tuples for eating combined into a single environment graph.

It was shown that the model for knowledge structures accommodates the duplication of the example Performances 1, 2, and 3 through the use of knowledge association, and knowledge association was accomplished by combining points of graphs. The model for knowledge structures also accommodates the duplication of Performance 4 of Section 1.1 through the use of knowledge association, but knowledge association, in this case, is accomplished by combining parallel lines instead of points. Performance 4 is

4. An observer sees and almost simultaneously hears the oral sound of an animal one day, hears and almost simultaneously smells (but does not see) that same animal at some later time, and still later, only smells that animal, and then draws a picture of that animal in answer to the question

   "What is that?"

The basis for combining parallel lines into single lines references the functions of the graph model. There is a graph function for each of the graphs, Processor, environment, and time graph, and the ordering of terms identifying two points and an adjacent line is not arbitrary but is a grouping of terms for a domain point and of terms for its image. Suppose an observation was made of a performing automaton at some moment. Assume the observation gives a domain point consisting of a group of terms (a, b, c, d) and a corresponding image point consisting of a group of terms $(e, -, g, -, i)$. The dashes, -, indicate unknown (unobserved) values. Suppose at some moment later, an observation was made giving a domain point consisting of the same group of terms (a, b, c, d). Then, if a corresponding image point consisting of terms $(-, f, -, h, i)$ was also observed, the two graphs (each of two points and an adjacent line) could be combined giving a single graph whose domain point is (a, b, c, d) and whose image point is now (e, f, g, h, i). This must be true based on the single-valueness of the function. This type of knowledge association and how it relates to similar performances are discussed in greater detail in Section 4.4.

1.2.4 Establishing Conclusions and Inferences

Conclusions and inferences are based on knowledge previously acquired, and knowledge association, discussed in the previous section, becomes paramount. Example Performances 5 and 6 described in Section 1.1 yield responses that are related to conclusions and inferences. It will be shown how the model for knowledge structures based on general automata accommodates the duplication of these example Performances 5 and 6.

Performance 5 is

5. A person hears the statement

   "Copper is a metal."

   and then responds with the statement

   "Copper is malleable."

   without having previously received that responding statement.

The two conversational statements are recognized as parts of the valid argument

"All metals are malleable.
Copper is a metal.
Therefore, copper is malleable."

The first two sentences of the argument are premises, and the last sentence is the conclusion.

Relate Performance 5 to operations on knowledge structures. The first premise

"All metals are malleable."

can be considered to have been previously stored as a Knowledge Structure described by the tuple

$$(e_i = (\text{metal, malleable, -, -}), -, -, -, -, -, -, -, -, -, (-, -, -, -, -), -, -, -, -, -)$$

When the second premise

"Copper is a metal."

is received, it is stored as a Knowledge Structure described by the tuple

$$(e_i = (\text{copper, -, -, metal}), -, -, -, -, -, -, -, -, -, (-, -, -, -, -), -, -, -, -, -)$$

The association of (metal, malleable, -, -) and (copper, -, -, metal) for identifying the environment e_i yields

$$e_i = ((\text{copper, -, -}), \text{metal, malleable, -, -})$$

and the graph tuple that contains the meaning of both premises is

$(e_i = ((\text{copper}, \text{-}, \text{-}), \text{metal}, \text{malleable}, \text{-}, \text{-}), \text{-}, \text{-}, \text{-}, \text{-}, \text{-}, \text{-}, \text{-}, \text{-}, \text{-}, (\text{-}, \text{-}, \text{-}, \text{-}, \text{-}), \text{-}, \text{-}, \text{-}, \text{-}, \text{-})$

Now, suppose an operation is performed to insert **malleable** in the position following the word **copper** in the above tuple for the two premises. Then the following tuple is obtained:

$(e_i = ((\text{copper}, \text{malleable}, \text{-}), \text{metal}, \text{malleable}, \text{-}, \text{-}), \text{-}, \text{-}, \text{-}, \text{-}, \text{-}, \text{-}, \text{-}, \text{-}, \text{-}, (\text{-}, \text{-}, \text{-}, \text{-}, \text{-}), \text{-}, \text{-}, \text{-}, \text{-}, \text{-})$

But this is the knowledge structure for the complete argument, and an operation on the structure yields the conclusion

"Copper is malleable."

which is the response in Performance 5. These operations on the knowledge structures are discussed in greater detail in Section 4.4.

The last performance, Performance 6 in Section 1.1, yields a response that involves an inference. To infer suggests the arriving at a decision or opinion by reasoning from known facts or evidence. It will now be shown how the model for knowledge structures accommodates the duplication of Performance 6, which is

6. A person hears the statement

   "Mary put the nut into her mouth."

   and then responds with the statement

   "Mary probably chewed the nut."

Assume the person had obtained knowledge about eating from observing or reading about individuals over a period of time eating various foods. Some of the knowledge the person obtained about eating can be described by the sentences

1. "*a* ate e_i."
2. "*a* put e_i into *a*'s mouth."
3. "*a* chewed e_i."
4. "*a* swallowed e_i."

where *a* and e_i are variables. Following this storage of knowledge, the person of Performance 6 hears the sentence

"Mary put the nut into her mouth."

and identifies the knowledge with that of Sentence 2. Sentence 3 describes what is likely to happen next, based on previous observations of people eating. So, with a=Mary and e_i=nut, the response that the person gives is

"Mary probably chewed the nut."

Now, relate the performance to the operations on knowledge structures. The association of knowledge over a period of time takes place in the manner discussed for Performance 3, and Figure 1.3 shows the four environment graphs for the above four sentences combined into a single graph. Operations on the structure locate the two points and adjacent line for the input sentence, and the output sentence

"Mary probably chewed the nut."

is synthesized from the next two points and adjacent line.

Section 4.5 covers the subject of inferences in detail. The inference of Performance 6 involves the knowledge of more than a single graph tuple. Inferences may involve the knowledge of only a single graph tuple.

EXERCISES

1.1 Read Glushkov (1966).

1.2 Read Wiener (1954).

1.3 Read Chapuis and Droz (1958).

1.4 Read George (1980).

The following publications deal with various aspects of sequential machines and computing automata:

1.5 Read Shannon (1938) [analysis of relay and switching circuits, 15 pages].

1.6 Read Moore (1956) [experiments on sequential machines, 24 pages].

1.7 Read Myhill (1957) [finite automata and the representation of events, 25 pages].

1.8 Read Ginsburg (1959) [minimal state sequential machines, 11 pages].

1.9 Read Gould (1959) [applications of graph theory, 48 pages].

1.10 Read Kohavi and Winograd (1973) [bounds concerning finite automata, 11 pages].

CHAPTER 2

A General Automaton

2.1 FORMAL ANALYSIS FOR A GENERAL AUTOMATON

To accomplish completeness and unification for the broad concept of a general automaton, the analysis applies the algebra of sets and functions and follows with the application of combinatorial mathematics for graph theory to define a general graph model. Various models called transition diagrams were established in the 1950s for the special case of sequential machines without the algebraic analysis required for a rigorous development of a graph model (Moore 1964, Shannon and McCarthy 1956), and an algebraic analysis, limited to the confines of sequential and digital computing machines, began in the early 1960s (Ginsburg 1962, Glushkov 1966, Hartmanis and Stearns 1966, Miller 1965).

Fundamental to the capability of the analysis to handle interactive automata is the development of a graph model consisting of three graphs, the processor graph, the environment graph, and the time graph. Each of these, in a hierarchical sense, reflects the action of the automaton as it performs internally and externally in space and time. Since interacting automata must be interfaced in space and time, the graph model is fundamental to the analysis of systems of automata.

2.1.1 General Analysis

The analysis for a general automaton was presented by Koenig and Frederick (1970a). In the broad concept, an automaton refers to the continuum of a stimulus processor that, when coupled with its environment, will continue to produce responses to an encountered series of stimuli from one or more sources over a period of time.

The operation of an automaton is described in a discrete manner in that it is viewed in a sequence of moments that are in a one-to-one correspondence

---

Knowledge Structures for Communications in Human-Computer Systems: General Automata-Based, by Eldo C. Koenig

with a naturally ordered subset of nonnegative integers. The correspondence serves to denote a current moment $m+i$, the next moment $m+(i+1)$, the next after that, $m+(i+2)$, and so forth for $i \geq 0$. Given a discrete moment $m+i$, an automaton is considered to be operating on one of a number of stimuli, to be in one of a number of states, and to be effecting one of a number of responses. The functions describing the stimulus, state, and response with respect to time between the moment $m+j$ and the next moment $m+(j+1)$ need not be discrete or continuous.

Then, for any automaton, the view may be taken that the stimulus operates on a set Q of internal states, receives a set S of stimuli from a set E of stimulus locations, called the automaton's environment, and produces a set R of responses. The processor P is treated as a black box other than to say that the internal state at a given time depends on the history of P, and the set of states Q represents equivalence classes of possible histories for P. A given automaton is said to affect its environment if a response serves as a subsequent stimulus for either itself or some other automaton.

Utilizing the notation of discrete moments of time, two equations describe the basic action of the automaton:

$$Q(m+(i+1))=\omega(Q(m+i),S(m+i)) \qquad 2.1E$$

$$R(m+(i+1))=\beta(Q(m+i),S(m+i)) \qquad 2.2E$$

That is, a next state (state at the next moment) $Q(m+(i+1))$ is a function ω of the state at a current moment $(m+i)$ and the stimulus at a current moment $S(m+i)$. Similarly, the response at the next moment $R(m+(i+1))$ is a function β of the current state $Q(m+i)$ and stimulus $S(m+i)$. It is clear from the second automaton Equation 2.2E that when the set of states and stimuli are finite sets, the set of responses is a finite set.

The notion of a response by the automaton is a ternary notion. First, there is a response that is interpreted as a stimulus if the automaton subsequently encounters the location of the response. The response is appropriately called a transformation response. Next, there is a response that causes the automaton to access another stimulus location or the current location at the next moment. This response is called a spatial change response. Thirdly, there is a response that indicates the elapsed time between the current moment and the next moment. This response is appropriately called the time change response. Transformation responses are denoted by a set F, and such responses can be interpreted by the automaton as stimuli if received at some later moment. The spatial change responses consist of a set D of vectors from R^n, where n is the dimension of the environment for the automaton, and the set D represents the allowable spatial changes at each moment. Finally, the time change responses consist of a set C of vectors from R^1, which represents the allowable time lapse between each moment. Thus, the set R of responses is described by a triadic relation on the sets F, D, and C by $R \subseteq F \times D \times C$, and the response is viewed as an ordered triple of the form (transformation response, spatial change response, time change response).

It follows from the definition of R that the second automaton Equation 2.2E is defined by three separate equations

$$F(m+(i+1))=\sigma(Q(m+i), S(m+i)) \quad 2.3E$$

$$D(m+(i+1))=\delta(Q(m+i), S(m+i)) \quad 2.4E$$

$$C(m+(i+1))=\tau(Q(m+i), S(m+i)) \quad 2.5E$$

so that $\beta(Q(m+i), S(m+i))=[\sigma(Q(m+i), S(m+i)), \delta(Q(m+i), S(m+i)), \tau(Q(m+i), S(m+i))]$. Here, as before, the transformation response at a given moment is a function σ of the previous state and stimulus. Similarly, the spatial change response and the time change response at the current moment are functions δ and τ, respectively, of the previous state and stimulus.

The environment E for a given automation is a set of accessible stimulus locations (environments) in n space encountered over a given set of moments and corresponds to a subset of vectors from the real n-dimensional vector space R^n. At any given moment, only one location is involved, and since the set of moments is a discrete set, the environment E is discrete. Similarly, time T is a set of absolute times that corresponds to the given set of moments and also corresponds to a subset of vectors (real numbers) from R^1. In the same sense that E is discrete, T is discrete. Note that if the given set of moments is an infinite set, environment E and T may not be infinite sets.

To complete the automaton equations, it is necessary to introduce the equations describing the environment and time for the automaton. There is a function γ', namely, ordinary addition of real vectors, which describes the environment or stimulus location at the next moment $E(m+(i+1))$ given the stimulus location $E(m+i)$ at the current moment and the spatial change response vector $D(m+(i+1))$. That is

$$E(m+(i+1))=\gamma'(E(m+i), D(m+(i+1))) \quad 2.6E$$

is defined by $E(m+(i+1))=E(m+i)+D(m+(i+1))$. Similarly, there is a function λ' that describes the absolute time at the next moment $T(m+(i+1))$ given the absolute time at the current moment $T(m+i)$ and the time change response $C(m+(i+1))$. That is

$$T(m+(i+1))=\lambda'(T(m+i), C(m+(i+1))) \quad 2.7E$$

and $T(m+(i+1))=T(m+i)+C(m+(i+1))$ can be interpreted by the automaton as stimuli if received at some later moment. If $F(m+i)$ denotes a transformation response at moment $m+i$, which is given to a stimulus location encountered again at moment $m+u$, $u>i$, then

$$S(m+u)=\alpha(F(m+i)) \quad 2.8E$$

α is not necessarily a one-to-one function; i.e., more than one transformation response may be interpreted as the same stimulus.

The diagram of Figure 2.1 describes the operation of a general automaton for the moments $m+i$ and $m+i+1$.

A transformation response by a given automaton at any given location serves as a subsequent stimulus if at some moment the given automaton or another automaton interprets the transformation response as a stimulus at this location. Clearly, this can occur with another automaton only if they have nonempty intersecting environments. Furthermore, if the transformation response is interpreted as a stimulus by another automaton at that same moment, only then can that automaton recognize the spatial change response and time change response of the given automaton. Whenever an automaton exchanges responses as stimuli with another automaton, they are said to be *interacting automata*.

Let the set of discrete moments used in describing the operation of an automaton throughout the *life of the automaton* be denoted by L. Let the automaton sets associated with L be denoted by

$$\overline{Q}, \overline{S}, \overline{R}, \overline{F}, \overline{D}, \overline{C}, \overline{E}, \overline{T}$$

which represent maximum sets for the given automaton. Any subset $M \subseteq L$ may serve as a *moment base* in describing the automaton. Denote the sets associated with M as *active sets* Q, S, R, F, D, C, E, T. Underlying the automaton equations is the notion that at any given moment in $M=\{m, m+1, ---, m+h\}$, excluding the initial moment m and final moment $m+h$ (if one exists), the automaton encounters only one stimulus location, receives only one

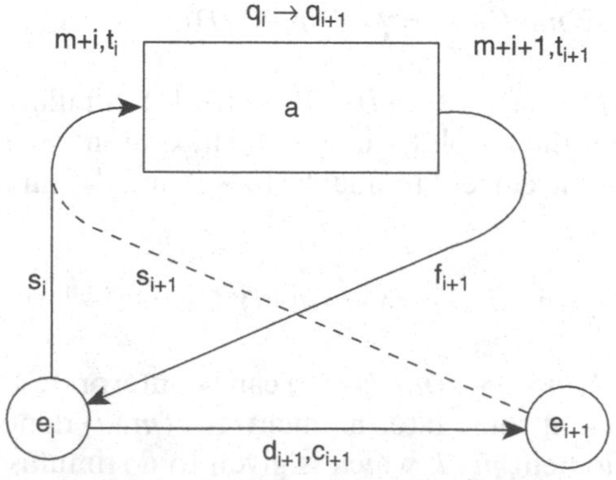

Figure 2.1 A general automaton performing during moments $m+i$ and $m+i+1$.

stimulus, makes only one response, and has only one absolute time associated with it. Stating this more precisely, there is a set of moment functions $\theta=\{\theta_Q, \theta_S, \theta_R, \theta_F, \theta_D, \theta_C, \theta_E, \theta_T\}$ mapping a subset of M into Q, S, F, D, C, E, and T.

$$\theta_Q : M \to \theta_Q(M) = Q \subseteq \bar{Q} \tag{2.1M}$$

$$\theta_S : M-\{m+h\} \to \theta_S(M-\{m+h\}) = S \subseteq \bar{S} \tag{2.2M}$$

$$\theta_R : M-\{m\} \to \theta_R(M-\{m\}) = R \subseteq \bar{R} \tag{2.3M}$$

$$\theta_F : M-\{m\} \to \theta_F(M-\{m\}) = F \subseteq \bar{F} \tag{2.4M}$$

$$\theta_D : M-\{m\} \to \theta_D(M-\{m\}) = D \subseteq \bar{D} \tag{2.5M}$$

$$\theta_C : M-\{m\} \to \theta_C(M-\{m\}) = C \subseteq \bar{C} \tag{2.6M}$$

$$\theta_E : M \to \theta_E(M) = E \subseteq \bar{E} \tag{2.7M}$$

$$\theta_T : M \to \theta_T(M) = T \subseteq \bar{T} \tag{2.8M}$$

The functions θ_i are onto functions but not necessarily one-to-one.

Let $M=\{m, m+1, m+2, ---, m+h\}$, $h\geq 1$, and $M\subseteq L$. Let a moment base M' be given such that M is properly contained in M', and M' is a subset of L. Let the active sets associated with M' be denoted by Q', S', R', F', D', C', E', T'. If $Q'-Q$, $S'-S$, $R'-R$, $F'-F$, $D'-D$, $C'-C$, $E'-E$, $T'-T$ are not empty sets, the elements in these sets are called *latent elements* with respect to the moment base M. Note that $Q\subseteq Q'\subseteq\bar{Q}$, $S\subseteq S'\subseteq\bar{S}$, $R\subseteq R'\subseteq\bar{R}$, $F\subseteq F'\subseteq\bar{F}$, $D\subseteq D'\subseteq\bar{D}$, $C\subseteq C'\subseteq\bar{C}$, $E\subseteq E'\subseteq\bar{E}$, $T\subseteq T'\subseteq\bar{T}$. It is clear that for L, the life of the automaton, the automaton sets $\bar{Q}$, $\bar{S}$, $\bar{R}$, $\bar{F}$, $\bar{D}$, $\bar{C}$, $\bar{E}$, $\bar{T}$ contain only active elements and are maximum sets for the given automaton.

For convenience, consider the following notation for the pertinent sets of an automaton. Let $Q=\{q_0, q_1, ---, q_m\}$ be the finite set of active states of the processor and $S=\{s_0, s_1, ---, s_n\}$ be a finite set of active stimuli over a given moment base $M=\{m, m+1, m+2, ---, m+h\}$, $M\subseteq L$. Then, for the automaton, there is a dyadic relation V on the sets Q and S; i.e., $V\subseteq Q\times S$, and V serves as the domain set for the functions ω, β, σ, τ. Let the respective range sets be denoted by $\omega(V)$, $\beta(V)$, $\sigma(V)$, $\delta(V)$, $\tau(V)$. Then $\omega(V)\subseteq Q\subseteq\bar{Q}$, $\beta(V)=R\subseteq\bar{R}$, $\sigma(V)=F\subseteq\bar{F}$, $\delta(V)=D\subseteq\bar{D}$, $\tau(V)=C\subseteq\bar{C}$, Denote $\beta(V)=R$ by $\{r_0, r_1, ---, r_p\}$, $\sigma(V)=F$ by $\{f_0, f_1, ---, f_u\}$, $\delta(V)=D$ by $\{d_0, d_1, ---, d_v\}$, $\tau(V)=C$ by $\{c_0, c_1, ---, c_w\}$. Finally, if E and T are finite sets, let $E=\{e_0, e_1, ---, e_x\}$ and $T=\{t_0, t_1, ---, t_y\}$.

The Equations 2.1E to 2.8E, describing an operational automaton, may now be written in an abstract mapping sense. For example, given abstract sets $\bar{Q}$, Q, $\bar{S}$, S, and V a dyadic relation on Q and S, then the automaton function ω is the abstract mapping

$$\omega : V \subseteq Q\times S \to \omega(V) \subseteq Q \subseteq \bar{Q} \tag{2.1F}$$

Similarly, given the sets $\bar{R}$, R, $\bar{F}$, F, $\bar{D}$, D, $\bar{C}$, C, $\bar{E}$, E, $\bar{T}$, T (where R is a triadic relation on F, D, C), the automaton functions β, σ, δ, τ, γ', λ', α are the following abstract mappings:

$$\beta : V \subseteq Q \times S \to \beta(V) = R \subseteq \bar{R} \qquad 2.2F$$

$$\sigma : V \subseteq Q \times S \to \sigma(V) = F \subseteq \bar{F} \qquad 2.3F$$

$$\delta : V \subseteq Q \times S \to \delta(V) = D \subseteq \bar{D} \qquad 2.4F$$

$$\tau : V \subseteq Q \times S \to \tau(V) = C \subseteq \bar{C} \qquad 2.5F$$

$$\gamma' : W' \subseteq E \times D \to \gamma'(W') \subseteq E \subseteq \bar{E} \qquad 2.6F$$

$$\lambda' : N' \subseteq T \times C \to \lambda'(N') \subseteq T \subseteq \bar{T} \qquad 2.7F$$

$$\alpha : F_0 \subseteq F \to \alpha(F_0) = S_0 \subseteq S \qquad 2.8F$$

The set F_0 consists of the transformation responses that are received as stimuli at some later moment.

Since γ' is defined on $W' \subseteq E \times D$, there is an interrelationship of the automaton functions δ with γ'. Considering $\bar{E}$, E, V, W as abstract sets, where W is a dyadic relation on E and V, define γ as the composition function of γ'' and γ'; i.e., $\gamma = \gamma' \circ \gamma''$ where $\gamma'' : W \subseteq E \times V \to \gamma''(W) \subseteq W'$ and $\gamma' : \gamma''(W) \to \gamma'(\gamma''(W)) \subseteq E$. There results

$$\gamma : W \subseteq E \times V \to \gamma(W) \subseteq E \subseteq \bar{E} \qquad 2.9F$$

Similarly, since λ' is defined on $N' \subseteq T \times C$, there is an interrelationship on the automaton functions τ and λ'. With $\bar{T}$, T, V, N as abstract sets, where N is a dyadic relation on T and V, define $\lambda = \lambda' \circ \lambda''$, where $\lambda'' : N \subseteq T \times V \to \lambda''(N) \subseteq N'$ and $\lambda' : \lambda''(N) \to \lambda'(\lambda''(N)) \subseteq T$. There results

$$\lambda : \mathrm{N} \subseteq \mathrm{T} \times \mathrm{V} \to \lambda(\mathrm{N}) \subseteq \mathrm{T} \subseteq \bar{\mathrm{T}} \qquad 2.10F$$

An interrelation of the functions γ and λ is used to describe the association of time with the environment. Let Y be a dyadic relation on $T \times W$, and define $\pi = \pi' \circ \pi''$ where $\pi'' : Y \subseteq T \times W \subseteq T \times E \times V \to \pi''(Y) \subseteq E \times T \times V$ and $\pi' : (\pi''(Y)) \to \pi'(\pi''(Y)) \subseteq E \times T$. There results

$$\pi : Y \subseteq T \times W \subseteq T \times E \times V \to \pi(Y) \subseteq \gamma(W) \times \lambda(N) \subseteq E \times T \qquad 2.11F$$

Clearly, there is a parallel function

$$\mu : Y' \subseteq E \times N \subseteq E \times T \times V \to \mu(Y') \subseteq \lambda(N) \times \gamma(W) \subseteq T \times E \qquad 2.12F$$

To determine what stimulus is received by the automaton at each moment, a set ϕ of environment functions is introduced that describes the assignment of stimuli to stimulus locations at each moment in a moment base. This set includes the function ϕ_0 for moment $m+0$ defined by ϕ_0: $E' \subseteq \bar{E} \to \phi_0(E') = S \subseteq \bar{S}$, where $E \subseteq E' \subseteq \bar{E}$. The set $E' = E$ if the environment is known, otherwise E' may be chosen as large as the environment set $\bar{E}$ for the life of the automaton. For a stimulus location element e_k in the domain, the stimulus image is $\phi_0(e_k) = s_j$. In general

$$\phi_i : E' \subseteq \bar{E} \to \phi_i(E') = S \qquad 2.13F$$

where i corresponds to the moment $m+i$. Thus, at an arbitrary moment $m+i \in M$, each stimulus location $e_k \in E'$ has associated with it a stimulus $s_j = \phi_i(e_k)$.

The following definitions pertain to the initial and final moments of a given finite moment base as they relate to the associated sets for the automaton and to special response conditions and stimuli:

Definition 2.1. By *initial stimulus*, *initial stimulus location*, *initial state*, and *initial time* is meant the stimulus, stimulus location, state, and time at the initial moment m; and the *initial response* is the response at the next moment $m+1$, following the initial moment (see Equation 2.2E).

Definition 2.2. For a given moment base $M = \{m, m+1, ---, m+h\}$, an absolute initial state is an initial state that occurs only at the moments $m+j$, $\forall j = 0, 1, ---, i$, where $i \in \{0, 1, ---, h\}$; if an initial stimulus is received only at moment m by an automaton in the absolute initial state, it is an *absolute initial stimulus*; an *absolute stimulus location* is an initial stimulus location that is encountered only at the moments $m+j$, $\forall j = 0, 1, ---, i$, where $i \in \{0, 1, ---, h\}$.

Definition 2.3. By *final state*, *final response*, *final stimulus location*, and *final time* is meant the state, response, location, and time at the final moment $m+h$ of a given moment base, and the *final stimulus* is the stimulus at the moment $m+(h-1)$ preceding the final moment.

Definition 2.4. For a given moment base $M = \{m, m+1, ---, m+h\}$, an *absolute final state* is a final state that occurs only at the moments $m+j$, $\forall_j = h, h-1, ---, h-i$, where $i \in \{0, 1, ---, h\}$; if a final stimulus is received only at the moment $m+(h-1)$ by an automaton in the absolute final state or takes the automaton into that state, it is an *absolute final stimulus*; an *absolute final stimulus location* is a final stimulus location that is encountered only at the moment $m+j$, $\forall j = h, h-1, ---, h-i$, where $i \in \{0, 1, ---, h\}$.

Definition 2.5. A *total null response* is a triple of the form (identity transformation response, null spatial change response, null time change response) as the stimulus, and the null spatial change response and null time change response are the respective zero vectors. A *null response* is a total null response that does not require the null time change. Finally, a *partial null response* is a response with the identity transformation response, the null spatial change response, and the null time change response as mutually exclusive events, but one must occur.

Definition 2.6. The *null stimulus* is interpreted as the absence of a stimulus but is considered a member of the set S of stimuli.

Let automaton sets Q and S be finite sets of active states and stimuli associated with a given moment base for the operation of a given automaton. Recall that a dyadic relation V on $Q \times S$ serves as a domain set for the automaton functions σ, δ, τ, and other pertinent automaton sets result from these mappings. For example, $\sigma(V)=F$ is the set of active transformation responses; $\delta(V)=D$ is the set of active spatial change responses; and $\tau(V)=C$ is the set of active time change responses. The following bounds necessarily follow for $|F|$, $|D|$, $|C|$:

Theorem 2.1. *Let* $|Q|=m+1$, $|S|=n+1$, $F=\{f_0, f_1, \text{---}, f_u\}$, $D=\{d_0, d_1, \text{---}, d_v\}$, *and* $C=\{c_0, c_1, \text{---}, c_w\}$. *Then* $|F|=u+1 \le |V| \le |Q|\ |S|=(m+1)\ (n+1)$; $|D|=v+1 \le |V| \le (m+1)\ (n+1)$; *and* $|C|=w+1 \le |V| \le (m+1)\ (n+1)$.

With $V \subseteq Q \times S$, then $|V| \le |Q \times S| = |Q|\ |S| = (m+1)\ (n+1)$. By the single-valueness of the functions σ, δ, τ, then

$$|\sigma(V)|=|F| \le |V|,\ |\delta(V)|=|D| \le |V|, \text{ and } |\tau(V)|=|C| \le |V|$$

Assuming automaton sets Q and S finite, R is a finite set, where R is the active responses of the form (transformation response, spatial change response, time change response). The following result is observed for $|R|$:

Theorem 2.2. *Let* $R=\{r_0, r_1, \text{---}, r_p\}$, $|F|=u+1$, $|D|=v+1$, *and* $|C|=w+1$. *Then* $max\ (u+1, v+1, w+1) \le p+1 \le |V|$.

By the single-valueness of β, $|\beta(V)|=|R| \le |V|$. But with $\beta(V)=R$, which is a set of ordered triples of the form $(\sigma(V), \delta(V), \tau(V))$, then $max\ (|\sigma(V)|, |\delta(V)|, |\tau(V)|) \le |R|$. Hence

$$max(u+1, v+1, w+1) \le |R| = p+1$$

The bounds placed on the number of elements in the finite sets for a given automaton are supplemented by the following relationship between the number of active states and active stimuli when the dyadic relation $V=Q \times S$:

Theorem 2.3. *Let* $V=Q \times S$, $|Q|=m+1$, $|S|=n+1$, *and put* $max\ (|F|, |D|, |C|)=k$. *Then* $|Q| \ge (k+x)/(n+1)$, *where* $k+x \equiv 0$ *modulo* $n+1$, *and* x *is the minimum nonnegative integer with this property.*

Clearly $k>0$ since $|F|>0$, $|D|>0$, and $|C|>0$. By Theorem 2.1, $k \leq |Q \times S| = |Q|(n+1)$, and $k/(n+1) \leq |Q|$. If $k/(n+1)$ is not an integer, $k/(n+1) < |Q|$. Then $(k+x)/(n+1)$ is the smallest integer with $(k+x)/(n+1) > k/(n+1)$ when $k+x \equiv 0$ modulo $n+1$, and x is the minimum nonnegative integer with this property. Hence

$$(k+x)/(n+1) \leq |Q|.$$

Corollary 2.3. *Given the hypothesis of Theorem 2.3, $|Q|$ can be replaced by $|S|$ and $n+1$ by $m+1$. Thus, $|S| \leq (k+x)/(m+1)$, where $k+x \equiv 0$ modulo $m+1$, and x is the minimum nonnegative integer with this property.*

A general classification of automata can be established based on whether transformation responses interpreted as stimuli occurred within or disjoint with a given moment base in describing the operation of an automaton.

Definition 2.7. An *internal stimulus* for a given automaton operating over a given moment base is a stimulus produced as a transformation response by that same automaton at some prior moment within the moment base.

Definition 2.8. An *external stimulus* for a given automaton operating over a given moment base is a stimulus produced as a transformation response either by that automaton at some moment prior to the given moment base or by another automaton whose moment base is not necessarily disjoint with the moment base of the given automaton.

An external stimulus for a given automaton is a stimulus that is not an internal stimulus. It arises whenever a stimulus location is encountered for the first time, or for the first time after another automaton provides a transformation response not interpreted identical to the transformation response left at that location by the given automaton.

Theorem 2.4. *For an automaton, the initial stimulus is an external stimulus.*

This follows immediately from the fact that the initial stimulus comes from the initial location, which is encountered for the first time at the initial moment. Alternatively, suppose $Q(m)$ and $S(m)$ denote the initial state and initial stimulus of the automaton. From 2.2E, the initial response occurs at the next moment $m+1$; i.e., $R(m+1) = \beta(Q(m), S(m)) = (F(m+1), D(m+1), C(m+1))$. Thus, $S(m)$ was not produced as a transformation response by the automaton during its moment base and is not an internal stimulus.

Let the moments of encounters of a single location that produce successive external stimuli be $m+i$ and $m+j$. All encounters of that location for moments between $m+i$ and $m+j$ must provide internal stimuli. Internal stimuli produced by successive encounters of a single location at successive moments correspond to a response containing a null spatial change response. The following result is observed:

Theorem 2.5. *Assuming no other automaton interacts with a given automaton, any null response not occurring with a change of state results in the given automaton operating internally in an infinite loop.*

Let $S(m+i)$ and $Q(m+i)$ denote the stimulus and state at moment $m+i$. From 2.1E and 2.2E, $Q(m+(i+1))=\omega(Q(m+i), S(m+i))$, and $R(m+(i+1))=\beta(Q(m+i), S(m+i))=(F(m+(i+1)), D(m+(i+1)), C(m+(i+1)))$. By hypothesis, the response is a null response, so $F(m+(i+1))$ is interpreted as $S(m+i)$, and $Q(m+(i+1))=Q(m+i)$. With $S(m+(i+1))$ the interpretation of $F(m+(i+1))$, then $S(m+(i+1))=S(m+i)$. Considering the moment $m+(i+2)$, $R(m+(i+2))=\beta(Q(m+(i+1)), S(m+(i+1)))=\beta(Q(m+i), S(m+i))=R(m+(i+1))$. Continuing without interaction with another automaton, then $R(m+(i+h))=R(m+i)$ $\forall h \geq 0$, and the automaton is operating internally in an infinite loop.

Definition 2.9. When all stimuli, excluding the initial stimulus, are internal stimuli within a given moment base, the automaton is said to operate internally over that moment base.

If over some moment base $M=\{m, m+1, ---, m+h\}$, u is defined recursively so that the moments $m+u$, for which there are external stimuli, are $m+1, m+2, ---, m+h$, then external stimuli are received at each moment. The automaton must always access a previously nonaccessed stimulus location if it is not properly interacting with another automaton.

Definition 2.10. When all stimuli of an automaton are external stimuli within a given moment base, the automation is said to operate externally over the moment base.

Definition 2.11. When stimuli of an automaton excluding the initial stimulus are both internal and external within a given moment base, the automaton is said to operate internally and externally over the moment base.

Note that when the latter happens, the moments are partitioned into two equivalence classes; namely those moments during which the automaton operates externally and those during which it operates internally.

2.1.2 Graph Model

A graph model of an automaton is a set of graphs. The notion of a general graph together with mapping functions define the graphs for the model. The mapping functions are based on the functions describing the automaton. A graph in its most general mathematical sense is a set of elements herein called points, some of which (possibly none or all) are paired together by a set of elements herein called lines. Every line has two of these points (not necessarily distinct) at each end, and if two lines intersect, it must be in one of these end points. Note that the lines are not to be mistaken as a continuum of points, as in the Euclidean sense, but merely as linkages. Furthermore, if one does not require uniqueness for the end points, such a line is said to be a loop. Also, any two lines (loops) having the same end points are said to be parallel. A graph that permits loops and parallel lines (loops) is said to be a general graph. Finally, if in the case of distinct end points, the end points are differentiated as beginning and ending points for all such lines, the lines are directed, and the graph is said to be a directed graph. In the case of loops, directness is a meaningless notion.

The above notion of a general graph together with a mapping function describes a particular type of graph. For example, Berge (1962) employed a function mapping a set of points X into X to establish a type of graph for which the outdegree of the points must be one for the function to be single-valued. Busacker and Saaty (1965) used a function mapping a set of lines E into V and V or $V \times V$, where V and V is the unordered product, and V is a set of points.

There are three types of graphs, processor, environment, and time, and for the environment and time graphs there are principal and alternate graphs. A graph model of an automaton is made up of the processor graph and a combination of principal and alternate environment and time graphs that provide total information about the automaton. The graphs defined here in the context of automaton modeling may be modified to allow the application of the theorems and algorithms of conventional graph theory (E.F. Koenig 1976).

Processor Graph. A function Ω is defined and employed in describing the first graph, called the processor graph of the graph model. Define Ω mapping a subset of the product space of internal states and stimuli, determined as a dyadic relation by the automaton, into the product space of responses and states. That is, Ω: $V \subseteq Q \times S \rightarrow R \times Q$, or more specifically

$$\Omega: V \subseteq Q \times S \rightarrow (\sigma(V) \times \delta(V) \times \tau(V) \times \omega(V)) \subseteq F \times D \times C \times Q \qquad 2.1\text{GF}$$

where $\Omega(q_i, s_i) = (\sigma(q_i, s_i), \delta(q_i, s_i), \tau(q_i, s_i), \omega(q_i, s_i))$. Hence, the image under Ω, denoted by $\Omega(V)$, is a tetradic relation on $F \times D \times C \times Q$; i.e., $\Omega(V) \subseteq F \times D \times C \times Q$.

A set of primitives for the processor graph follows:

Primitive P1. A set Q' of elements called points.

Primitive P2. A family of sets $\{S', F', D', C'\}$ and a set U of elements called lines, where $U \subseteq S' \times (F' \times D' \times C')$.

Primitive P3. A set of functions $\{\sigma, \delta, \tau, \omega\}$ and a function Ω whose domain is $V \subseteq Q \times S$, $Q \subseteq Q'$, $S \subseteq S'$ and whose range is $\Omega(V) \subseteq (\sigma(V) \times \delta(V) \times \tau(V) \times \omega(V)) \subseteq F \times D \times C \times Q$, where σ: $V \subseteq Q \times S \rightarrow \sigma(V) = F \subseteq F'$,

δ: $(V) \rightarrow \delta(V) = D \subseteq D'$, τ: $V \rightarrow \tau(V) = C \subseteq C'$, ω: $V \rightarrow \omega(V) \subseteq Q$. Furthermore, $Q = \{q_i \mid (q_i, s_i) \in V\} \cup \{q_{i+1} \mid q_{i+1} \in \omega(V), (q_{i+1}, s_i) \notin V\}$, and Ω is defined by $\Omega(q_i, s_i) = (\sigma(q_i, s_i), \delta(q_i, s_i), \tau(q_i, s_i), \omega(q_i, s_i)) = (f_{i+1}, d_{i+1}, c_{i+1}, q_{i+1})$. The set of lines U is all the $(s_i, (f_{i+1}, d_{i+1}, c_{i+1}))$ in the sextuples $(q_i, s_i, f_{i+1}, d_{i+1}, c_{i+1}, q_{i+1})$, such that $\Omega(q_i, s_i) = (f_{i+1}, d_{i+1}, c_{i+1}, q_{i+1})$, and if $Q' - Q$ is not empty, the $q_0 \in Q' - Q$ are points with no lines.

Primitive P4. A subset $U_0 \subseteq U$ is a set of directed lines if for $u \in U_0$, $u = (s_i, (f_{i+1}, d_{i+1}, c_{i+1}))$, the point q_i associated with the element (q_i, s_i) from the domain is not equal to the point q_{i+1} in the image quadruple $\Omega(q_i, s_i) = (f_{i+1}, d_{i+1}, c_{i+1}, q_{i+1})$. The line u is said to be directed from point q_i to point q_{i+1}.

The details of the processor graph are presented more specifically in terms of the automaton. An element (q_i, s_i) from the domain and the image element $\Omega(q_i, s_i) = (f_{i+1}, d_{i+1}, c_{i+1}, q_{i+1})$ form a sextuple $(q_i, s_i, f_{i+1}, d_{i+1}, c_{i+1}, q_{i+1})$, which is represented in the graph by two points q_i and q_{i+1}, $(q_i \neq q_{i+1})$, and a connecting line directed from q_i to q_{i+1} identified by s_i at its origin and f_{i+1}, d_{i+1}, c_{i+1}, at its terminal. The graph belongs to the graph model of a general automaton because of its ability to reflect some of the integral parts of the automaton as it operates over any moment base within its life. These integral parts include the stimulus set, both active and latent states in the set of internal states of the processor, and the response set. The set of points Q' for the graph correspond to a set of internal states for the processor. There is a directed line $(s_i, (f_{i+1}, d_{i+1}, c_{i+1}))$ in the graph from a point representing state q_i to another point representing state q_{i+1} if, whenever the processor is in state q_i and receives a stimulus s_i, it responds a moment later with $r_{i+1} = (f_{i+1}, d_{i+1}, c_{i+1})$ and changes to state q_{i+1}. An action of the automaton, not resulting in a change of state, is reflected in the graph by a nondirected line called a loop; i.e., when $q_i = q_{i+1}$. Also, if the processor is in state q_i at different moments and therein receives different stimuli before changing to state q_{i+1}, this is represented in the graph by parallel lines (or parallel loops).

Let A denote an automaton and $G_L(P)$ the processor graph of P defined over its respective sets for the life of the automaton. Let M be an arbitrary moment base, $M \subseteq L$, and denote the corresponding processor graph by $G_M(P)$ for automaton A. Then $G_M(P)$ is a subgraph of $G_L(P)$ since the points and lines in $G_M(P)$ are also points and lines in $G_L(P)$. A significant aspect of this follows: Let $G_{M'}(P)$ be the subgraph for moment base M', $M \subseteq M' \subseteq L$. Comparing $G_{M'}(P)$ with $G_M(P)$, any additional points in $G_{M'}(P)$ reflect latent states with respect to the moment base M. Thus, the subgraph $G_M(P)$ may be viewed as having the same number of points as $G_{M'}(P)$, where the latent states are represented by isolate points in $G_M(P)$. In a similar manner, additional lines in $G_{M'}(P)$ may reflect latent stimuli and responses for moment base M.

Consider the following example of an abstract automaton A as reflected by its pertinent sets, functions, and processor graph. Let the moment base of A be $M' = \{m, m+1, ---, m+12\}$. The associated sets for this moment base are observed to be $S'=\{s_0, s_1, ---, s_4\}\subseteq\bar{S}$, $F'=\{f_0=s_0, f_1=s_1, f_2=s_2\}=\sigma(V)\subseteq\bar{F}$, $Q'=\{q_0, q_1, ---, q_5\}\subseteq\bar{Q}$, $D'=\{d_0, +d_1, -d_1, +d_2, -d_2, +d_3, -d_3\}=\delta(V)\subseteq\bar{D}$, $C'=\{c_0, c_1, c_2\}=\tau(V)\subseteq\bar{C}$, and $V=\{(q_0, s_4)\}\cup\{q_1\times(S'-\{s_3, s_4\})\}\cup\{q_2\times(S'-\{s_1, s_3\})\}\cup\{q_3\times(\{s_0, s_2\})\}\cup\{q_4, s_1)\}$. Note from the partitioning that the automaton leaves states q_1, q_2, q_3 upon receiving any one of two or more stimuli. The state that the automaton enters depends on which stimulus is at the encountered stimulus location. The automaton functions σ, δ, τ, ω of Table 2.1(a) further define the operation of the automaton over the given base M'. The graph function Ω, defined on $V\subseteq Q'\times S'$ (Table 2.1(b)), and the graph primitives determine the processor graph $G_{M'}(P)$ of Figure 2.2(a). Observe that Q', S', F', D', C' are active sets for the moment base M'. From Definitions 2.2 and 2.4, q_0 is an absolute initial state, and q_5 is an absolute final state for the given moment base. From the same definitions, s_3 is an absolute initial stimulus, and s_4 is an absolute final stimulus.

Now, if a moment base M is taken to be $M=\{m, m+1, ---, m+4\}$, properly contained in M', and if the associated state set is $Q=\{q_0, q_1, q_3\}$, the corresponding graph $G_M(P)$, shown in Figure 2.2(b), is a proper subgraph of $G_{M'}(P)$. The states q_2, q_4, q_5 are latent states, stimuli s_1, s_4 are latent stimuli, and responses (f_2, d_3, c_2), $(f_2, -d_3, c_2)$, (f_1, d_2, c_2), and $(f_1, -d_1, c_1)$ are latent responses with respect to the moment base M.

TABLE 2.1. Functions defining an example automaton and its processor graph $G_M(P)$

(a) Automaton Functions σ, δ, τ, ω

$\sigma: V\subseteq Q'\times S' \to\sigma(V)=F'\subseteq\bar{F}$	$\delta: V\to\delta(V) =D'\subseteq\bar{D}$	$\tau: V\to\tau(V) =C'\subseteq\bar{C}$	$\omega: V\to\omega(V) \subseteq Q'\subseteq\bar{Q}$
$\sigma(q_0, s_3)=f_0$	$\delta(q_0, s_3)=d_1$	$\tau(q_0, s_3)=c_1$	$\omega(q_0, s_3)=q_1$
$\sigma(q_1, s_0)=f_2$	$\delta(q_1, s_0)=-d_2$	$\tau(q_1, s_0)=c_2$	$\omega(q_1, s_0)=q_3$
$\sigma(q_1, s_1)=f_2$	$\delta(q_1, s_1)=d_3$	$\tau(q_1, s_1)=c_2$	$\omega(q_1, s_1)-q_2$
$\sigma(q_1, s_2)=f_0$	$\delta(q_1, s_2)=d_0$	$\tau(q_1, s_2)=c_0$	$\omega(q_1, s_2)=q_1$
$\sigma(q_2, s_0)=f_0$	$\delta(q_2, s_0)=d_0$	$\tau(q_2, s_0)=c_0$	$\omega(q_2, s_0)=q_3$
$\sigma(q_2, s_4)=f_0$	$\delta(q_2, s_4)=d_0$	$\tau(q_2, s_4)=c_0$	$\omega(q_2, s_4)=q_5$
$\sigma(q_2, s_2)=f_2$	$\delta(q_2, s_2)=-d_3$	$\tau(q_2, s_2)=c_2$	$\omega(q_2, s_2)=q_3$
$\sigma(q_3, s_0)=f_1$	$\delta(q_3, s_0)=d_2$	$\tau(q_3, s_0)=c_2$	$\omega(q_3, s_0)=q_4$
$\sigma(q_3, s_2)=f_1$	$\delta(q_3, s_2)=d_0$	$\tau(q_3, s_2)=c_0$	$\omega(q_3, s_2)=q_1$
$\sigma(q_4, s_1)=f_1$	$\delta(q_4, s_1)=-d_1$	$\tau(q_4, s_1)=c_1$	$\omega(q_4, s_1)=q_2$

(b) Graph Function Ω for Defining the Graph

$$\Omega: V\subseteq Q'\times S'\to\sigma(V)\times\delta(V)\times\tau(V)\times\omega(V)$$

$(q_0, s_3)\to(f_0, d_1, c_1, q_1)$	$(q_2, s_4)\to(f_0, d_0, c_0, q_5)$
$(q_1, s_0)\to(f_2, -d_2, c_2, q_3)$	$(q_2, s_2)\to(f_2, -d_3, c_2, q_3)$
$(q_1, s_1)\to(f_2, d_3, c_2, q_2)$	$(q_3, s_0)\to(f_1, d_2, c_2, q_4)$
$(q_1, s_2)\to(f_0, d_0, c_0, q_1)$	$(q_3, s_2)\to(f_1, d_0, c_0, q_1)$
$(q_2, s_0)\to(f_0, d_0, c_0, q_3)$	$(q_4, s_1)\to(f_1, -d_1, c_1, q_2)$

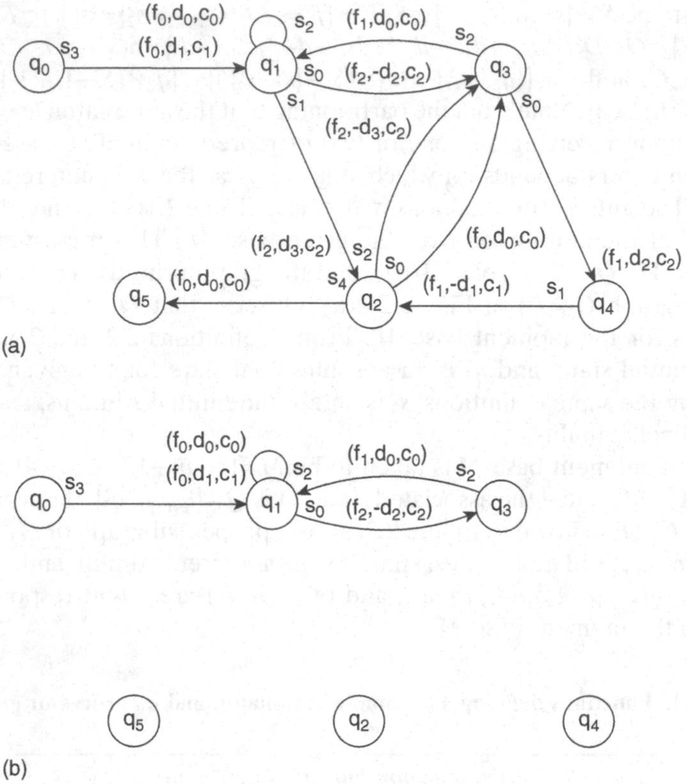

Figure 2.2 Processor graph for the example abstract automaton. (a) Processor graph $G_{M'}(P)$ determined by the function Ω of Table 2.1 (b) and the primitives P. (b) Processor graph $G_M(P)$ for $M \subseteq M'$ of $G_{M'}(P)$ of (a) appropriately denoted as $G_{M' \setminus M}(P)$.

Environment Graph. Of special importance to the study of interactive automata, which necessarily have sets of nonempty intersecting environments, is the environment graph. In contrast with the processor graph, whose points represent the internal states of the processor, the points of this graph represent the automaton's environment. An alternate environment graph is noted in a later discussion.

Define Γ mapping a subset of the product space of the environment, internal states, and stimuli into the product space of responses, states, and environment. That is, Γ: $W \subseteq E \times V \rightarrow \Omega(V) \times E$. More specifically

$$\Gamma: W \subseteq E \times V \rightarrow (\sigma(V) \times \delta(V) \times \tau(V) \times \omega(V) \times \gamma(W)) \subseteq F \times D \times C \times Q \times E \quad 2.2\text{GF}$$

where $\Gamma(e_i, (q_i, s_i)) = ((\sigma(q_i, s_i), \delta(q_i, s_i), \tau(q_i, s_i), \omega(q_i, s_i)), \gamma(e_i, q_i, s_i)), V \subseteq Q \times S$, $\Omega(V) \subseteq F \times D \times C \times Q$, and where $\sigma, \delta, \tau, \omega, \Omega$ refer to Functions 2.3F, 2.4F, 2.5F, 2.1F, 2.1GF, respectively. The image under Γ, $\Gamma(W)$, is a pentadic relation on $F \times D \times C \times Q \times E$, i.e., $\Gamma(W) \subseteq \Omega(V) \times E$.

Primitive E_p1. A set E' of elements called points.

Primitive E_p2. A family of sets $\{Q', S', F', D', C'\}$ and a set X of elements called lines, where $X \subseteq (Q' \times S') \times (F' \times D' \times C' \times D')$.

Primitive E_p3. A set of functions $\{\sigma, \delta, \tau, \omega, \gamma\}$ and a function Γ whose domain is $W \subseteq E \times V$, where $E \subseteq E'$, $V \subseteq Q \times S$, $Q \subseteq Q'$, $S \subseteq S'$ and whose range is $\Gamma(W) \subseteq (\sigma(V) \times \delta(V) \times \tau(V) \times \omega(V) \times \gamma(W)) \subseteq F \times D \times C \times Q \times E$, where σ: $V \subseteq Q \times S \rightarrow \sigma(V) = F \subseteq F'$, δ: $V \rightarrow \delta(V) = D \subseteq D'$, τ: $V \rightarrow \tau(V) = C \subseteq C'$, ω: $V \rightarrow \omega(V) \subseteq Q \subseteq Q'$, and γ: $W \rightarrow \gamma(W) \subseteq E \subseteq E'$. Furthermore, $Q = \{q_i \mid (q_i, s_i) \in V\} \cup \{q_{i+1} \mid q_{i+1} \in \omega(V), (q_{i+1}, s_i) \notin V\}$, $E = \{e_i \mid (e_i, (q_i, s_i)) \in W\} \cup \{e_{i+1} \mid e_{i+1} \in \gamma(W), (e_{i+1}, (q_i, s_i)) \notin W\}$, and Γ is defined by $\Gamma(e_i, (q_i, s_i)) = (\sigma(q_i, s_i), \delta(q_i, s_i), \tau(q_i, s_i), \omega(q_i, s_i), \gamma(e_i, q_i, s_i))$. The set of lines X is all the $((q_i, s_i), (f_{i+1}, d_{i+1}, c_{i+1}, q_{i+1}))$ in the octuples $(e_i, q_i, s_i, f_{i+1}, d_{i+1}, c_{i+1}, q_{i+1}, e_{i+1})$ such that $\Gamma(e_i, (q_i, s_i)) = (f_{i+1}, d_{i+1}, c_{i+1}, q_{i+1}, e_{i+1})$, and if $E' - E$ is not empty, the $e_0 \in E' - E$ are points with no lines.

Primitive E_p4. A subset $X_0 \subseteq X$ is a set of directed lines if for $x \in X_0$, $x = ((q_i, s_i), (f_{i+1}, d_{i+1}, c_{i+1}, q_{i+1}))$, the point e_i associated with the element $(e_i, (q_i, s_i))$ from the domain is not equal to the point e_{i+1} in the image quintuple $\Gamma(e_i, (q_i, s_i)) = (f_{i+1}, d_{i+1}, c_{i+1}, q_{i+1}, e_{i+1})$. The line x is said to be directed from point e_i to point e_{i+1}.

The primitives and construction details for the graph may be stated in terms of the processor and its environment. An element $(e_i, (q_i, s_i))$ from the domain and the image element $\Gamma(e_i, (q_i, s_i)) = (f_{i+1}, d_{i+1}, c_{i+1}, q_{i+1}, e_{i+1})$ form an octuple $(e_i, q_i, s_i, f_{i+1}, d_{i+1}, c_{i+1}, q_{i+1}, e_{i+1})$, which is represented in the graph by two points e_i and e_{i+1}, $(e_i \neq e_{i+1})$, and a connecting line directed from e_i to e_{i+1} identified by (q_i, s_i) at its origin and $(f_{i+1}, d_{i+1}, c_{i+1}, q_{i+1})$ at its terminal. If $e_i = e_{i+1}$, e_i and e_{i+1} are the same point and the line is a nondirected loop with the line identification established on the basis of an assumed line direction. Now, in terms of the processor and its environment, the octuple associated with two points and a connecting line means that at some moment $m+i$ the environment e_i supplies a stimulus s_i to the processor in state q_i, and at the next moment $m+(i+1)$ the processor is in a state q_{i+1} and gives a transformation response f_{i+1}, a spatial change response d_{i+1}, and a time change response c_{i+1}. d_{i+1} provides the processor with a new environment $e_{i+1} = e_i + d_{i+1}$ as the location of a next stimulus. For the case when d_{i+1} is the zero vector d_0, i.e., when the response is a partial null response (a total null response if in addition $f_{i+1} = s_{i+1}$ and c_{i+1} is a null time change response), the environment e_i at moment $m+i$ and e_{i+1} at the next moment are the same. This condition corresponds to a nondirected loop on the graph.

Let $G_L(E_p)$ denote the principal environment graph of a given automaton defined by its respective sets for the life of the automaton. Let M be an arbitrary moment base, $M \subseteq L$, and denote the corresponding environment graph by $G_M(E_p)$. Then, $G_M(E_p)$ is a subgraph of $G_L(E_p)$ since the points and lines in $G_M(E_p)$ are also points and lines in $G_L(E_p)$. Now, let $G_{M'}(E_p)$ be the subgraph for the moment base M', $M \subseteq M' \subseteq L$. Comparing $G_{M'}(E_p)$ with $G_M(E_p)$, any additional points in $G_{M'}(E_p)$ reflect environments (stimulus locations) with respect to the moment base M, which are represented by isolated points in $G_M(E_p)$. Additional lines in $G_{M'}(E_p)$ may reflect latent states, stimuli, and responses for moment base M.

The environment graph whose points represent the locations of the stimuli in space provides a convenient means for describing an automaton when information associated with the stimulus locations is known. For the example automaton, assume the observed information over the moment base $M' = \{m, m+1, ---, m+12\}$ also include the external stimuli $s_3, s_2, s_2, s_0, s_1, s_2, s_1, s_4$ at the location elements of the environment set $E' = \{e_{0,0,0}, e_{1,0,0}, e_{1,-1,0}, e_{1,-1,1}, e_{1,0,1}, e_{0,0,1}, e_{0,1,0}, e_{-1,1,0}\}$, respectively, at the initial moment m. E' is a subset of the integer module of R^3, and the subscripts of the elements identify the stimulus locations given by integer vectors with these entries. Similarly, the integer vector representation of the elements of the spatial change set D are $d_0 = \mathrm{d}_{0,0,0}$, $\pm d_1 = d_{\pm 1,0,0}$, $\pm d_2 = d_{0,\pm 1,0}$, $\pm d_3 = d_{0,0,\pm 1}$. The only external stimuli are those at the initial moment m. The initial stimulus and the initial stimulus location is s_3 and $e_{0,0,0}$. The automaton's operation is further defined by the functions γ' and γ'' as composition functions for obtaining γ (Function 2.9F) as shown in Table 2.2(a).

With the description of the automaton completed, the graph function Γ can now be defined and is shown in Table 2.2(b). This function Γ and the graph primitives establish the environment graph $G_{M'}(E_p)$ of Figure 2.3(a). Note that the final stimulus location $e_{-1,1,0}$ is an absolute final stimulus location (Definition 2.4).

For the same example automaton, impose the additional condition of a change of environment of the automaton by a second automaton. Assume that between the two successive encounters of $e_{0,0,0}$ by the given automaton, a second automaton is observed to encounter that same location and replaces the internal stimulus s_0 with an external stimulus s_3=the initial stimulus. Then, for the same moment base M', the sets are observed to be $E' = \{e_{0,0,0}, e_{1,0,0}, e_{1,-1,0}, e_{1,-1,1}, e_{1,0,1}, e_{0,0,1}\}$ (reduced by two elements), $Q' = \{q_0, q_1, q_2, q_3, q_4\}$ (reduced by one element), $S' = \{s_0, s_1, s_2, s_3\}$ (reduced by one element), and F', D', and C' remain the same. After corresponding changes are made in the functions γ' and γ'' and Γ of Table 2.2(a) and (b), the environment graph of Figure 2.3(b) is established. The processor graph $G_{M'}(P)$ of Figure 2.2(a) can be altered to reflect this imposed condition by removing the point q_5 and the adjacent line and by making appropriate changes in line identifications.

TABLE 2.2. Functions defining an example automaton and its environment graph $G_M(E_p)$

(a) Automaton Function $\gamma=\gamma'\circ\gamma''$

$\gamma'': W\subseteq E\times V\rightarrow\gamma''(W)\subseteq E\times\delta(V)$	$\gamma': \gamma''(W)\rightarrow\gamma'(\gamma''(W))\subseteq E$
$\gamma''(e_{0,0,0}, (q_0, s_3)) = (e_{0,0,0}, d_1)$	$\gamma'(e_{0,0,0}, d_1) = e_{0,0,0} + d_1 = e_{1,0,0}$
$\gamma''(e_{0,0,0}, (q_3, s_0)) = (e_{0,0,0}, d_2)$	$\gamma'(e_{0,0,0}, d_2) = e_{0,1,0}$
$\gamma''(e_{1,0,0}, (q_1, s_0)) = (e_{1,0,0}, -d_2)$	$\gamma'(e_{1,0,0}, -d_2) = e_{1,-1,0}$
$\gamma''(e_{1,0,0}, (q_1, s_2)) = (e_{1,0,0}, d_0)$	$\gamma'(e_{1,0,0}, d_0) = e_{1,0,0}$
$\gamma''(e_{1,-1,0}, (q_1, s_1)) = (e_{1,-1,0}, d_3)$	$\gamma'(e_{1,-1,0}, d_3) = e_{1,-1,1}$
$\gamma''(e_{1,-1,0}, (q_3, s_2)) = (e_{1,-1,0}, d_0)$	$\gamma'(e_{1,-1,0}, d_0) = e_{1,-1,0}$
$\gamma''(e_{1,-1,1}, (q_2, s_0)) = (e_{1,-1,1}, d_0)$	$\gamma'(e_{1,-1,1}, d_0) = e_{1,-1,1}$
$\gamma''(e_{1,-1,1}, (q_3, s_0)) = (e_{1,-1,1}, d_2)$	$\gamma'(e_{1,-1,1}, d_2) = e_{1,0,1}$
$\gamma''(e_{1,0,1}, (q_4, s_1)) = (e_{1,0,1}, -d_1)$	$\gamma'(e_{1,0,1}, -d_1) = e_{0,0,1}$
$\gamma''(e_{0,0,1}, (q_2, s_2)) = (e_{0,0,1}, -d_3)$	$\gamma'(e_{0,0,1}, -d_3) = e_{0,0,0}$
$\gamma''(e_{0,1,0}, (q_4, s_1)) = (e_{0,1,0}, -d_1)$	$\gamma'(e_{0,1,0}, -d_1) = e_{-1,1,0}$
$\gamma''(e_{-1,1,0}, (q_2, s_4)) = (e_{-1,1,0}, d_0)$	$\gamma'(e_{-1,1,0}, d_0) = e_{1,1,0}$

(b) Graph Function Γ for Defining the Graph

$(e_{0,0,0}, (q_0, s_3)) \rightarrow (f_0, d_1, c_1, q_1, (e_{1,0,0}))$	$(e_{1,-1,1}, (q_2, s_0)) \rightarrow (f_0, d_0, c_0, q_3, (e_{1,-1,1}))$
$(e_{0,0,0}, (q_3, s_0)) \rightarrow (f_1, d_2, c_2, q_4, (e_{0,1,0}))$	$(e_{1,-1,1}, (q_3, s_0)) \rightarrow (f_1, d_2, c_2, q_4, (e_{1,0,1}))$
$(e_{1,0,0}, (q_1, s_0)) \rightarrow (f_2, -d_2, c_2, q_3, (e_{1,-1,0}))$	$(e_{1,0,1}, (q_4, s_1)) \rightarrow (f_1, -d_1, c_1, q_2, (e_{0,0,1}))$
$(e_{1,0,0}, (q_1, s_2)) \rightarrow (f_0, d_0, c_0, q_1, (e_{1,0,0}))$	$(e_{0,0,1}, (q_2, s_2)) \rightarrow (f_2, -d_3, c_2, q_3, (e_{0,0,0}))$
$(e_{1,-1,0}, (q_1, s_1)) \rightarrow (f_2, d_3, c_2, q_2, (e_{1,-1,1}))$	$(e_{0,1,0}, (q_4, s_1)) \rightarrow (f_1, -d_1, c_1, q_2, (e_{-1,1,0}))$
$(e_{1,-1,0}, (q_3, s_2)) \rightarrow (f_1, d_0, c_0, q_1, (e_{1,-1,0}))$	$(e_{-1,1,0}, (q_2, s_4)) \rightarrow (f_0, d_0, c_0, q_5, (e_{-1,1,0}))$

Time Graph. The time graph is a third graph contained in the graph model and provides for the study of automata in the time domain, where the points of the graph represent time. An alternate time graph is noted in a later discussion. The time graph chosen to be the principal time graph is determined by a function based on all of the developed automaton functions, and, therefore, it can give complete information on the automaton (under the assumed detail).

The function employed is Λ mapping a subset of the product space of the time, environment, internal states, and stimuli into the product space of responses, states, environment and time, i.e., Λ: $Y\subseteq T\times W\rightarrow\Gamma(W)\times T$. More specifically

$$\Lambda = Y\subseteq T\times W\rightarrow(\sigma(V)\times\delta(V)\times\tau(V)\times\omega(V)\times\pi(Y)) \subseteq F\times D\times C\times Q\times E\times T \qquad 2.4\text{GF}$$

where $\Lambda(t_i, (e_i, (q_i, s_i))) = (\sigma(q_i, s_i), \delta(q_i, s_i), \tau(q_i, s_i), \omega(q_i, s_i), \gamma(e_i, (q_i, s_i)), \lambda(t_i, (q_i, s_i)))$, $W\subseteq E\times V$, $V\subseteq Q\times S$, Γ $(W)\subseteq\Omega(V)\times E$, $\Omega(V)\subseteq F\times D\times C\times Q$, and where σ, δ, τ, ω, γ, λ, π, Ω, Γ refer to Functions 2.3F, 2.4F, 2.5F, 2.1F, 2.9F, 2.10F, 2.11F, 2.1GF, 2.2GF, respectively. The image under Λ,

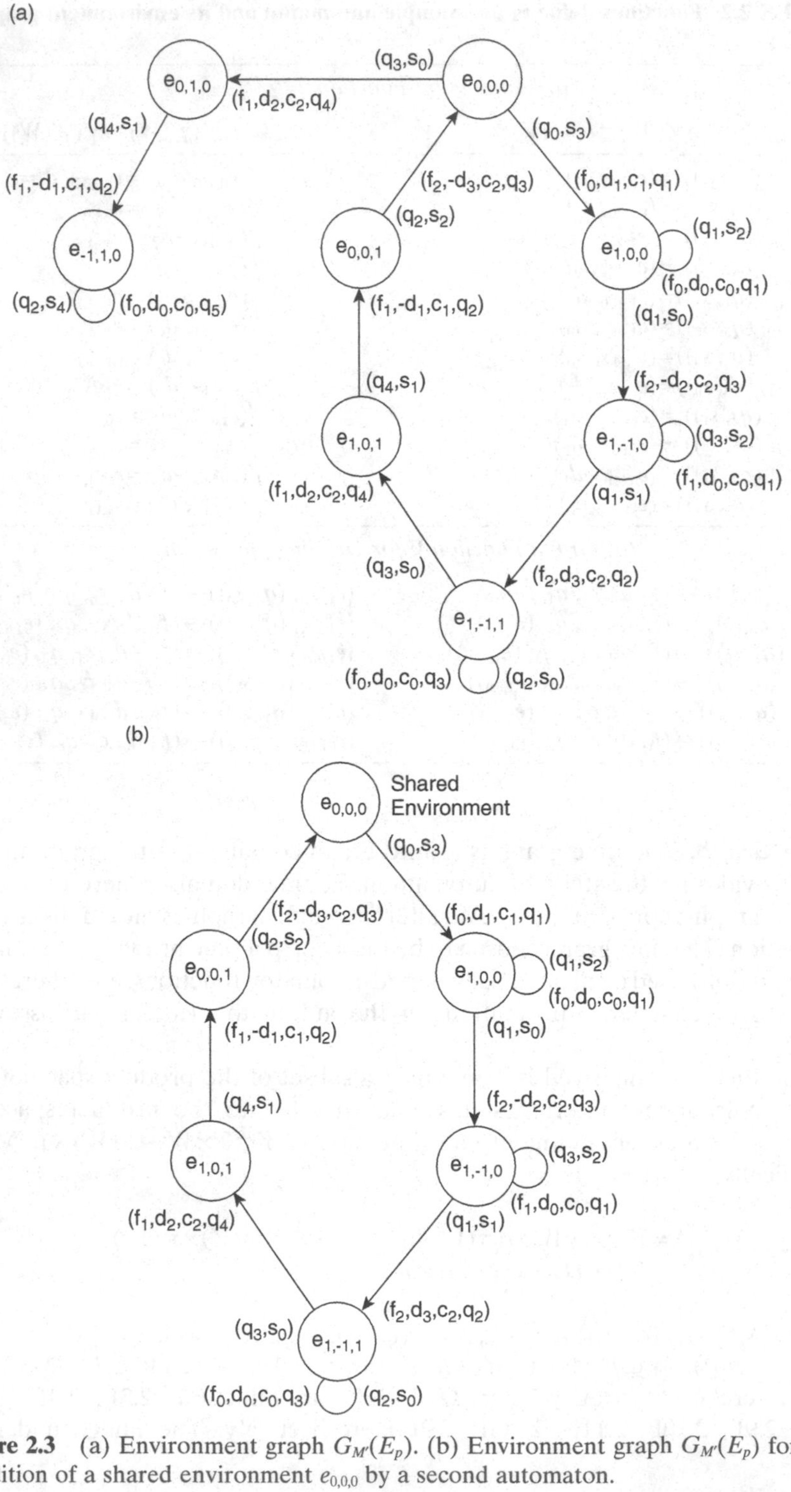

Figure 2.3 (a) Environment graph $G_{M'}(E_p)$. (b) Environment graph $G_{M'}(E_p)$ for the condition of a shared environment $e_{0,0,0}$ by a second automaton.

$\Lambda(Y)$, is a hexadic relation on $F \times D \times C \times Q \times E \times T$; i.e., $\Lambda(Y) \subseteq \Gamma(W) \times T \subseteq \Omega(V) \times E \times T \subseteq F \times D \times C \times Q \times E \times T$.

The set of primitives for the time graph follows:

Primitive $\mathbf{T_p 1}$. A set T' of elements called points.

Primitive $\mathbf{T_p 2}$. A family of sets $\{E', Q', S', F', D', C'\}$ and a set Z of elements called lines, where $Z \subseteq (E' \times Q' \times S') \times (F \times D \times C \times Q \times E)$.

Primitive $\mathbf{T_p 3}$. A set of $\{\sigma, \delta, \tau, \omega, \gamma, \lambda, \pi\}$ and a function Λ whose domain is $Y \subseteq T \times W$, where $W \subseteq E \times V$, $E \subseteq E'$, $T \subseteq T'$, and whose range is $\Lambda(Y) \subseteq (\sigma(V) \times \delta(V) \times \tau(V) \times \omega(V) \times \pi(Y)) \subseteq (\sigma(V) \times \delta(V) \times \tau(V) \times \omega(V) \times \gamma(W) \times \lambda(N)) \subseteq F \times D \times C \times Q \times E \times T$, where $\sigma: V \subseteq Q \times S \rightarrow \sigma(V) = F \subseteq F'$, $\delta: V \rightarrow \delta(V) = D \subseteq D'$, $\tau: V \rightarrow \tau(V) = C \subseteq C'$, $\omega: V \rightarrow \omega\ V) \subseteq Q \subseteq Q'$, $\gamma: W \rightarrow \gamma(W) \subseteq E \subseteq E'$, $\lambda: N \rightarrow \lambda(N) \subseteq T \subseteq T'$, and $\pi: Y \rightarrow \pi(Y) \subseteq \gamma(W) \times \lambda(N) \subseteq E \times T$. Furthermore, $Q = \{q_i \mid (q_i, s_i) \in V\} \cup \{q_{i+1} \mid q_{i+1} \in \omega(V), (q_{i+1}, s_i) \notin V\}$, $E = \{e_i \mid (e_i, (q_i, s_i)) \in W\} \cup \{e_{i+1} \mid e_{i+1} \in \gamma(W), (e_{i+1}, (q_i, s_i)) \notin W\}$, $T = \{t_i \mid (t_i, (e_i, (q_i, s_i))) \in Y\} \cup \{t_{i+1} \mid t_{i+1} \in \lambda(N), (t_{i+1}, (e_i, (q_i, s_i))) \notin Y\}$, and Λ is defined by $\Lambda(t_i, (e_i, (q_i, s_i))) = (\sigma(q_i, s_i), \delta(q_i, s_i), \tau(q_i, s_i), \omega(q_i, s_i), \gamma(e_i, (q_i, s_i)), \lambda(t_i, (q_i, s_i)))$. The set of lines Z is all the $((e_i, q_i, s_i), (f_{i+1}, d_{i+1}, c_{i+1}, q_{i+1}, e_{i+1}))$ in the 10-tuple $(t_i, e_i, q_i, s_i, f_{i+1}, d_{i+1}, c_{i+1}, q_{i+1}, e_{i+1}, t_{i+1})$ such that $\Lambda(t_i, (e_i, (q_i, s_i))) = (f_{i+1}, d_{i+1}, c_{i+1}, q_{i+1}, e_{i+1}, t_{i+1})$, and if $T' - T$ is not empty, the $t_0 \in T' - T$ are points with no lines.

Primitive $\mathbf{T_p 4}$. A subset $Z_0 \subseteq Z$ is a set of directed lines if for $z \in Z_0$, $z = (e_i, q_i, s_i, f_{i+1}, d_{i+1}, c_{i+1}, q_{i+1}, e_{i+1})$, the point t_i associated with the element $(t_i, (e_i, (q_i, s_i)))$ from the domain is not equal to the point t_{i+1} in the image sextuple $\Lambda(t_{i+1}, (e_{i+1}, (q_{i+1}, s_{i+1}))) = (f_{i+1}, d_{i+1}, c_{i+1}, q_{i+1}, e_{i+1}, t_{i+1})$. The line z is said to be directed from point t_i to point t_{i+1}.

To describe these primitives in terms of the physical system, consider the 10-tuple $(t_i, (e_i, q_i, s_i), (f_{i+1}, d_{i+1}, c_{i+1}, q_{i+1}, e_{i+1}), t_{i+1})$ associated with two points and a connecting line, where $(t_i, (e_i, q_i, s_i))$ is an element from the domain and $(f_{i+1}, d_{i+1}, c_{i+1}, q_{i+1}, e_{i+1}, t_{i+1})$ is the image element. The 10-tuple is represented in the graph by two points t_i and t_{i+1}, $(t_i \neq t_{i+1})$, and the connecting line carries the identification of the octuple of the Λ function where (e_i, q_i, s_i) identifies its origin at point t_i and $(f_{i+1}, d_{i+1}, c_{i+1}, q_{i+1}, e_{i+1})$ identifies its terminal at point t_{i+1}. If $t_i = t_{i+1}$, the time change response is the zero vector and there are loops in the time graph $G(T_p)$. Let $G_L(T_p)$ denote the time graph of an automaton defined by its respective sets for the life of the automaton, and let $G_M(T_p)$ be the graph for an arbitrary moment base, $M \subseteq L$. Then $G_M(T_p)$ is a subgraph of $G_L(T_p)$ since the points and lines in $G_M(T_p)$ are also points and lines in $G_L(T_p)$.

To demonstrate the time graph, consider the example automaton previously introduced. The observed information on the automaton described by functions σ, δ, τ, ω of Table 2.1(a) (required to establish $G_{M'}(P)$) and by $\gamma = \gamma' \circ \gamma''$ of Table 2.2(a) (required as additional information for $G_{M'}(E_p)$) is necessary to establish the time graph $G_{M'}(T_p)$. Also required for $G_{M'}(T_p)$ is the observed information described by the automaton function $\lambda = \lambda' \circ \lambda''$ of Table 2.3(a). From the information of this table, the time set is determined as $T' = \{t_0, t_1, t_3, t_5, t_7, t_8, t_{10}, t_{12}, t_{13}\}$. T' and C' are each a subset of the integer module

TABLE 2.3.

(a) Automaton Function $\lambda=\lambda'\circ\lambda''$

$\lambda'': N\subseteq T\times V\rightarrow\lambda''(N)\subseteq T\times\tau(V)$	$\lambda': \lambda''(N)\subseteq T\times\tau(V)\rightarrow\lambda'(\lambda''(N))\subseteq T$
$\lambda''(t_0, (q_0, s_3)) = (t_0, c_1)$	$\lambda'(t_0, c_1) = t_0+c_1=t_1$
$\lambda''(t_{10}, (q_3, s_0)) = (t_{10}, c_2)$	$\lambda'(t_{10}, c_2) = t_{12}$
$\lambda''(t_1, (q_1, s_0)) = (t_1, c_2)$	$\lambda'(t_1, c_2) = t_3$
$\lambda''(t_1, (q_1, s_2)) = (t_1, c_0)$	$\lambda'(t_1, c_0) = t_1$
$\lambda''(t_3, (q_1, s_1)) = (t_3, c_2)$	$\lambda'(t_3, c_2) = t_5$
$\lambda''(t_3, (q_3, s_2)) = (t_3, c_0)$	$\lambda'(t_3, c_0) = t_3$
$\lambda''(t_5, (q_2, s_0)) = (t_5, c_0)$	$\lambda'(t_5, c_0) = t_5$
$\lambda''(t_5, (q_3, s_0)) = (t_5, c_2)$	$\lambda'(t_5, c_2) = t_7$
$\lambda''(t_7, (q_4, s_1)) = (t_7, c_1)$	$\lambda'(t_7, c_1) = t_8$
$\lambda''(t_8, (q_2, s_2)) = (t_8, c_2)$	$\lambda'(t_8, c_2) = t_{10}$
$\lambda''(t_{12}, (q_4, s_1)) = (t_{12}, c_1)$	$\lambda'(t_{12}, c_1) = t_{13}$
$\lambda''(t_{13}, (q_2, s_4)) = (t_{13}, c_0)$	$\lambda'(t_{13}, c_0) = t_{13}$

(b) Automaton Function π

$\pi: Y\subseteq T\times E\times V\rightarrow\pi(Y)\subseteq\gamma(W)\times\lambda(N)\subseteq E\times T$
$\pi(t_0, e_{0,0,0}, (q_0, s_3)) = e_{1,0,0}, t_1$
$\pi(t_{10}, e_{0,0,0}, (q_3, s_0)) = e_{0,1,0}, t_{12}$
$\pi(t_1, e_{1,0,0}, (q_1, s_0)) = e_{1,-1,0}, t_3$
$\pi(t_1, e_{1,0,0}, (q_1, s_2)) = e_{1,0,0}, t_1$
$\pi(t_3, e_{1,-1,0}, (q_1, s_1)) = e_{1,-1,1}, t_5$
$\pi(t_3, e_{1,-1,0}, (q_3, s_2)) = e_{1,-1,0}, t_3$
$\pi(t_5, e_{1,-1,1}, (q_2, s_0)) = e_{1,-1,1}, t_5$
$\pi(t_5, e_{1,-1,1}, (q_3, s_0)) = e_{1,0,1}, t_7$
$\pi(t_7, e_{1,0,1}, (q_4, s_1)) = e_{0,0,1}, t_8$
$\pi(t_8, e_{0,0,1}, (q_2, s_2)) = e_{0,0,0}, t_{10}$
$\pi(t_{12}, e_{0,1,0}, (q_4, s_1)) = e_{-1,1,0}, t_{13}$
$\pi(t_{13}, e_{-1,1,0}, (q_2, s_4)) = e_{-1,1,0}, t_{13}$

(c) Graph Function Λ *for the Principle Time Graph of an Example Automaton*

$\Lambda: Y\subseteq T\times E\times V\rightarrow\sigma(V)\times\delta(V)\times\tau(V)\times\omega(V)\times\gamma(W)\times\lambda(N)$
$(t_0, e_{0,0,0}, (q_0, s_3)) \rightarrow ((f_0, d_1, c_1, q_1), e_{1,0,0}, t_1)$
$(t_{10}, e_{0,0,0}, (q_3, s_0)) \rightarrow ((f_1, d_2, c_2, q_4), e_{0,1,0}, t_{12})$
$(t_1, e_{1,0,0}, (q_1, s_0)) \rightarrow ((f_2, -d_2, c_2, q_3), e_{1,-1,0}, t_3)$
$(t_1, e_{1,0,0}, (q_1, s_2)) \rightarrow ((f_0, d_0, c_0, q_1), e_{1,0,0}, t_1)$
$(t_3, e_{1,-1,0}, (q_1, s_1)) \rightarrow ((f_2, d_3, c_2, q_2), e_{1,-1,1}, t_5)$
$(t_3, e_{1,-1,0}, (q_3, s_2)) \rightarrow ((f_1, d_0, c_0, q_1), e_{1,-1,0}, t_3)$
$(t_5, e_{1,-1,1}, (q_2, s_0)) \rightarrow ((f_0, d_0, c_0, q_3), e_{1,-1,1}, t_5)$
$(t_5, e_{1,-1,1}, (q_3, s_0)) \rightarrow ((f_1, d_2, c_2, q_4), e_{1,0,1}, t_7)$
$(t_7, e_{1,0,1}, (q_4, s_1)) \rightarrow ((f_1, -d_1, c_1, q_2), e_{0,0,1}, t_8)$
$(t_8, e_{0,0,1}, (q_2, s_2)) \rightarrow ((f_2, -d_3, c_2, q_3), e_{0,0,0}, t_{10})$
$(t_{12}, e_{0,1,0}, (q_4, s_1)) \rightarrow ((f_1, -d_1, c_1, q_2), e_{-1,1,0}, t_{13})$
$(t_{13}, e_{-1,1,0}, (q_2, s_4)) \rightarrow ((f_0, d_0, c_0, q_5), e_{-1,1,0}, t_{13})$

of R^1 and the identifying subscripts of t_i and t_{i+1} are considered as integer vectors with these entries. The information of Table 2.2(a) and 2.3(a) defines the function π of Table 2.3(b), and the graph function Λ for the time graph is defined next (Table 2.3(c)) from the information of Tables 2.1(a) and 2.3(b). As before, the graph function and the corresponding primitives establish the graph $G_{M'}(T_p)$ shown in Figure 2.4. Those points and lines drawn with solid lines are for the described conditions; those drawn with broken lines are for an imposed condition of an interacting automaton to be described later. When the time between successive moments corresponding to a time change response c_0 is zero, the condition is described by a loop in the graph.

The time graph provides a convenient means for describing a given automaton when other automata change its environment within the operating moment base. This will be demonstrated by imposing an interacting condition on the example automaton. Let any two successive encounters of a stimulus location

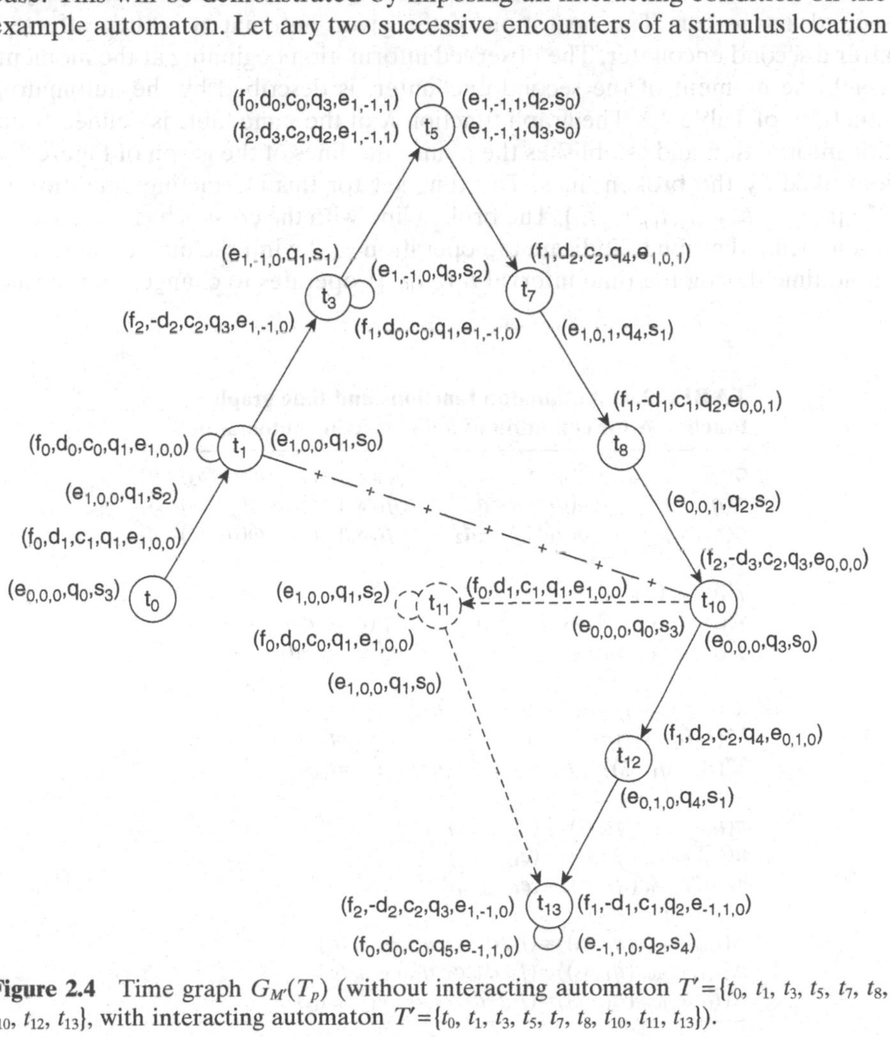

Figure 2.4 Time graph $G_{M'}(T_p)$ (without interacting automaton $T' = \{t_0, t_1, t_3, t_5, t_7, t_8, t_{10}, t_{12}, t_{13}\}$, with interacting automaton $T' = \{t_0, t_1, t_3, t_5, t_7, t_8, t_{10}, t_{11}, t_{13}\}$).

e_k by the given automaton occur at moments $m+i$ and $m+j$, $i \leq j-2$, within a given moment base M'. A second automaton changes the given automaton's environment when, between $m+i$ and $m+j$, at least one nonidentity transformation response, $f \neq s$, is produced by the second automaton at the stimulus location e_k, i.e., when, between $m+i$ and $m+j$, at least one internal stimulus is changed to a nonidentity external stimulus by the second automaton. This type of condition was imposed on the example automaton in the discussion of the environment graph, and the effect was illustrated by comparing the graphs in Figure 2.3(b) and 2.3(a) with and without the interacting automaton. This same condition will be imposed on the example automaton to illustrate the effect on the time graph.

Recall that between the two encounters of location $e_{0,0,0}$, the internal stimulus s_0 was observed to have been replaced by an external stimulus s_3 by a second automaton. The operation of the given automaton is then affected after a second encounter. The observed information beginning at the moment $m+9$, the moment of the second encounter, is described by the automaton functions of Table 2.4. The graph function Λ of the same table is defined from this information and establishes the points and lines of the graph of Figure 2.4 identified by the broken lines. The time set for this interacting condition is $T' = \{t_0, t_1, t_3, t_5, t_7, t_8, t_{10}, t_{11}, t_{13}\}$. The broken line with the crosses has been added to the defined graph to indicate the operation of the interacting automaton at some time during the time interval t_1 to t_{10}. It operates to change the internal

TABLE 2.4. Automaton functions and time graph function Λ for condition of an interacting automaton

$\sigma(q_1, s_3) = f_0$	$\delta(q_2, s_3) = d_1$	$\tau(q_0, s_3) = c_1$	$\omega(q_0, s_3) = q_1$
$\sigma(q_1, s_2) = f_0$	$\delta(q_1, s_2) = d_0$	$\tau(q_1, s_2) = c_0$	$\omega(q_0, s_2) = q_1$
$\sigma(q_1, s_0) = f_2$	$\delta(q_1, s_0) = -d_2$	$\tau(q_1, s_0) = c_2$	$\omega(q_1, s_0) = q_3$

$\gamma''(e_{0,0,0}, (q_0, s_3)) = (e_{0,0,0}, d_1)$	$\gamma'(e_{0,0,0}, d_1) = e_{1,0,0}$
$\gamma''(e_{0,0,0}, (q_1, s_2)) = (e_{1,0,0}, d_0)$	$\gamma'(e_{1,0,0}, d_0) = e_{1,0,0}$
$\gamma''(e_{1,0,0}, (q_1, s_0)) = (e_{1,0,0}, -d_2)$	$\gamma'(e_{1,0,0}, -d_2) = e_{1,-1,0}$

$\lambda''(t_{10}(q_0, s_3)) = (t_{10}, c_1)$	$\lambda'(t_{10}, c_1) = t_{11}$
$\lambda''(t_{11}(q_1, s_2)) = (t_{11}, c_0)$	$\lambda'(t_{11}, c_0) = t_{11}$
$\lambda''(t_{11}, (q_1, s_0)) = (t_{11}, c_2)$	$\lambda'(t_{11}, c_2) = t_{13}$

$\pi(t_{10}, e_{0,0,0}, (q_0, s_3)) = (e_{1,0,0}, t_{11})$
$\pi(t_{11}, e_{1,0,0}, (q_1, s_2)) = (e_{1,0,0}, t_{11})$
$\pi(t_{11}, e_{1,0,0}, (q_1, s_0)) = (e_{1,-1,0}, t_{13})$

$\Lambda(t_{10}, e_{0,0,0}, (q_0, s_3)) = (f_0, d_1, c_1, q_1, e_{1,0,0}, t_{11})$
$\Lambda(t_{11}, e_{1,0,0}, (q_1, s_2)) = (f_0, d_0, c_0, q_1, e_{1,0,0}, t_{11})$
$\Lambda(t_{11}, e_{1,0,0}, (q_1, s_0)) = (f_2, -d_2, c_2, q_3, e_{1,-1,0}, t_{13})$

stimulus s_3 required no later than moment $m+9$. Interactive automata will be studied in greater detail in a subsequent chapter.

Alternate Environment Graph. There is an alternate environment graph $G_M(E_a)$ similar to the principal time graph in that it also gives complete information on the automaton. This results from the parallelism of the functions π and μ of 2.11F and 2.12F. If Λ′ is taken as the function, then

$$\Lambda': Y' \subseteq E \times N \rightarrow (\sigma(V) \times \delta(V) \times \tau(V) \times \omega(V) \times \mu(Y')) \subseteq F \times D \times C \times Q \times T \times E \qquad \text{2.5GF}$$

(compare with Function 2.4GF), and the parallelism follows through the established primitives of the time graph. An element $(e_i, (t_i, (q_i, s_i)))$ from the domain and the image element $\Lambda'(e_i, (t_i, (q_i, s_i))) = (f_{i+1}, d_{i+1}, c_{i+1}, q_{i+1}, t_{i+1}, e_{i+1})$ form a 10-tuple $(e_i, t_i, q_i, s_i, f_{i+1}, d_{i+1}, c_{i+1}, q_{i+1}, t_{i+1}, e_{i+1})$, which is represented in the graph by two points e_i and e_{i+1}, $(e_i \neq e_{i+1})$, and a connecting line directed from e_i to e_{i+1} identified by t_i, q_i, s_i at its origin and $(f_{i+1}, d_{i+1}, c_{i+1}, q_{i+1}, t_{i+1})$ at its terminal. If $e_i = e_{i+1}$, e_i and e_{i+1} are the same point and the line is a nondirected loop with the line identification established on the basis of an assumed line direction.

Alternate Time Graph. There is an alternate time graph $G_M(T_a)$ that is similar to the principal environment graph in that the line identification is the same. This results from the parallelism of the functions λ′ defining time and γ′ defining environment (2.7F and 2.6F). In the case of time, a subset of the product space of time (instead of environment), internal states, and stimuli is mapped into the product space of response, states, and time (instead of environment); i.e., if Γ′ is taken as the function, then

$$\Gamma': N \subseteq T \times V \rightarrow (\sigma(V) \times \delta(V) \times \tau(V) \times \omega(V) \times \lambda(N)) \subseteq F \times D \times C \times Q \times T \qquad \text{2.3GF}$$

(compare with 2.2GF), and the parallelism follows through the established primitives of the environment graph. An element $(t_i, (q_i, s_i))$ from the domain and the image element $\Gamma'(t_i, (q_i, s_i)) = (f_{i+1}, d_{i+1}, c_{i+1}, q_{i+1}, t_{i+1})$ form an octuple $(t_i, q_i, s_i, f_{i+1}, d_{i+1}, c_{i+1}, q_{i+1}, t_{i+1})$ which is represented in the graph by two points t_i and t_{i+1}, $(t_i \neq t_{i+1})$, and a connecting line directed from t_i to t_{i+1} identified by (q_i, s_i) at its origin and $(f_{i+1}, d_{i+1}, c_{i+1}, q_{i+1})$ at its terminal. If $t_i = t_{i+1}$, then t_i and t_{i+1} are the same point and the line is a nondirected loop with the line identification established on the basis of an assumed line direction.

2.1.3 Select Properties of the Graph Model

A select number of properties of the graph model of a general automaton are presented dealing with the model as it relates to the automaton's sets, functions, and characteristics (Koenig and Frederick 1972). The properties are stated in the form of theorems and corollaries, which are labeled either a **C**, **P**, $\mathbf{E_p}$, or $\mathbf{T_p}$, in parentheses. The **C** denotes properties common to all three graphs, while **P**, $\mathbf{E_p}$, $\mathbf{T_p}$ denote properties for the processor graph, principal environment graph, and principal time graph, respectively. Some of the proofs are not given and may be assigned as problems.

Connective Properties: The following two theorems and corollary indicate the connectivity of the graphs in the graph model. Connectivity is used in the usual graph sense.

Theorem 2.6 (C). *The processor graph $G_L(P)$, environment graph $G_L(E_p)$, and time graph $G_L(T_p)$ for the life of a given automaton L are connected graphs.*

Recall that all associated sets with the life of an automaton contain only active elements. In particular for $G_L(P)$, all points represent active states. If $G_L(P)$ were not connected, there would exist an active state for which the processor could not enter and leave, which is contrary to the definition of active states. Equivalently, since the moment base is L, the life of the automaton, $\bar{Q}=Q$, and by primitive P3, $Q=\{q_i \mid (q_i, s_i)\in V\}\cup\{q_{i+1} \mid q_{i+1}\in \omega(V), (q_{i+1}, s_{i+1})\notin V\}$. So all points $\bar{Q}$ are either represented in ω (V), or in V as elements from equivalence classes in the natural partitioning of $\bar{Q}\times\bar{S}$ by the members of $\bar{Q}$. Thus, $G_L(P)$ is connected. Next, suppose $G_L(E_p)$ is not connected. Then, there are stimulus locations in the environment $\bar{E}$ that the automaton cannot encounter during L. But this contradicts the definition of $\bar{E}$, which is taken to be all the stimulus locations that the automaton encounters during its life. Expressed mathematically, since $\bar{E}=E$, by Primitive E_p3, $E=\{e_i \mid (e_i, (q_i, s_i))\in W\}\cup\{e_{i+1} \mid e_{i+1}\in \gamma(W), (e_{i+1}, (q_i, s_i))\notin W\}$. So all points $\bar{E}$ are either represented in Γ (W) or in W as elements from equivalence classes in the natural partitioning of $\bar{E}\times V$ by the members of $\bar{E}$. Thus, $G_L(E_p)$ is connected. Finally, $G_L(T_p)$ is connected, for the points correspond to absolute times from the beginning of the automaton's operation throughout its life, and lines must necessarily join these points to reflect the passing from one time to the next. Since $\bar{T}=T$, by Primitive T3, $T=\{t_i \mid (t_i, (e_i, (q_i, s_i)))\in Y\}\cup\{t_{i+1} \mid t_{i+1}\in \lambda(N), (t_{i+1}, (e_i, (q_i, s_i)))\notin Y\}$. All points $\bar{T}$ are either represented in Λ (Y), or in Y as elements from equivalence classes in the natural partitioning of $\bar{T}\times W$ by the members of $\bar{T}$. Thus, G_L (T_p) is also connected.

Corollary 2.6 (C). *Let M be a moment base, and $M\subseteq L$. Suppose no information is given for any other moment base M′ properly containing M. Then, G_M (P), G_M (E_p), G_M (T_p) are connected graphs.*

This follows immediately from Theorem 2.6. Since the graphs are unknown for any moment base M' properly containing M, then M can be treated as L in the sense of Theorem 2.6.

Theorem 2.7 (C). *Let M' be a moment base, $m \in M'$, $M' \subseteq L$. Suppose $G_{M'}(P)$, $G_{M'}(E_p)$, $G_{M'}(T_p)$ for an automaton are given. If a moment base M, properly contained in M' with $m \in M'$, is chosen so that there are latent states, latent stimulus locations, and latent times, the resulting graphs $G_{M'|M}(P)$, $G_{M'|M}(E_p)$, and $G_{M'|M}(T_p)$, which denote the graphs for M' restricted to M, are disconnected proper subgraphs of $G_{M'}(P)$, $G_{M'}(E_p)$, $G_{M'}(T_p)$, respectively.*

Let Q' and S' denote the sets of active states and active stimuli for the given automaton during moment base M'. Denote by V' the dyadic relation on $Q' \times S'$ determined by the automaton. More specifically, there exists functions θ_1 and θ_2 mapping a subset of M' onto Q' and S'; i.e., $\theta_1 : M' \to Q$ and $\theta_2 : M' - \{m+h\} \to S$, and $m+h$ is the last moment if one exists. Then there is a new function $\theta_v : M' - \{m+h\} \subseteq M' \to V$ by $\theta_v(m+i) = (\theta_1(m+i), \theta_2(m+i))$. With M a proper subset of M', then $M \subseteq M' - \{m+h\} \subseteq M'$. Denote the restrictions of θ_1, θ_2, θ_v to M by $\theta_1 \mid M$, $\theta_2 \mid M$, and $\theta_v \mid M$. Then, $\theta_1 \mid M(M) = Q \subseteq Q'$, $\theta_2 \mid M(M) = S \subseteq S'$, $\theta_v \mid M(M) = V \subseteq V'$. In particular, Q and S denote the sets of active states and stimuli for the given automaton during the moment base $M \subseteq M'$. Now, by hypothesis, there are latent states so $Q' - Q \neq \phi$, where ϕ is the empty set. Using the map $\theta_v \mid M$, $V' - V \neq \phi$. Then, applying the graph function Ω, $\Omega(V') \neq \Omega(V)$. By Primitive P4, the $q_0 \in Q' - Q$ are points with no lines; i.e., isolate points. Thus, the processor graph $G_{M'|M}(P)$, read as "the processor graph for M' restricted to M," is disconnected. By Corollary 2.6, $G_{M'}(P)$ is a connected graph, so $G_{M'|M}(P)$ is a proper subgraph of $G_{M'}(P)$; i.e., there are lines in $G_{M'}(P)$ not found in $G_{M'|M}(P)$. Similarly, the argument is used to show $G_{M'|M}(E_p)$ and $G_{M'|M}(T_p)$ are disconnected proper subgraphs of $G_{M'}(E_p)$ and $G_{M'}(T_p)$.

The last two results are illustrated by the example automaton (Figures 2.2, 2.3, 2.4). Recall that the moment base was $M' = \{m, m+1, ---, m+12\}$, and all sets contained only active elements. The processor graph $G_{M'}(P)$ (Figure 2.2(a)) is a connected graph, whereas the graph $G_{M'|M}(P)$ (Figure 2.2(b)) corresponding to the moment base $M = \{m, m+1, ---, m+4\}$ is a disconnected proper subgraph of $G_{M}'(P)$. Note that the states q_2, q_4, q_5 are latent states; s_1 and s_4 are latent stimuli; and (f_2, d_3, c_2), $(f_2, -d_3, c_2)$, (f_1, d_2, c_2), $(f_1, -d_1, c_1)$ are latent responses with respect to moment base M. With the environment in 3-space and with external stimuli $s_3, s_2, s_2, s_0, s_1, s_2, s_1, s_4$ in locations respectively referenced by the vectors (0, 0, 0), (1, 0, 0), (1, –1, 0), (1, –1, 1), (1, 0, 1), (0, 0, 1), (0, 1, 0), (–1, 1, 0), the corresponding environment graph $G_{M''}(E_p)$ (Figure 2.3(a)) is also a connected graph. Although $G_{M'|M}(E_p)$ is not presented, it is clear that it would be a disconnected proper subgraph of $G_{M'}(E_p)$ with isolate points $e_{1,-1,1}$, $e_{1,0,1}$, $e_{0,0,1}$, $e_{0,1,0}$, $e_{-1,1,0}$, representing latent stimulus locations. Similarly, the time graph $G_{M'}(T_p)$ (Figure 2.4) is a connected graph, and although $G_{M'|M}(T_p)$ is not presented, it would be a disconnected proper subgraph of $G_{M'}(T_p)$ with isolate points $t_5, t_7, t_8, t_{10}, t_{12}, t_{13}$ representing latent times.

The graphs in the graph model are either finite graphs or infinite graphs. Knowing that the graphs are finite does not imply that the moment base used to describe the operation of an automaton is a finite set. However, if the moment base is a finite set, the corresponding graphs are finite. For the purpose of discussion of the remaining properties, it is assumed the graphs are finite graphs.

Bounds for the Number of Points and Lines, In-degree and Out-degree: The following theorems and corollaries deal with the bounds for the graphs of the automaton model.

Theorem 2.8 (P). *Let* $|\bar{Q}|=m+1$ *and* $|\bar{S}|=n+1$ *denote the maximum number of states and stimuli for a given automaton. The number of points in* $G(P)$ *is* $|Q'|$ *and* $|Q'|\leq m+1$. *The number of lines in* $G\ (P)$ *is* $|U|$, *and* $|U|=|V|\leq(m+1)(n+1)$. *Furthermore, if* $o\ (q_i)$ *denotes the out-degree of point* q_i, *then* $o\ (q_i)\leq n+1$.

Since the points of $G\ (P)$ correspond directly to the internal states of the processor, it is clear from Primitive P1 that the number of points in the graph is $|Q'|$, and $|Q'|\leq|\bar{Q}|=m+1$. By definition of the graph function Ω, the number of lines is $|U|=|V|$ since V is the domain for Ω. With $V\subseteq\bar{Q}\times\bar{S}$, then $|U|=|V|\leq|\bar{Q}|\ |\bar{S}|=(m+1)\ (n+1)$. It follows immediately from the natural partitioning by states that for any arbitrary equivalence class $q_i\times\bar{S}$ if $q_i\times S\in V$, where $S\subseteq\bar{S}$, then $|q_i\times S|\leq|q_i\times\bar{S}|$. Hence

$$o(q_i)\leq|q_i\times\bar{S}|=n+1$$

Corollary 2.8.1 (P). *Let* Q *and* S *denote the sets of active states and active stimuli associated with a given moment base* M *for a given automaton. Then the number of lines is* $|U|\leq|Q|\ |S|$. *Furthermore,* $|U|<|Q|\ |S|$ *if the processor* P *cannot receive all stimuli in every state.*

Corollary 2.8.2 (P). *If within a given moment base* M *every active state receives all possible active stimuli, then* $V=Q\times S$. *The number of lines in* $G_M(P)$ *is* $|U|=|Q|\ |S|$; *the out-degree of every connected point is* $o\ (q_i)=|S|$; *and* $|M|\geq|Q|\ |S|$.

Theorem 2.9 (P). *Let moment base* $M=\{m, m+1, ---, m+h\}$ *for an operating automaton* $\boldsymbol{a}$. *The connected set of points in* $G_M(P)$ *corresponds to the active states* Q *of* $\boldsymbol{a}$ *during* M, *and* $|Q|\leq h+1$. *The set* S *of active stimuli is* $|S|\leq h$. *The number of lines in* $G_M(P)$ *is* $|U|=|V|\leq h$. *Moreover, if* $|Q|=h+1$, *there is an absolute initial state and an absolute final state in* $\boldsymbol{a}$. *If* $|S|=h$ *and if there is an absolute initial state and an absolute final state, there is an absolute initial stimulus and an absolute final stimulus for the automaton.*

By previous remarks, only active states correspond to connected points. Since there are $h+1$ moments in M, there can be at most $h+1$ active states and h active stimuli. Thus $|Q|\leq h+1$, $|S|=h$, and the number of points in $G_M(P)$ is, at most, $h+1$. There can be no line at the final moment $m+h$, so $|U|\leq h$.

Note that $|U|<h$ if one or more lines are traversed more than once during M. If $|Q|=h+1$, the number of lines is h, and every point has out-degree one except the point representing the final state and in-degree one except the point representing the initial state. Since the point representing the final state has out-degree zero, the final state is an absolute final state, and since the point representing the initial state has in-degree zero, the initial state is an absolute initial state. Finally, if $|S|=h$, there are h lines, and every stimulus is received only once. Thus, the existence of the absolute states implies the existence of absolute stimuli.

Theorem 2.10 (P). *Assuming a given automaton **a** has more than one state, an absolute initial state is represented in $G_M(P)$ by a point with i loops and one directed line originating from it, where $0 \leq j < |S|$. Similarly, an absolute final state is represented by a point with j loops, and one directed line terminating on it, where $0 \leq j < |S|$.*

By definition, an absolute initial state can never be re-entered once left. From the automaton function ω, the initial state must enter another state. Thus, there exists one directed line originating from the point representing the absolute initial state in $G_M(P)$. This, however, does not preclude loops which are bounded in number by $|S|-1$ by the single-valueness of ω. Similarly, the result for the absolute final state follows:

Corollary 2.10 (P). *In Theorem 2.10, if S contains an absolute initial stimulus and an absolute final stimulus, then $0 \leq i < |S|-2$ and $0 \leq j < |S|-2$.*

Theorem 2.11 (P). *Let ω (V) and Ω (V) denote the range sets under the automaton function ω and the graph function Ω. Then ω (V) represents the set of points in G (P) with in-degree greater than zero and $|\omega(V)| \leq |\Omega(V)| \leq |V|$.*

By the single-valueness of ω and Ω, $|\omega(V)| \leq |V|$, $|\Omega(V)| \leq |V|$. But $\Omega(V)$ is a tetradic relation on $\sigma(V) \times \delta(V) \times \tau(V) \times \omega(V)$ of the form $((f_{i+1}, d_{i+1}, c_{i+1}), \omega(V))$, hence $|\Omega(V)| \geq |\omega(V)|$. The fact that $\omega(V)$ represents the set of points with in-degree greater than zero follows directly from Ω; i.e., $\Omega(q_i, s_i) = (\sigma(q_i, s_i), \delta(q_i, s_i), \tau(q_i, s_i), \omega(q_i, s_i))$, so that the set of all $\omega(q_i, s_i)$, namely $\omega(V)$, have in-degree greater than zero. Note that although a loop is undirected, a point with only a loop is said to have in-degree one and out-degree one, and direction may be arbitrarily assumed.

Corollary 2.11 (P). *The automaton function ω maps $V \subseteq Q \times S$ into the set Q, hence $\omega(V) \subseteq Q$. If $|\omega(V)| < |Q|$, then $|\omega(V)| = |Q|-1$, and the automaton has an absolute initial state.*

The example automaton (see Table 2.1 and Figure 2.2) can be used to illustrate the results dealing with the processor graph G (P). In the example, $|Q|=6$, $|S|=5$, and since Q represents only active states, the number of points in $G_{M'}(P)$ is $|Q'|=|Q|=6$. The number of lines is $|U|=|V|=10$ (Table 2.1(b)), and the inequality of Corollary 2.8 is trivially satisfied since $|V|=10<|Q|\,|S|=30$. With $|M'|=13$, the inequality expressed in Theorem 2.9 holds for the example in that $|U|=10<|M'|-1=12$. In addition, the maximum out-degree occurs at point q_1 and q_2, where $o(q_1)=o(q_2)=3<|S|=5$. The set of points with in-degree greater than zero is $\omega(V)=\{q_1, q_2, q_3, q_4, q_5\}=Q-\{q_0\}$ (Table 2.1 (a)), $|\omega(V)|=|Q|-1$, and the point $q_0 \notin \omega(V)$ represents an absolute initial state.

If the principal environment graph is restricted to a finite graph, the following properties correspond to those of the processor graph. Essentially, all that is necessary is to replace the sets given in the primitives for the processor graph by sets in the primitives for the environment graph. Theorems 2.12–2.15 and corollaries for the principal environment graph correspond to Theorems 2.8–2.11 and corollaries for the processor graph. Their proofs are left to the exercises.

Theorem 2.12 ($\mathbf{E_p}$). *Let $\bar{E}=\{e_0, e_1, ---, e_x\}$, so $|\bar{E}|=x+1$. The number of points in $G(E_p)$ is $|E'|$, and $|E'|\leq x+1$. The number of lines in $G(E_p)$ is $|X|$, and $|X|=|W|\leq(x+1)\,(m+1)\,(n+1)$. Furthermore, if $o\,(e_i)$ denotes the out-degree of point e_i, then $o\,(e_i)\leq(m+1)\,(n+1)$.*

Corollary 2.12.1 ($\mathbf{E_p}$). *Let E, Q, S denote the sets of active stimulus locations, active states, and active stimuli associated with a given moment base $M\subseteq L$ for a given automaton. Then the number of lines is $|X|\leq|E|\,|V|$. Furthermore, $|X|<|E|\,|V|$ if the processor P doesn't encounter all stimuli at every location during M, or P is not in all states when receiving all possible stimuli from every location in E during M.*

Corollary 2.12.2 ($\mathbf{E_p}$). *If within a given moment base M, every active stimulus location contains all possible active stimuli and the processor is in all states when receiving all stimuli from every possible location, then $W=E\times V$. The number of lines in $G_M(E_p)$ is $|X|=|E|\,|V|$, the out-degree of every connected point is $|V|$, and $|M|\geq|E|\,|V|$.*

Theorem 2.13 ($\mathbf{E_p}$). *Let a moment base $M=\{m, m+1, ---, m+h\}$ for an operating automaton **a**. The connected set of points in $G_M(E_p)$ correspond to the active stimulus locations E for **a** during M, and $|E|\leq h+1$. The set of active pairs of states and stimuli is $|V|\leq h$, and the number of lines in $G_M(E_p)$ is $|X|=|W|\leq h$. Moreover, if $|E|=h+1$, there is an absolute initial environment and an absolute final environment of **a** during M.*

Theorem 2.14 ($\mathbf{E_p}$). *Assuming the environment for a given automaton **a** contains more than one location, an absolute initial stimulus location is represented in $G_M(E_p)$ by a point with i loops and one directed line originating from it, where $0\leq i<|V|$. Similarly, an absolute final stimulus location is represented by a point with j loops and one directed line terminating on it, where $0<j<|V|$.*

Corollary 2.14 ($\mathbf{E_p}$). *If the automaton **a** has an absolute initial state and stimulus and an absolute final state and stimulus, the bounds expressed in Theorem 2.14 are* $0\leq i<|V|-2$ and $0\leq j<|V|-2$.

Theorem 2.15 (E_p). *Let $\gamma(W)$ and $\Gamma(W)$ denote the range sets under the automaton function γ and the graph function Γ. Then $\gamma(W)$ represents the set of points in $G(E_p)$ with in-degree greater than zero and $|\gamma(W)| \leq |\Gamma(W)| \leq |W|$.*

Corollary 2.15 (E_p). *The automaton function γ maps $W \subseteq E \times V$ into the set E, hence $\gamma(W) \subseteq E$. If $|\gamma(W)| < |E|$, then $|\gamma(W)| = |E| - 1$, and the automaton has an absolute initial environment.*

Again, referring to the example automaton (Table 2.2 and Figure 2.3(a)), the results stated herein dealing with the principal environment graph can be illustrated by the example. The number of stimulus locations is $|E| = 8$, and the number of state-stimulus pairs is $|V| = 10$. Since E represents only active states for $M' = \{m, m+1, ---, m+12\}$, the number of points in $G_{M'}(E_p)$ is $|E'| = |E| = 8$. The number of lines is $|X| = |W| = 12$ (Table 2.2(b)), and the inequality of Corollary 2.12.1 holds as $|W| = 12 < |E|\,|V| = 80$. With $|M'| = 13$, the inequality expressed in Theorem 2.13, namely $|X| \leq |M'| - 1$, is satisfied since $|X| = 12 = |M'| - 1$. The maximum out-degree occurs at several points and is 2, which is less than $|V| = 10$. $\gamma(V) = E$, so $|\gamma(V)| = |E| = 8$. And all the points have in-degree greater than zero. Point $e_{-1,1,0}$ does represent an absolute final stimulus location with one loop, and the bound given in Theorem 2.14, namely $0 \leq i < |V|$, holds.

If the time graph is a finite graph, the following properties are observed:

Theorem 2.16 (T_p). *Let $\bar{T} = \{t_0, t_1, ---, t_y\}$, so $|\bar{T}| = y+1$. The number of points in the time graph $G(T_p)$ is $|T'| \leq y+1$. Let $T \subseteq T' \subseteq \bar{T}$, and let T denote the set of active times. Then the number of lines is $|Z| = |Y|$, and $|T| - 1 \leq |Z| \leq |T|\,|W| \leq (y+1)\ (x+1)\ (m+1)\ (n+1)$. Furthermore, each point of $G\ (T_p)$ not representing an initial or final time is a point with one terminating line, one originating line, and n loops where $0 \leq n \leq |Z| - |T| + 1$.*

The points of $G(T_p)$ correspond to absolute times associated with the moments used in describing the action of the automaton, so $|T'| \leq |\bar{T}| = y+1$. Every point associated with an active time, excluding the final time, is joined by a line to a point representing the next active time. Thus, there must be at least $|T| - 1$ lines, where T denotes the set of points corresponding to active times. From the definition of the graph function Δ, the number of lines is the same as the cardinality of the domain set Y, so $|Z| = |Y| \leq |T|\,|W| \leq (y+1)\ (x+1)\ (m+1)\ (n+1)$. The null time change response is reflected in $G(T_p)$ by the presence of loops on the points. Since there are $|T| - 1$ directed lines, the number of loops on any arbitrary point is $0 \leq n \leq |Z| - |T| + 1$.

Theorem 2.17 (T_p). *Let a moment base $M = \{m, m+1, ---, m+h\}$ be used in describing an operating automaton. The connected set of points in $G_M(T_p)$ correspond to the active absolute times T, and $|T| \leq h+1$. The set of active triples of environments, states, and stimuli is $|W| \leq h$, and the number of lines is $|Z| = |Y| \leq h$.*

With $h+1$ moments in M, there are at most $h+1$ active absolute times associated with M, so $|T| \leq h+1$. From Theorem 2.13, $|W| \leq h$, hence the set of active triples of environments, states, and stimuli is bounded by h. With no line at the last moment $m+h$ by definition, the number of lines is bounded by h; i.e., $|Z| \leq h$.

Again, the example automaton (see Table 2.3 and Figure 2.4) can illustrate some properties of the time graph. The number of points is $|T'|=|T|=9$. Since $|M'|=13$, the number of lines in $G'_M(T_p)$, namely $|Z|=|Y|=12$, satisfies the inequality $|T|-1 \leq |Z| \leq h$ as $8<12=12$. Note that four points each contain a single loop with no point containing more than one loop. Thus, the inequality $0 \leq n \leq |Z|-|T|+1$ is satisfied as $0<1<4$.

A natural extension of the analysis for a general automaton is the development and application of recursive methods for determining the effective operation of a general automaton through select properties of these established graphs. The material follows the work of Frederick and Koenig (1971a) and is presented in Appendix A.

2.2 AN APPLICATION OF THE DISCIPLINES TO THE MODELING OF NATURAL AUTOMATA

With the disciplines of 2.1, it is proposed that natural automata can be modeled for certain recognized mental abilities and that these abilities can be imparted to artificial automata (Koenig 1973–74, 1978, 1984).

Automata are observed to observe automata performing over time intervals, and over later time intervals, are observed to perform like automata they observed over previous time intervals. From this observation, three inherited abilities are identified.

1. The ability to structure in nonshared environments (memory) knowledge about automata and histories of automata from observations of the performing automata and of environments.
2. The ability to operate on knowledge structures in nonshared environments (memory) and to produce in shared environments like responses of automata described by the knowledge structures using the effector organs for locomotion and manipulation.
3. The ability to modify and extend knowledge structures with a minimal of additional knowledge from shared environments.

The principles involved in the modeling of these abilities are demonstrated by a case study, and a theorem is presented for establishing a set of states for an observed automaton when the states are not observable and are not included in an environment graph obtained by observation. There will be further discussion of the modeling for these abilities in later chapters.

2.2.1 A Case Study

There are given a teacher, a student and a chalkboard. Also given are two 2-digit numbers on the chalkboard, one above the other. Over a time interval, the teacher is to teach the student how to multiply two 2-digit numbers by performing the multiplication on the numbers on the board. The top number is the multiplicand and the lower number is the multiplier. Both teacher and student have knowledge for the multiplication and addition of two 1-digit numbers.

Let the teacher with a piece of chalk be described as an automaton a_1, and let the student be described as an observing automaton a_0. The chalkboard is the set of shared environments in which a_1 operates receiving stimuli and recording responses. a_1 is included in the environment of a_0. The principal and auxiliary receptors and effectors are integral parts of the structures of the automaton; e.g., the eyes and eyeglasses (if they are worn) are integral parts of a_1 and a_0, and the hand and the chalk held by the hand are integral parts of a_1. Each digit of the two numbers is identified with an environment location.

a_0 observes a_1 performing in the shared environments multiplying the numbers over a moment base M. At each moment $m+i \in M$ an important observation is made, and at the end of the time interval, the student has stored in his nonshared environments multiple patterns similar to the frames of a movie film. He is able to visualize the teacher as he operated in the shared environments. The two numbers that appeared on the chalkboard at the beginning of the time interval (at moment $m \in M$), and those that are on the board at the end of the period (at the final moment $m+h \in M$) appear in the student's book. What happened during the intervening moments could not be determined from the book. Over some later time interval, the student performs like the teacher in the multiplication of two 2-digit numbers (not necessarily the same numbers) using effector organs for locomotion and manipulation.

What a_0 observed of a_1 operating over the time interval and what he has stored in his nonshared environments can be described by the principal environment graph in application of the disciplines of Section 2.1. This is done by substituting specific sets of elements for the universal sets in the primitives that define the principal environment graph of the general automaton. The graph function (2.2GF) is Γ and an element $(e_i, (q_i, s_i))$ from the domain and the image $\Gamma(e_i, (q_i, s_i))=(f_{i+1}, d_{i+1}, c_{i+1}, q_{i+1}, e_{i+1})$ form a graph octuple $(e_i, q_i, s_i, f_{i+1}, d_{i+1}, c_{i+1}, q_{i+1}, e_{i+1})$. The automaton a_1 was included as part of the environment of a_0 during the first time interval, but in this case, the states q_i of a_1 could not be observed. So the environment graph without the states represents what a_0 observed of a_1 performing on the chalkboard. The graph is shown in Figure 2.5, and the Table 2.5 shows the graph octuples with the state elements blank, where the line numbers given in the table correspond to the numbers associated with the graph of the figure. The environment functions

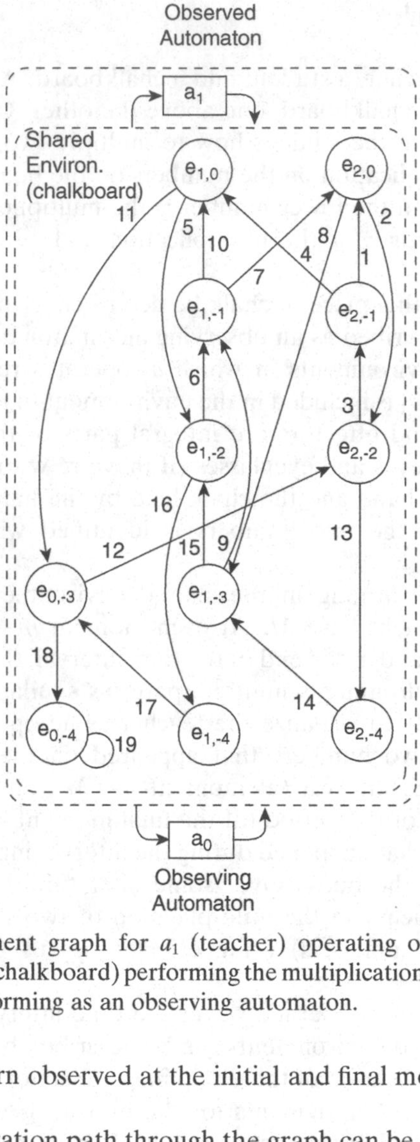

Figure 2.5 Environment graph for a_1 (teacher) operating over a time interval in shared environments (chalkboard) performing the multiplication of two 2-digit numbers with a_0 (student) performing as an observing automaton.

describing the pattern observed at the initial and final moments are shown in Table 2.6.

An effective operation path through the graph can be described as a point, line, point sequence corresponding to moments $(m, m+1)$, $(m+1, m+2)$, ---, $(m+(h-1), m+h)$. The calculation of this sequence is a mathematical representation of a_0 operating on the knowledge structure in his nonshared environments to produce in his shared environments like responses of a_1 using his effector organs of manipulation.

The mathematical development of the set of disciplines of 2.1 allows for the modeling of the inherited ability to modify and extend knowledge structures with a minimal of additional knowledge from shared environments. This is demonstrated in part for this case study. Recall that the states of a_1 could

TABLE 2.5. Graph octuples for principle environment graph of Figure 2.5 with undetermined state elements

Line No.	Moments	Graph Octuples
1	$(m, m+1)$	$e_{2,-1}$, __, s_4, f_4, $d_{0,1}$, c_1, __, $e_{2,0}$
2	$(m+1, m+2)$	$e_{2,0}$, __, s_2, f_2, $d_{0,-2}$, c_2, __, $e_{2,-2}$
3	$(m+2, m+3)$	$e_{2,-2}$, __, s_0, f_6, $d_{0,1}$, c_1, __, $e_{2,-1}$
4	$(m+3, m+4)$	$e_{2,-1}$, __, s_4, f_4, $d_{-1,1}$, c_1, __, $e_{1,0}$
5	$(m+4, m+5)$	$e_{1,0}$, __, s_1, f_1, $d_{0,-2}$, c_2, __, $e_{1,-2}$
6	$(m+5, m+6)$	$e_{1,-2}$, __, s_0, f_5, $d_{0,1}$, c_1, __, $e_{1,-1}$
7	$(m+6, m+7)$	$e_{1,-1}$, __, s_3, f_3, $d_{1,1}$, c_1, __, $e_{2,0}$
8	$(m+7, m+8)$	$e_{2,0}$, __, s_2, f_2, $d_{-1,-3}$, c_3, __, $e_{1,-3}$
9	$(m+8, m+9)$	$e_{1,-3}$, __, s_0, f_8, $d_{0,2}$, c_2, __, $e_{1,-1}$
10	$(m+9, m+10)$	$e_{1,-1}$, __, s_3, f_3, $d_{0,1}$, c_1, __, $e_{1,0}$
11	$(m+10, m+11)$	$e_{1,0}$, __, s_1, f_1, $d_{-1,-3}$, c_3, __, $e_{0,-3}$
12	$(m+11, m+12)$	$e_{0,-3}$, __, s_0, f_7, $d_{2,1}$, c_2, __, $e_{2,-2}$
13	$(m+12, m+13)$	$e_{2,-2}$, __, s_6, f_6, $d_{0,-2}$, c_2, __, $e_{2,-4}$
14	$(m+13, m+14)$	$e_{2,-4}$, __, s_0, f_{11}, $d_{-1,1}$, c_1, __, $e_{1,-3}$
15	$(m+14, m+15)$	$e_{1,-3}$, __, s_8, f_8, $d_{0,1}$, c_1, __, $e_{1,-2}$
16	$(m+15, m+16)$	$e_{1,-2}$, __, s_5, f_5, $d_{0,-2}$, c_2, __, $e_{1,-4}$
17	$(m+16, m+17)$	$e_{1,-4}$, __, s_0, f_{10}, $d_{-1,1}$, c_1, __, $e_{0,-3}$
18	$(m+17, m+18)$	$e_{0,-3}$, __, s_7, f_7, $d_{0,-1}$, c_1, __, $e_{0,-4}$
19	$(m+18, m+19)$	$e_{0,-4}$, __, s_0, f_9, $d_{0,0}$, c_0, __, $e_{0,-4}$

TABLE 2.6. Environment functions

m	$m+h$
$\phi: E \to \phi_0(E)$	$\phi_h: E \to \phi_h(E)$
$\phi_0(e_{1,0}) = s_1$	$\phi_h(e_{1,0}) = s_1$
$\phi_0(e_{2,0}) = s_2$	$\phi_h(e_{2,0}) = s_2$
$\phi_0(e_{1,-1}) = s_3$	$\phi_h(e_{1,-1}) = s_3$
$\phi_0(e_{2,-1}) = s_4$	$\phi_h(e_{2,-1}) = s_4$
$\phi_0(e_{1,-2}) = s_0$	$\phi_h(e_{1,-2}) = s_5$
.	$\phi_h(e_{2,-2}) = s_6$
.	$\phi_h(e_{0,-3}) = s_7$
.	$\phi_h(e_{1,-3}) = s_8$
.	$\phi_h(e_{0,-4}) = s_9$
.	$\phi_h(e_{1,-4}) = s_{10}$
$\phi_0(e_{2,-4}) = s_0$	$\phi_h(e_{2,-4}) = s_{11}$

not be observed by a_0. The hierarchical development of the three types of graphs of the model in the sequence, processor, environment, and time requires that if the states are to be determined, they must be determined through the mathematical development of the processor graph whose points are states. Basic to the primitives that define the processor graph is the processor graph function Ω. It is the single-valueness of this function that enables the establishment of the processor graph. First, it is established from Theorem 2.6 that the processor graph is a connected graph. Before a previous state is reentered for a current moment, it must be determined whether the rule for single-valueness of the graph function will be violated at some later moment.

2.2.2 Required State Changes

The following theorem presents conditions when changes of state are required. It is then applied to the case study of Section 2.2.1.

Theorem 2.18. *Given an observed effective operation path of a principal environment graph $G_M(E_p)$ for an automaton that has performed over a moment base $M=\{m, m+1, ---, m+h\}$ and without a determined set of states. Let the observed set of stimuli and triadic responses over M be denoted by S and R and let the undetermined set of states be denoted by Q. Let the construction of the processor graph proceed from the sextuples (__, s_i, (f_{i+1}, d_{i+1}, c_{i+1}), __) of the lines of the environment graph taken in sequence following the observed effective operation path, and let the set of lines be denoted by U. Let the set of points already established for the processor graph for moments (m, $m+1$, ---, $m+i$) be denoted by $Q_i \subseteq Q$, and let $U_i \subseteq U$ denote the set of lines for moments (m, $m+1$), ($m+1$, $m+2$), ---, ($m+i$, $m+(i+1)$) directed from the points of the set Q_i. Let a quadruple (s_i, (f_{i+1}, d_{i+1}, c_{i+1})) for moments ($m+i$, $m+(i+1)$) denote a current line to be drawn from point $q_i \in Q_i$ for moment $m+i$ to some next undetermined point q_{i+1} for moment $m+(i+1)$, and let the processor graph, to be completed, be described by the effective operation path defined by a point, line, point sequence with undetermined points __, (s_{i+1}, $(f, d, c)_{i+2}$), __, (s_{i+2}, $(f, d, c)_{i+3}$), __, ---, __, (s_{h+1}, $(f, d, c)_h$), __ corresponding to moments ($m+(i+1)$, $m+(i+2)$), ($m+(i+2)$, $m+(i+3)$), ---, ($m+(h-1)$, $m+h$). Let r_{i+n} denote a triadic response $(f, d, c)_{i+n}$. For $0 \leq i \leq (h-2)$, $2 \leq n \leq (h-1)$*

Case 1: $A_{11}, A_{12} \vdash B$.

A_{11}: *Assume $s_{i+1} \in S$ of the quadruple (s_{i+1}, $(f, d, c)_{i+2}$) for moments ($m+(i+1)$, $m+(i+2)$) appears in a quadruple for a line directed from each of the points of the set $Q_i \subseteq Q$, and let U_{i+1} denote the set of these lines, $|U_{i+1}| = |Q_i|$, $U_{i+1} = \{s_{i+1} \times R_{i+1}\} \subseteq U_i \subseteq U$, where $R_{i+1} = \{r_k \mid r_k$ is a triadic response $(f, d, c)_k$, and r_k appears with s_{i+1} in a quadruple for a line directed from each of the points of the set $Q_i \subseteq Q\} \subseteq R$.*

A_{12}: *Furthermore, assume $r_{i+2} \notin R_{i+1}$.*

B: *Then the choice must be to direct a current line (s_i, $(f, d, c)_{i+1}$) from $q_i \in Q_i$ for moment $m+i$ to a new point $q_{i+1} \in Q - Q_i$ for moment $m+(i+1)$.*

Case 2: $A_{11}, A_{21}, A_{22}, A_{23} \vdash B$.

A_{11}: *See Case 1.*

A_{21}: *Assume $r_{i+2} \in R_{i+1}$, and let the set of lines for $r_{i+2} \in R_{i+1}$ be denoted by $U'_{i+1} \subseteq U_{i+1}$. Let the set of points* ***from*** *which these lines are directed be denoted by $Q'_{i+1} \subseteq Q_i$, and the set of points* ***to*** *which these lines are directed be $Q_{i+2} \subseteq Q_i$. For $q_i \in Q'_{i+1}$, let $q_i \in Q_{i+2}$.*

A_{22}: *Assume $s_{i+2} \in S$ of the quadruple $(s_{i+2}, (f, d, c)_{i+3})$ for moments $(m+(i+2), m+(i+3))$ appears in a quadruple for a line directed from each of the points of the set $Q_{i+2} \subseteq Q_i$, and let U_{i+2} denote the set of these lines, $|U_{i+2}| = |Q_{i+2}|$, $U_{i+2} = \{s_{i+2} \times R_{i+2}\} \subseteq U_i$, where $R_{i+2} = \{r_k \mid r_k = (f, d, c)_k$ appears with s_{i+2} in a quadruple for a line directed from each of the points of the set $Q_{i+2} \subseteq Q_i\} \subseteq R$.*

A_{23}: *Furthermore, assume $r_{i+3} \notin R_{i+2}$.*

B: *See Case 1.*

Case n: A_{11}, A_{21}, A_{22}, ---, $A_{n1}, A_{n2}, A_{n3} \vdash B$.

A_{11}: *See Case 1.*

A_{21}: *See Case 2.*

A_{22}: *See Case 2.*

A_{n1}: *Assume $r_{i+n} \in R_{i+(n-1)}$, and let the set of lines for $r_{i+n} \in R_{i+(n-1)}$ be denoted by $U'_{i+(n-1)} \subseteq U_{i+(n-1)}$. Let the set of points* ***from*** *which these lines are directed be denoted by $Q'_{i+(n-1)} \subseteq Q_{i+(n-1)} \subseteq Q_i$, and the set of points* ***to*** *which these lines are directed be $Q_{i+n} \subseteq Q_i$. For $q_i \in Q'_{i+(n-1)}$, let $q_i \in Q_{i+n}$.*

A_{n2}: *Assume $s_{i+n} \in S$ of the quadruple $(s_{i+n}, (f, d, c)_{i+(n+1)})$ for moments $(m+(i+n), m+(i+(n+1)))$ appears in a quadruple for a line directed from each of the points of the set $Q_{i+n} \subseteq Q_i$,* and let *U_{i+n} denote the set of these lines, $|U_{i+n}| = |Q_{i+n}|$, $U_{i+n} = \{s_{i+n} \times R_{i+n}\} \subseteq U_i$, where $R_{i+n} = \{r_k \mid r_k = (f, d, c)_k$ appears with s_{i+n} in a quadruple for a line directed from each of the points $Q_{i+n} \subseteq Q_i\} \subseteq R$.*

A_{n3}: *Furthermore, assume $r_{i+(n+1)} \notin R_{i+n}$.*

B: *See Case 1.*

Case h−1: A_{11}, A_{21}, A_{22}, ---, A_{n1}, A_{n2}, ---, $A_{(h-1)1}, A_{(h-1)2}, A_{(h-1)3} \vdash B$.

From the primitives of the processor graph of 2.1, there is the processor graph function Ω whose domain is $V \subseteq Q \times S$ and whose range is $\Omega(V) \subseteq (\sigma(V) \times \delta(V) \times \tau(V) \times \omega(V)) \subseteq F \times D \times C \times Q$. The function Ω is defined by $\Omega(q_i, s_i) = (\sigma(q_i, s_i), \delta(q_i, s_i), \tau(q_i, s_i), \omega(q_i, s_i)) = ((f, d, c)_{i+1}, q_{i+1})$. The set Q of elements is the set of points, and the set of lines U is all the $(s_i, (f, d, c)_{i+1})$ in the sextuples $(q_i, s_i, (f, d, c)_{i+1}, q_{i+1})$ such that $\Omega(q_i, s_i) = ((f, d, c)_{i+1}, q_{i+1})$. A subset $U_0 \subseteq U$ is a set of directed lines if for $u \in U_0$, $u=(s_i, (f, d, c)_{i+1})$, the point q_i associated with the element (q_i, s_i) from the domain is not equal to the point q_{i+1} in the image quadruple $\Omega\ (q_i, s_i)=((f, d, c)_{i+1}, q_{i+1})$. The line u is said to be directed from point q_i to point q_{i+1}. Consider moments $(m, m+1)$, $(m+1, m+2)$, ---, $(m+i, m+(i+1))$ for the effective operation path of the processor graph and let $V_i \subseteq Q_i \times S_i \subseteq Q \times S$ be the set of elements in the domain of the processor graph function Ω corresponding to these moments. Then $\Omega\ (V_i) \subseteq F \times D \times C \times Q$ is the image set. Point $q_i \in Q_i$ is the last point in the path already established and $(s_i, (f, d, c)_{i+1}) \in U_i$ is the line to be directed from q_i to some point $q_{i+1} \in Q_i$ or $q_{i+1} \in Q-Q_i$. The domain element $(q_i, s_i) \in V_i$ has an image $\Omega\ (q_i, s_i)=((f, d, c)_{i+1}, q_{i+1})$. Consider Case 1. With $i=0$, the point, line, point sequence for consideration is q_0, $(s_0, (f, d, c)_1)$, __, $(s_1, (f, d, c)_2)$, __. The point q_0 at moment m is the only point in the established graph and $Q_0=\{q_0\}$. The current line $(s_0, (f, d, c)_1)$ at moments $(m, m+1)$ is the only line directed from the point q_0. From the Assumption A_{11}, $s_1 \in S$ of the quadruple $(s_1, (f, d, c)_2)$ for moments $(m+1, m+2)$ appears in a quadruple for a line directed from each of the points of the set $Q_0=\{q_0\}$. It follows that $s_i=s_0$ and $\{(q_0, s_0)\}=V_0$. From the second Assumption A_{12}, $r_2 \notin R_1$. It follows that $(f, d, c)_2 \neq (f, d, c)_1$, so that $((f, d, c)_2, q_2) \notin \Omega\ (V_0)$, $q_2=q_0$ or $q_2 \in Q-\{q_0\}$. Suppose for a next point q_1 of the graph for moment $m+(i+1)$, $q_1 \in Q_0$; i.e., $q_1=q_0$. With $\{(q_0, s_0)\}=V_0$, the domain element (q_0, s_0) at moment $m+1$ is also the domain element at moment m. With $((f, d, c)_2, q_2) \notin \Omega\ (V_0)$, the image $((f, d, c)_2, q_2)$ of (q_0, s_0) for moment $m+1$ is a different image of that same domain element at moment m, and the requirement for single-valueness of the function Ω is not satisfied. For $q_1 \in Q-\{q_0\}$ for a point of the graph at moment $m+1$, the corresponding domain element $(q_1, s_1) \notin V_i$, and it has a unique image when the line $(s_1, (f, d, c)_2)$ is directed to some point q_2 for moment $m+2$. Therefore, the choice must be to direct the current line $(s_0, (f, d, c)_1)$ to a new point $q_1 \in Q-\{q_0\}$. Now consider $i+1$. From the Assumptions A_{11} and A_{12}, it follows that $\{Q_i \times s_{i+1}\} \subseteq V_1$ and $(f, d, c)_{i+2} \notin R_{i+1}$, so that $((f, d, c)_{i+2}, q_{i+2}) \notin \Omega\ (\{Q_i \times s_{i+1}\})$, $q_{i+2} \in Q_i$ or $q_{i+2} \in Q-Q_i$. Suppose for a next point q_{i+1} of the graph for moment $m+(i+1)$, $q_{i+1} \in Q_i$. With $\{Q_i \times s_{i+1}\} \subseteq V_i$, the domain element (q_{i+1}, s_{i+1}) at moment $m+(i+1)$ is also the domain element at some moment $(m+k)$, $k \leq i$. With $((f, d, c)_{i+2}, q_{i+2}) \notin \Omega\ (\{Q_i \times s_{i+1}\})$, the image $((f, d, c)_{i+2}, q_{i+2})$ of (q_{i+1}, s_{i+1}) for moment $m+(i+1)$ is a different image than the image of that same domain element (q_{i+1}, s_{i+1}) for moment $(m+k)$, $k \leq i$ and the requirement for single-valueness of the function Ω is not satisfied. For $q_{i+1} \in Q-Q_i$ for a point of the graph at moment $m+(i+1)$, the corresponding domain element $(q_{i+1}, s_{i+1}) \notin V_i$, and it has a unique image when the line $(s_{i+1}, (f, d, c)_{i+2})$ is directed to some point q_{i+2} for moment $m+(i+2)$. Therefore, the choice must be to direct the current line $(s_i, (f, d, c)_{i+1})$ to a new point $q_{i+1} \in Q-Q_i$. For $i=h-2$, a current line $(s_{h-2}, (f, d, c)_{h-1})$ is to be directed from q_{h-2} to a next point q_{h-1} and from the Assumptions A_{11}, A_{12}, $\{Q_i \times s_{h-1}\} \subseteq V_i$ at moment $m+(h-2)$, and $(f, d, c)_h \notin R_{h-1}$. In a similar manner, it can be shown that the choice must be to direct the current line to a new point $q_{h-1} \in Q-Q_i$. Consider Case 2. With $i=0$, the point, line, point sequence for consideration is q_0, $(s_0, (f, d, c)_1)$, __, $(s_1, (f, d, c)_2)$, __, $(s_2, (f, d, c)_3)$, __ corresponding to moments $(m, m+1)$, $(m+1, m+2)$, $(m+2, m+3)$. The current line $(s_0, (f, d, c)_1)$ for moments $(m, m+1)$ is the only line directed from the established point $q_0 \in Q_0 = \{q_0\}$. Recall, by Assumption A_{11}, $s_1 \subseteq S$ of the quadruple $(s_1, (f, d, c)_2)$ for moments $(m+1, m+2)$ appears in a quadruple for a line directed from each of the points of the set $Q_0 = \{q_0\}$. It follows that $\{Q_0 \times s_1\} \subseteq V_0 = \{(q_0, s_0)\}$ and $s_1 = s_0$. From the Assumption A_{21}, $r_2 \in R_1 = \{(f, d, c)_1\}$ and it follows that $(f, d, c)_2 = (f, d, c)_1$. The set of lines containing $r_2 \in R_1$ is $U_1' = U_1$ directed from the points of the set $Q_1' = Q_0 = \{q_0\}$ to the set Q_2, and with $q_0 \in Q_1'$, $q_0 \in Q_2 = \{q_0\}$. Suppose $q_1 \in Q_1' = \{q_0\}$. With $\{Q_1' \times s_1\} \subseteq \{Q_0 \times s_1\} \subseteq V_0 = \{(q_0, s_0)\}$, the domain element (q_0, s_0) at moment $m+1$ is the same domain element at moment m. With $q_2 \in Q_2 = \{q_0\}$ and $(f, d, c)_2 \in R$, by A_{21}, $((f, d, c)_2, q_2) \in \Omega(\{Q_1' \times s_1\}) = \Omega(\{(q_0, s_0)\})$, and the image $((f, d, c)_2, q_2)$ of $(q_1, s_1) = (q_0, s_0)$ for moment $m+1$ is also the image of that same domain element (q_0, s_0) for moment m. The point, line, point, q_1, $(s_1, (f, d, c)_2)$, q_2, for moments $(m+1, m+2)$ is the same point, line, point, q_0, $(s_0, (f, d, c)_1)$, q_0, at moments $(m, m+1)$; i.e., the point, line, point is repeated as a loop to point q_0 when the current line $(s_0, (f, d, c)_1)$ is directed from q_0 to a point $q_1 \in Q_0 = \{q_0\}$. From Assumption A_{22}, $s_2 \in S$ of the quadruple $(s_2, (f, d, c)_3)$ for moments $(m+2, m+3)$ appears in a quadruple for a line directed from each of the points of the set $Q_2 \subseteq Q_0 = \{q_0\}$. It follows that $\{Q_2 \times s_2\} \subseteq V_t = \{(q_0, s_0)\}$ and $s_2 = s_0$. From Assumption A_{23}, $(f, d, c)_3 \notin R_2$ so that $((f, d, c)_3, q_3) \notin \Omega(\{Q_2 \times s_2\}) = \Omega(\{(q_0, s_2)\}) = ((f, d, c)_2, q_0)$, $q_3 \in Q_0 = \{q_0\}$ or $q_3 \in Q - Q_0$. It follows that the domain element $(q_2, s_2) = (q_0, s_0)$ at moment $m+2$ is also the domain element at moment m and $m+1$, and with $((f, d, c)_3, q_3) \notin \Omega(\{Q_2 \times s_2\}) = ((f, d, c)_2, q_0) = ((f, d, c)_1, q_0)$, the domain element has two images and the single-valueness of the function Ω is not satisfied at moments $(m+2, m+3)$ with $q_1 \in Q_i = \{q_0\}$ as a point of the graph at moment $m+1$. From the proof of Case 1 with $q_1 \in Q - Q_i = Q - \{q_0\}$, the image of the domain element $(q_1, s_1) \notin V_i$ is unique when the line $(s_1, (f, d, c)_2)$ is directed to some point q_2 for moment $m+2$. For Case 2, consider now the point, line, point sequence q_i, $(s_i, (f, d, c)_{i+1})$, __, $(s_{i+1}, (f, d, c)_{i+2})$, __, $(s_{i+2}, (f, d, c)_{i+3})$, __ corresponding to moments $(m+i, m+(i+1))$, $(m+(i+1), m+(i+2))$, $(m+(i+2), m+(i+3))$. By the Assumption A_{11}, $s_{i+1} \in S$ for the quadruple $(s_{i+1}, (f, d, c)_{i+2})$ for moments $(m+(i+1), m+(i+2))$ appears in a quadruple for a line directed from each of the points of the set $Q_i \subseteq Q$, and $\{Q_i \times s_{i+1}\} \subseteq V_i$. From the Assumption A_{21}, $r_{i+2} \in R_{i+1}$, and the set of lines containing $r_{i+2} \in R_{i+1}$ is $U_{i+1}' \subseteq U_{i+1}$ directed from the points of the set $Q_{i+1}' \subseteq Q_i$ and to the points of the set $Q_{i+2} \subseteq Q_i$. Suppose for a next point q_{i+1} of the graph for moment $m+(i+1)$, $q_{i+1} \in Q_{i+1}' \subseteq Q_i$. For $q_i \in Q_{i+1}'$, q_i is included in the set of points Q_{i+2} and the current line $(s_i, (f, d, c)_{i+1})$, $s_{i+1} = s_i$, can be directed from q_i to q_i; i.e., the line can be a loop on a point. With $\{Q_{i+1}' \times s_{i+1}\} \subseteq \{Q_i \times s_{i+1}\} \subseteq V_i$, the domain element (q_{i+1}, s_{i+1}) at moment $m+(i+1)$ is also the domain element at some moment $(m+w_2)$, $w_2 \in W_2 \subseteq I = \{0, 1, 2, ---, i\}$. With $q_{i+1} \in Q_{i+2} \subseteq Q$, and $(f, d, c)_{i+2} \in R_{i+1}$, $((f, d, c)_{i+2}, q_{i+2}) \in \Omega(\{Q_{i+1}' \times s_{i+1}\})$ and the image $((f, d, c)_{i+2}, q_{i+2})$ of (q_{i+1}, s_{i+1}) for moment $m+(i+1)$ is also the image of that same domain element (q_{i+1}, s_{i+1}) for moment $(m+p_2)$,

$p_2 \in P_2 \subseteq I$, $P_2 \cap W_2 \neq \phi$. And the point, line, point, q_{i+1}, $(s_{i+1}, (f, d, c)_{i+2})$, q_{i+2}, for moments $(m+(i+1), m+(i+2))$ is the same point, line, point for moments $(m+p_2, m+(p_2+1))$, $p_2 \in P_2 \cap W_2$; i.e., the point, line, point is repeated. From Assumptions A_{22} and A_{23}, it follows that $\{Q_{i+2} \times s_{i+2}\} \subseteq V_i$ and $(f, d, c)_{i+3} \notin R_{i+2}$ so that $((f, d, c)_{i+3}, q_{i+3}) \notin \Omega$ $(\{Q_{i+2} \times s_{i+2}\})$, $q_{i+3} \in Q_i$, or $q_{i+3} \in Q - Q_i$. Suppose $q_{i+2} \in Q_{i+2} \subseteq Q_i$, the domain element (q_{i+2}, s_{i+2}) at moment $m+(i+2)$ is also the domain element at some moment $m+z_2$, $z_2 \in Z_2 \subseteq I$ and with $((f, d, c)_{i+3}, q_{i+3}) \notin \Omega$ $(\{Q_{i+1} \times s_{i+2}\})$, the domain element (q_{i+2}, s_{i+2}) has two images and the single-valueness of the function Ω is not satisfied at moments $(m+(i+2), m+(i+3))$ with $q_{i+1} \in Q_i$ as a point of the graph at moment $m+(i+1)$. From the proof of Case 1, with $q_{i+1} \in Q - Q_i$, the image of the domain element $(q_{i+1}, s_{i+1}) \notin V_i$ is unique when the line $(s_{i+1}, (f, d, c)_{i+2})$ is directed to some point q_{i+2} for moment $m+(i+2)$. Therefore, B. For Case 2, $i=h-3$, the point, line, point sequence is q_{h-3}, $(s_{h-3}, (f, d, c)_{h-2})$, __, $(s_{h-2}, (f, d, c)_{h-1})$, __, $(s_{h-1}, (f, d, c)_h)$, __ corresponding to moments $(m+(h-3), m+(h-2))$, $(m+(h-2), m+(h-1))$, $(m+(h-1), m+h)$, and Assumptions A_{11}, A_{21}, A_{22}, A_{23} yield B in a similar manner. Consider Case n. The point, line, point sequence is q_i, $(s_i, (f, d, c)_{i+1})$, __, $(s_{i+1}, (f, d, c)_{i+2})$, __, $(s_{i+2}, (f, d, c)_{i+3})$, __, ---, __, $(s_{i+(n-2)}, (f, d, c)_{i+(n-1)})$, __, $(s_{i+(n-1)}, (f, d, c)_{i+n})$, __, $(s_{i+n}, (f, d, c)_{i+(n+1)})$, __ corresponding to moments $(m+i, m+(i+1))$, $(m+(i+1), m+(i+2))$, $(m+(i+2), m+(i+3))$, ---, $(m+(i+(n-2)), m+(i+(n-1)))$, $(m+(i+(n-1)), m+(i+n))$, $(m+(i+n), m+(i+(n+1)))$. Proceeding with Assumption A_{n1}, $r_{i+n} \in R_{i+(n-1)}$, and the set of lines containing $r_{i+n} \in R_{i+(n-1)}$ is $U'_{i+(n-1)} \subseteq U_{i+(n-1)}$ directed from the points of the set $Q'_{i+(n-1)} \subseteq Q_i$ and to the points of the set $Q_{i+n} \subseteq Q_i$. Suppose for a next point $q_{i+(n-1)}$ of the graph for moment $m+(i+(n-1))$, $q_{i+(n-1)} \in Q'_{i+(n-1)} \subseteq Q_i$. For $q_i \in Q'_{i+(n-1)}$, q_i is included in the set of points Q_{i+n} and the current line, $(s_i, (f, d, c)_{i+1})$, $s_{i+(n-1)} = s_i$, can be directed from q_i to q_i; i.e., the line can be a loop on the point. With $\{Q'_{i+(n-1)} \times s_{i+(n-1)}\} \subseteq \{Q_i \times s_{i+(n-1)}\} \subseteq V_i$ the domain element $(q_{i+(n-1)}, s_{i+(n-1)})$ at moment $m+(i+(n-1))$ is also the domain element at some moment $(m+w_n)$, $w_n \in W_n \subseteq I = \{0, 1, 2, ---, i\}$. With $q_{i+n} \in Q_{i+n} \subseteq Q$, and $(f, d, c)_{i+n} \in R_{i+(n-1)}$, $((f, d, c)_{i+n}, q_{i+n}) \in \Omega$ $(\{Q'_{i+(n-1)} \times s_{i+(n-1)}\})$ and the image, $((f, d, c)_{i+n}, q_{i+n})$ of $(q_{i+(n-1)}, s_{i+(n-1)})$, for moment $m+(i+(n-1))$ is also the image of that same domain element $(q_{i+(n-1)}, s_{i+(n-1)})$ for moment $(m+p_n)$, $p_n \in P_n \subseteq I$, $P_n \cap W_n \neq \phi$, and the point, line, point, $q_{i+(n-1)}$, $(s_{i+(n-1)}, (f, d, c)_{i+n})$, q_{i+n} for moments $(m+(i+(n-1)), m+(i+n))$ is the same point, line, point for moments $(m+p_n, m+(p_n+1))$, $p_n \in P_n \cap W_n$; i.e., the point, line, point is repeated. From Assumption A_{n2} and A_{n3}, it follows that $\{Q_{i+n} \times s_{i+n}\} \subseteq V_i$ and $(f, d, c)_{i+(n+1)} \notin R_{i+n}$ so that $((f, d, c)_{i+(n+1)}, q_{i+(n+1)}) \notin \Omega$ $(\{Q_{i+n} \times s_{i+n}\})$, $q_{i+(n+1)} \in Q_i$, or $q_{i+(n+1)} \in Q - Q_i$. Suppose $q_{i+n} \in Q_{i+n} \subseteq Q_i$, the domain element (q_{i+n}, s_{i+n}) at moment $m+(i+n)$ is also the domain element at some moment $m+z_n$, $z_n \in Z_n \subseteq I$, and with $((f, d, c)_{i+(n+1)}, q_{i+(n+1)}) \in \Omega$ $(\{Q_{i+(n-1)} \times s_{i+n}\})$, the domain element (q_{i+n}, s_{i+n}) has two images and the single-valueness of the function Ω is not satisfied at moments $(m+(i+n), m+(i+(n+1)))$ with $q_{i+1} \in Q_i$ as a point of the graph at moment $m+(i+1)$. With $q_{i+1} \in Q - Q_i$, the image of the domain element $(q_{i+1}, s_{i+1}) \notin V_i$ is unique when the line $(s_{i+1}, (f, d, c)_{i+2})$ is directed to some point q_{i+2} for moment $m+(i+2)$. Therefore, the choice must be to direct a current line $(s_i, (f, d, c)_{i+1})$ from $q_i \in Q_i$ for moment $m+i$ to a new point $q_{i+1} \in Q-Q_i$ for moment $(m+(i+1)$.

Table 2.7 was established from the line sextuples of Table 2.5 and Theorem 2.18. The corresponding processor graphs are shown in Figures 2.6 and 2.7. The line numbers of the table correspond to the line numbers of the graphs. The differences in the two graphs is that in the graph in Figure 2.6, a state once left is never reentered while in the graph of Figure 2.7, state q_0 is reentered a maximum number of times, with states q_1 and q_2 having second and third priorities. The theorem allows this, and both graphs are valid. Note that for both graphs, states are entered at the same moments. The effective operation path is a point, line, point sequence and for the moments $(m, m+1)$, $(m+1, m+2)$, $(m+2, m+3)$, the effective operation path for the graph of Figure 2.6 is q_0, (s_4, r_1), q_0, (s_2, r_2), q_0, (s_0, r_3), q_1.

Only Case 1 of Theorem 2.18 applies in establishing the states for the above study. Table 2.8 shows an arbitrary set of graph sextuples with the blanks for the states established by Theorem 2.18 requiring Cases 1, 2, and 3.

Since Theorem 2.18 represents an algorithm, an artificial automaton of the digital class receiving the sextuples from an observed automaton with undetermined state elements can establish the states and when the states are first entered.

TABLE 2.7. Graph sextuples for processor graphs of Figures 2.6 and 2.7 obtained from line sextuples of Table 2.5 for the observed environment graph of Figure 2.5 with states determined by Theorem 2.18

		Graph Sextuples, $(r_i=(f, d, c)_i)$	
Line No.	Moments	Figure 2.6	Figure 2.7
1	$(m, m+1)$	$\underline{q_0}, s_4, r_1, \underline{q_0}$	$\underline{q_0}, s_4, r_1, \underline{q_0}$
2	$(m+1, m+2)$	$\underline{q_0}, s_2, r_2, \underline{q_0}$	$\underline{q_0}, s_2, r_2, \underline{q_0}$
3	$(m+2, m+3)$	$\underline{q_0}, s_0, r_3, \underline{q_1}$	$q_0, s_0, r_3, \underline{q_1}$
4	$(m+3, m+4)$	$\underline{q_1}, s_4, r_4, \underline{q_1}$	$\underline{q_1}, s_4, r_4, \underline{q_0}$
5	$(m+4, m+5)$	$\underline{q_1}, s_1, r_5, \underline{q_1}$	$\underline{q_0}, s_1, r_5, \underline{q_1}$
6	$(m+5, m+6)$	$\underline{q_1}, s_0, r_6, \underline{q_1}$	$\underline{q_1}, s_0, r_6, \underline{q_0}$
7	$(m+6, m+7)$	$\underline{q_1}, s_3, r_7, \underline{q_1}$	$\underline{q_0}, s_3, r_7, \underline{q_1}$
8	$(m+7, m+8)$	$\underline{q_1}, s_2, r_8, \underline{q_2}$	$\underline{q_1}, s_2, r_8, \underline{q_2}$
9	$(m+8, m+9)$	$\underline{q_2}, s_0, r_9, \underline{q_2}$	$\underline{q_2}, s_0, r_9, \underline{q_1}$
10	$(m+9, m+10)$	$\underline{q_2}, s_3, r_{10}, \underline{q_2}$	$\underline{q_1}, s_3, r_{10}, \underline{q_2}$
11	$(m+10, m+11)$	$\underline{q_2}, s_1, r_{11}, \underline{q_3}$	$\underline{q_2}, s_1, r_{11}, \underline{q_3}$
12	$(m+11, m+12)$	$\underline{q_3}, s_0, r_{12}, \underline{q_3}$	$\underline{q_3}, s_0, r_{12}, \underline{q_0}$
13	$(m+12, m+13)$	$\underline{q_3}, s_6, r_{13}, \underline{q_4}$	$\underline{q_0}, s_6, r_{13}, \underline{q_4}$
14	$(m+13, m+14)$	$\underline{q_4}, s_0, r_{14}, \underline{q_4}$	$\underline{q_4}, s_0, r_{14}, \underline{q_0}$
15	$(m+14, m+15)$	$\underline{q_4}, s_8, r_{15}, \underline{q_4}$	$\underline{q_0}, s_8, r_{15}, \underline{q_0}$
16	$(m+15, m+16)$	$\underline{q_4}, s_5, r_{16}, \underline{q_5}$	$\underline{q_0}, s_5, r_{16}, \underline{q_5}$
17	$(m+16, m+17)$	$\underline{q_5}, s_0, r_{17}, \underline{q_5}$	$\underline{q_5}, s_0, r_{17}, \underline{q_0}$
18	$(m+17, m+18)$	$\underline{q_5}, s_7, r_{18}, \underline{q_6}$	$\underline{q_0}, s_7, r_{18}, \underline{q_6}$
19	$(m+18, m+19)$	$\underline{q_6}, s_0, r_{19}, \underline{q_6}$	$\underline{q_6}, s_0, r_{19}, \underline{q_0}$

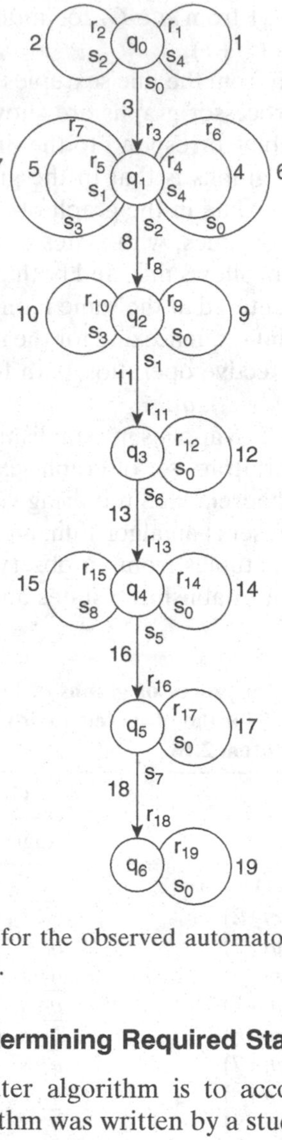

Figure 2.6 Processor graph for the observed automaton of Figure 2.5, where states once left are never reentered.

2.2.3 Algorithm for Determining Required State Changes

The object of the computer algorithm is to accomplish what is stated in Theorem 2.18. (The algorithm was written by a student in computer sciences, John J. Schauf, as a problem assignment.)

Given an observed effective operation path of a principal environment graph G_M (E_p) for an automaton that has performed over a moment base $M=\{m, m+1, ---, m+h\}$ and without a determined set of states. Let the construction of the processor graph proceed from the sextuples $(\_\_, s_i, (f, d, c)_{i+1}, \_\_)$ of the environment graph taken in sequence following the observed effective operation path.

Two processing strategies could be employed.

1. Always return to state q_0 whenever possible.
2. Remain in a current state until forced to change to a new state.

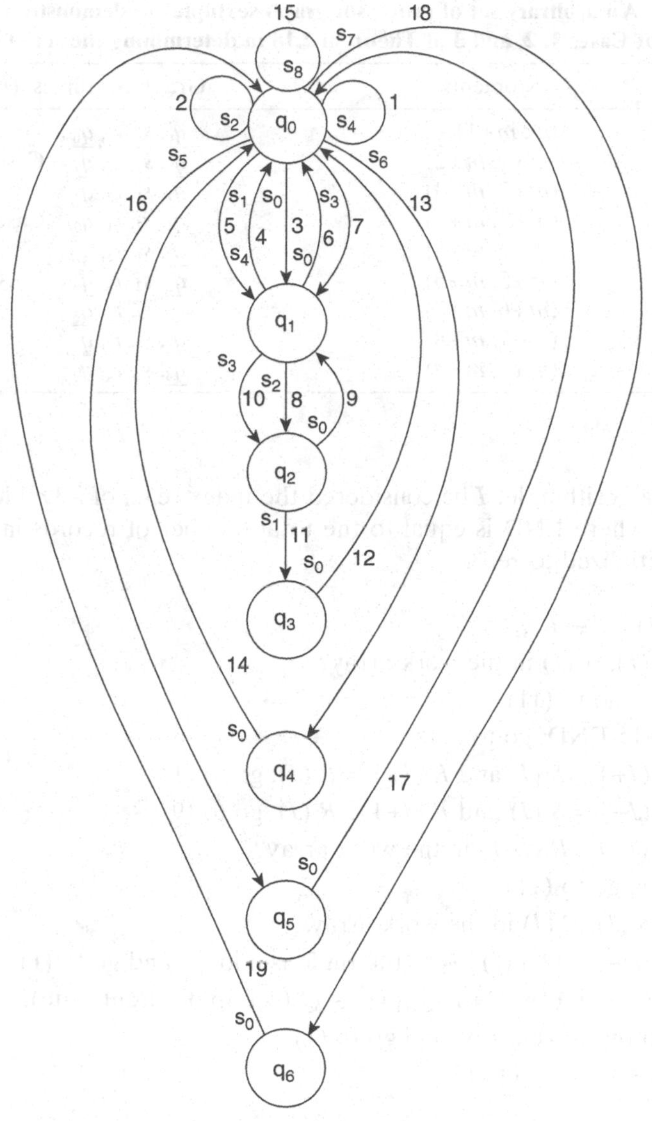

Figure 2.7 Processor graph for the observed automaton of Figure 2.5, where state q_0 is reentered a maximum number of times, with states q_1 and q_2 having second and third priorities.

A basic approach is to look ahead in the input array, and utilizing the criteria set forth in Theorem 2.18, determine whether the current s_i, $(f, d, c)_{i+1}$ will produce a loop or whether it will require a different state. A secondary work array may be used to record loops established for a current state to facilitate the look ahead procedure.

TABLE 2.8. An arbitrary set of processor graph sextuples to demonstrate the application of Cases 1, 2, and 3 of Theorem 2.18 in determining the set of states

Line No.	Moments	Graph Sextuples, $(r_i=(f, d, c)_i)$	
1	$(m, m+1)$	$\underline{q_0}, s_1, r_1, \underline{q_0}$	
2	$(m+1, m+2)$	$\underline{q_0}, s_2, r_2, \underline{q_1}$	Case 1
3	$(m+2, m+3)$	$\underline{q_1}, s_1, r_3, \underline{q_1}$	
4	$(m+3, m+4)$	$\underline{q_1}, s_2, r_1, \underline{q_2}$	Case 2
5	$(m+4, m+5)$	$\underline{q_2}, s_1, r_1, \underline{q_2}$	
6	$(m+5, m+6)$	$\underline{q_2}, s_2, r_3, \underline{q_3}$	Case 3
7	$(m+6, m+7)$	$\underline{q_3}, s_1, r_1, \underline{q_3}$	
8	$(m+7, m+8)$	$\underline{q_3}, s_2, r_2, \underline{q_4}$	Case 1
9	$(m+8, m+9)$	$\underline{q_4}, s_1, r_4, \underline{q_4}$	

For the algorithm, let I be considered the index (e.g., of a DO loop) from 1 to END, where END is equal to the total number of records in a current set. Q is initialized to zero.

1. $I \leftarrow I+1$, $J \leftarrow I$.
2. Is $S(J)$, $R(J)$ in the work array?
   If yes, go to (11).
3. If $J+1 > \text{END}$, go to (7).
4. If $S(J+1)=S(J)$ and $R(J+1)=R(J)$, go to (11).
5. If $S(J+1)=S(J)$ and $R(J+1)\neq R(J)$, go to (9).
6. Is $S(J+1)$, $R(J+1)$ in the work array?
   If yes, go to (11).
7. Put $S(J)$, $R(J)$ in the work array.
8. $Q_i(I) \leftarrow Q$, $Q_{i+1}(I) \leftarrow Q$, (the tuple is a loop) and go to (1).
9. $Q_i(I) \leftarrow Q$, $Q \leftarrow Q+1$, $Q_{i+1}(I) \leftarrow Q$ (set up different state).
10. Zero the work array, and go to (1).
11. $J \leftarrow J+1$, and go to (2).

EXERCISES

2.1 Establish in detail the four primitives for the (a) alternate time graph $G_M(T_a)$; (b) alternate environment graph $G_M(E_a)$.

2.2 Give the proof to (a) Corollaries 2.8.1 and 2.8.2; (b) Corollaries 2.10 and 2.11.

2.3 Give the proofs to (a) Theorems 2.12 and 2.13; (b) Theorems 2.14 and 2.15.

2.4 Write a program using the algorithm given in 2.2.3 for determining the required state changes and check the results for the example automaton of 2.2.2.

2.5 Parallel the case study of 2.2.1 with a case study for adding two 2-digit numbers.

CHAPTER 3

A General Automaton: Detailed Analysis

The analysis for a general automaton was presented in Section 2.1 and primitives were established for a graph model consisting of the processor, environment, and time graphs. An application of this work was given in Section 2.2, and recursive methods for operating on the graph model for determining effective operation paths and graph properties were presented in Appendix A. A number of assumptions were made to simplify the initial analyses. In this chapter, attention will be given to generality by removing a number of the assumptions. New equations and functions will be added and modifications made on the equations and functions of Chapter 2 reflecting the detailed analysis. Finally, summaries of the modified graph primitives are given.

First, the receptors and effectors for the simplified automaton of the previous analysis were considered an integral part of the automaton. Here, the receptors and effectors are made distinguishable. Second, in Chapter 2, the automaton was assumed to perform in homogeneous environments. In this chapter, nonhomogeneous environments are considered. Third, the transformation response of Chapter 2 will now be considered to have five components that provide stimuli for the five senses of an observing automaton. Fourth, nonshared environments (memory) were considered an integral part of the automaton for the previous analysis. An analysis is made in this chapter to study the performance of the automaton when certain nonshared environments are made distinguishable from the processor of the automaton.

3.1 DISTINGUISHABLE RECEPTORS AND EFFECTORS

In the work of Chapter 2, a general automaton was treated as a black box containing a stimulus processor and receiving stimuli and producing responses. In this section, the automaton is broken down into greater structural detail,

Knowledge Structures for Communications in Human-Computer Systems: General Automata-Based, by Eldo C. Koenig

making the receptors and effectors distinguishable from the stimulus processor (Koenig 1975a). The sets of stimuli and triadic responses remain observed sets over a moment base. For a natural automaton, there are the natural receptors and effectors. Auxiliary receptors and effectors may also operate as part of the automaton, such as eyeglasses, microscopes, and tools in general.

Figure 3.1a shows a diagram of an automaton a and the associated sets for the receptors and effectors nondistinguishable from the processor p, which represents the conditions for analysis in Chapter 2. Q, S, R denote the sets of states, stimuli, and responses for a operating over a moment base M. Recall, the set R of responses is described by a triadic relation on the sets F, D, C by $R \subseteq F \times D \times C$ where F, D, C denote sets of transformation, spatial change, and time change responses over M.

Figure 3.1b shows the automaton a with receptors and effectors distinguishable from the processor p^* and is our concern here. Let $\bar{X}$ and $\bar{Z}$ be the set of receptors and effectors and Q^* the set of states of p^* over M. $\bar{X}$ and $\bar{Z}$ are described by dyadic relations on the sets X'' and X', and Z' and Z'', i.e., $\bar{X} \subseteq X'' \times X'$, $\bar{Z} \subseteq Z' \times Z''$, where X' and X'' are the sets of principal and auxiliary receptors and Z' and Z'' are the sets of principal and auxiliary effectors over M. At any moment of observation, a in a performance role can be represented by a quintuple $(x'', x', p^*, z', z'') \in X'' \times X' \times P^* \times Z' \times Z''$. Sets Z'' and X'' each contain null elements.

Consider now the effect of distinguishable receptors and effectors on the automaton equations. The sets S and $R \subseteq F \times D \times C$ remain the sets of observed stimuli and responses of a operating over the moment base M. Recall, Q is the set of states over M when the receptors and effectors are nondistinguishable, and Q^* is the set of states when they are distinguishable. Let S^* be the set of stimuli that enters the stimulus processor p^* over M. A stimulus entering p^* at any moment $m+i$, $S^*(m+i)$, is a function of the observed stimulus, $S(m+i)$, and the receptors, $\bar{X}(m+i)$, at moment $m+i$, that is

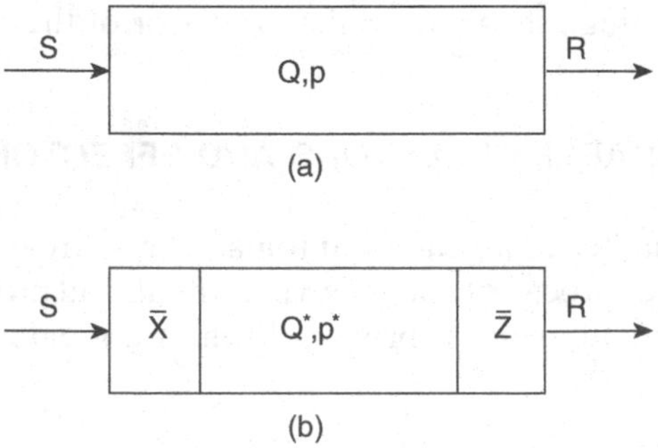

Figure 3.1 An automaton for (a) nondistinguishable receptors and effectors (b) distinguishable receptors and effectors.

$$S^*(m+i)=f(S(m+i), \bar{X}(m+i)) \quad 3.1E$$

and the Equation 2.1E, $Q(m+(i+1))=\omega(Q(m+i), S(m+i))$, for the nondistinguishable receptors becomes

$$Q^*(m+(i+1))=\omega'(Q^*(m+i), S^*(m+i)) \quad 3.2E$$

when the receptors are distinguishable. Let K be a dyadic relation on the sets $\bar{X}$ and S. Then, from Equation 3.1E, there is the function

$$f:K\subseteq\bar{X}\times S\rightarrow f(K)\subseteq S^*$$

Let V''' be a dyadic relation on the sets Q^* and S^*. Then, from Equation 3.2E, there is the function

$$\omega':V''\subseteq Q^*\times S^*\rightarrow\omega'(V'')\subseteq Q^*$$

Since ω' is defined on $V'''\subseteq Q^* \times S^*$, there is an interrelationship of the automaton functions f with ω'. Let V' be a dyadic relation on sets Q^* and K and define $\omega^*=\omega'\circ\omega''$ where ω'': $V'\subseteq Q^*\times K\rightarrow\omega''(V')\subseteq V'''$ and ω': $\omega''(V')\rightarrow\omega'(\omega''(V'))\subseteq Q^*\subseteq\bar{Q}^*$ where $\bar{Q}^*$ is the set of states for the life of the automaton. There results

$$\omega^*:V'\subseteq Q^*\times K\subseteq Q^*\times\bar{X}\times S\rightarrow\omega^*(V')\subseteq Q^*\subseteq\bar{Q}^* \quad 3.1F$$

For a given stimulus received at two consecutive moments by an automaton with nondistinguishable receptors and effectors, if a change of receptors and/or a change of effectors actually occurred and the responses were different, a change of state had to be assumed. The apparent state at any moment $(m+i)$, $Q(m+i)$, of p is a function of the state of p^*, $Q^*(m+i)$, at $m+i$, a function of the receptors, $\bar{X}(m+i)$ at $m+i$, and a function of the effectors, $\bar{Z}(m+i)$, at $m+i$, that is

$$Q(m+i)=f'(Q^*(m+i), \bar{X}(m+i), \bar{Z}(m+i)) \quad 3.3E$$

Let H be a triadic relation on the sets Q^*, $\bar{X}$, $\bar{Z}$ of states of p^*, receptors, effectors defined over the moment base M. Then the function f' results

$$f':H\subseteq Q^*\times\bar{X}\times\bar{Z}\rightarrow f'(H)\subseteq Q$$

The response Equation 2.2E and Function 2.2F for the automaton with the nondistinguishable receptors and effectors are

$$R(m+(i+1))=\beta(Q(m+i), S(m+i)) \quad 2.2E$$

$$\beta: V \subseteq Q \times S \rightarrow \beta(V) = R \subseteq \bar{R} \qquad 2.2\text{F}$$

Since β is defined on $V \subseteq Q \times S$, there is an interrelationship of the automaton functions f' with β. Let V^* be a dyadic relation on H and S and define $\beta^* = \beta \circ \beta'$, where $\beta': V^* \subseteq H \times S \rightarrow \beta'(V^*) \subseteq V$ and β: $\beta'(V^*) \rightarrow \beta(\beta'(V^*)) = R \subseteq \bar{R}$. There results

$$\beta^*: V^* \subseteq H \times S \subseteq Q^* \times \bar{X} \times \bar{Z} \times S \rightarrow \beta^*(V^*) = R \subseteq \bar{R} \qquad 3.2\text{F}$$

In a similar manner, define functions σ^*, δ^*, τ^* for the case of automaton a with distinguishable receptors and effectors

$$\sigma^*: V^* \rightarrow \sigma^*(V^*) = F \subseteq \bar{F} \qquad 3.3\text{F}$$

$$\delta^*: V^* \rightarrow \delta^*(V^*) = D \subseteq \bar{D} \qquad 3.4\text{F}$$

$$\tau^*: V^* \rightarrow \tau^*(V^*) = C \subseteq \bar{C} \qquad 3.5\text{F}$$

There are also the functions

$$\gamma': W' \subseteq E \times D \rightarrow \gamma'(W') \subseteq E \subseteq \bar{E} \qquad 3.6\text{F}$$

$$\lambda': N' \subseteq T \times C \rightarrow \lambda'(N') \subseteq T \subseteq \bar{T} \qquad 3.7\text{F}$$

and these are unchanged from Functions 2.6F, 2.7F for the automaton with nondistinguishable receptors and effectors.

A modification in Function 2.8F is deferred until the next section when additional considerations must be made dealing with nonhomogeneous environments.

The functions resulting from the compositions remain to be modified to reflect the detailed analysis of the automaton. First, consider for modification Function 2.9F, γ: $W \subseteq E \times V \rightarrow \gamma(W) \subseteq E \subseteq \bar{E}$. Since γ' is defined on $W' \subseteq E \times D$, there is an interrelation of δ^* with γ'. Let W^* be a dyadic relation on E and V^*, and define $\gamma^* = \gamma' \circ \gamma''$, where γ'': $W^* \subseteq E \times V^* \rightarrow \gamma''(W^*) \subseteq W^*$ and γ': $\gamma''(W^*) \rightarrow \gamma'(\gamma''(W^*)) \subseteq E$. There results

$$\gamma^*: W^* \subseteq E \times V^* \rightarrow \gamma^*(W^*) \subseteq E \subseteq \bar{E} \qquad 3.8\text{F}$$

There are similar modifications in the compositions of functions for obtaining the functions λ, π, μ of 2.10F, 2.11F, 2.12F. The corresponding new functions are

$$\lambda^*: N^* \subseteq T \times V^* \rightarrow \lambda^*(N^*) \subseteq T \subseteq \bar{T} \qquad 3.9\text{F}$$

$$\pi^*: Y^* \subseteq T \times W^* \subseteq T \times E \times V^* \rightarrow \pi^*(Y^*) \subseteq \gamma^*(W^*) \times \lambda^*(N^*) \subseteq E \times T \qquad 3.10\text{F}$$

$$\mu^*: Y^{*\prime} \subseteq E \times N^* \subseteq E \times T \times V^* \rightarrow \mu^*(Y^{*\prime}) \subseteq \lambda^*(N^*) \times \gamma^*(W^*) \subseteq T \times E \qquad 3.11\text{F}$$

There is also the environment function

$$\phi_i : E' \subseteq \bar{E} \rightarrow \phi_i(E') = S \qquad 3.12F$$

and it is unchanged from Function 2.13F for the automaton with nondistinguishable receptors and effectors.

The change in the moment functions must also be considered. There is a moment function θ^*_Q mapping a set M into the set of active states Q^* associated with the processor p^* for the automaton with distinguishable receptors and effectors

$$\theta^*_Q : M \rightarrow \theta^*_Q(M) = Q^* \subseteq \bar{Q}^* \qquad 3.1M$$

The remaining moment functions are

$$\theta_S : M - \{m+h\} \rightarrow \theta_S(M - \{m+h\}) = S \subseteq \bar{S} \qquad 3.2M$$

$$\theta_R : M - \{m\} \rightarrow \theta_R(M - \{m\}) = R \subseteq \bar{R} \qquad 3.3M$$

$$\theta_F : M - \{m\} \rightarrow \theta_F(M - \{m\}) = F \subseteq \bar{F} \qquad 3.4M$$

$$\theta_D : M - \{m\} \rightarrow \theta_D(M - \{m\}) = D \subseteq \bar{D} \qquad 3.5M$$

$$\theta_C : M - \{m\} \rightarrow \theta_C(M - \{m\}) = C \subseteq \bar{C} \qquad 3.6M$$

$$\theta_E : M \rightarrow \theta_E(M) = E \subseteq \bar{E} \qquad 3.7M$$

$$\theta_T : M \rightarrow \theta_T(M) = T \subseteq \bar{T} \qquad 3.8M$$

They are unchanged from those of 2.2M–2.8M for the automaton with nondistinguishable receptors and effectors.

3.2 NONHOMOGENEOUS ENVIRONMENTS

The condition of a nonhomogeneous environment media effect is considered and results in two additional equations (Koenig, Frederick 1971). For this condition, a transformation response produced at two different environments by an automaton will be recorded differently, and, hence, the two recorded responses will be interpreted as two different stimuli. Let B denote the set of recorded responses for the automaton over the moment base $M=\{m, m+1, m+2, ---, m+h\}$. Let $B(m+(i+1))$ be the recorded response at moment $m+(i+1)$. $B(m+(i+1))$ is a function of the transformation response produced at moment $m+(i+1)$, $F(m+(i+1))$, and the environment accessed at moment $m+i$, $E(m+i)$, which is the location where the transformation response is recorded. Then

$$B(m+(i+1)) = \eta'(E(m+i), F(m+(i+1))) \qquad 3.4E$$

An arbitrarily recorded response $B(m+i)$ of the automaton has potential of being interpreted as a stimulus by the automaton (or any other automaton) at any time during the period, beginning but not including $T(m+i)$ and ending at time $T(m+(h-1))$. The potential stimulus for the automaton is described by the equation

$$S(m+(i+p))=\alpha^*(B(m+i)) \qquad 3.5E$$

restricted in time to

$$((T(m+i)<T(m+(i+p))|i\in I-\{0,h\}|)\, p\in P)$$

where $P=\{0, 1, 2, ---, h-(i+1)\}$. Equation 3.5E now replaces Equation 2.8E, $S(m+j)=\alpha(F(m+i))$.

Let B be the set of recorded responses for the automaton over the moment base M and define η' on W'' a subset of the automaton sets $E\times F$. Then the mapping corresponding to Equation 3.4E is

$$\eta':W''\subseteq E\times F\rightarrow\eta'(W'')=B$$

Since η' is defined on $W''\subseteq E\times F$, there is an interrelationship of functions σ^* with η'. Let W^* be a dyadic relation on sets E and V^* and define $\eta=\eta'\circ\eta''$ where $\eta'':W^*\subseteq E\times V^*\rightarrow\eta''(W^*)\subseteq W''$ and $\eta':\eta''(W^*)\rightarrow\eta'(\eta''(W^*))=B\subseteq\bar{B}$, where $\bar{B}$, denotes the set of recorded responses for the life of the automaton. There results

$$\eta:W^*\subseteq E\times V^*\rightarrow\eta(W^*)=B\subseteq\bar{B} \qquad 3.13F$$

And the mapping corresponding to Equation 3.5E is

$$\alpha^*:B\rightarrow\alpha^*(B)=S\subseteq\bar{S} \qquad 3.14F$$

Modifications for the five graph functions 2.1GF–2.5GF are required to reflect the detailed analysis. All five of the functions are affected by the detailed analysis of the automaton of Section 3.1, but only the principal environment graph function (2.2GF), principal time graph function (2.4GF), and alternate environment graph function (2.5GF) involve the environment set and are affected by the analysis for nonhomogeneous environments. Define a processor graph function Ω^* for the automaton with the distinguishable receptors and effectors mapping $V^*\subseteq Q^*\times\bar{X}\times\bar{Z}\times S$ into the product space of responses and states. That is Ω^*: $V^*\rightarrow R\times Q^*$, or more specifically

$$Q^*:V^*\subseteq Q^*\times\bar{X}\times\bar{Z}\times S\rightarrow(\sigma^*(V^*)\times\delta^*(V^*)\times\tau^*(V^*)\times\omega^*(V'))$$
$$\subseteq F\times D\times C\times Q^* \qquad 3.1GF$$

This function corresponds to 2.1GF for the automaton with the simplified analysis, $\Omega: V \subseteq Q \times S \rightarrow (\sigma(V) \times \delta(V) \times \tau(V) \times \omega(V)) \subseteq F \times D \times C \times Q$. Similarly, graph functions Γ^*, $\Gamma^{*\prime}$, Λ^*, $\Lambda^{*\prime}$ are defined for the detailed analysis to correspond to the graph functions Γ, Γ', Λ, Λ' of the simplified analysis (Functions 2.2GF–2.5GF)

$$\Gamma^*: W^* \subseteq E \times V^* \rightarrow (\eta(W^*) \times \sigma^*(V^*) \times \delta^*(V^*) \times \tau^*(V^*) \times \omega^*(V') \times \gamma^*(W^*)) \subseteq B \times F \times D \times C \times Q^* \times E \quad \text{3.2GF}$$

$$\Gamma^{*\prime}: N^* \subseteq T \times V^* \rightarrow (\sigma^*(V^*) \times \delta^*(V^*) \times \tau^*(V^*) \times \omega^*(V') \times \lambda^*(N^*)) \subseteq F \times D \times C \times Q^* \times T \quad \text{3.3GF}$$

$$A^*: Y^* \subseteq T \times W^* \rightarrow (\eta(W^*) \times \sigma^*(V^*) \times \delta^*(V^*) \times \tau^*(V^*) \times \omega^*(V') \times \pi^*(Y^*)) \subseteq B \times F \times D \times C \times Q^* \times E \times T \quad \text{3.4GF}$$

$$\Lambda^{*\prime}: Y^{*\prime} \subseteq E \times N^* \rightarrow (\eta(W^*) \times \sigma^*(V^*) \times \delta^*(V^*) \times \tau^*(V^*) \times \omega^*(V') \times \mu^*(Y^{*\prime})) \subseteq B \times F \times D \times C \times Q^* \times T \times E \quad \text{3.5GF}$$

A brief summary of the primitives of an automaton for the analysis completed thus far will now be presented. The set of four primitives P1–P4 defining the processor graph gives a graph 10-tuple associated with two points (not necessarily distinct) and a connecting line

$$TP^*(m+i, m+i+1) = (q_i^*, x_i'', x_i', z_i', z_i'', s_i, f_{i+1}, d_{i+1}, c_{i+1}, q_{i+1}^*) \quad \text{3.1GT*}$$

such that $\Omega^*(q_i^*, x_i'', x_i', z_i', z_i'', s_i) = (f_{i+1}, d_{i+1}, c_{i+1}, q_{i+1}^*)$, and the line is directed from q_i^* of the domain element to q_{i+1}^* of the image.

The set of four primitives E_p1–E_p4 defining the principal environment graph gives a graph 13-tuple associated with two points (not necessarily distinct) and a connecting line

$$TE_p^*(m+i, m+i+1) = (e_i, q_i^*, x_i'', x_i', z_i', z_i'', s_i, b_{i+1}, f_{i+1}, d_{i+1}, c_{i+1}, q_{i+1}^*, e_{i+1}) \quad \text{3.2GT*}$$

such that $\Gamma^*(e_i, q_i^*, x_i'', x_i', z_i', z_i'', s_i) = (b_{i+1}, f_{i+1}, d_{i+1}, c_{i+1}, q_{i+1}^*, e_{i+1})$, and the line is directed from e_i of the domain element to e_{i+1} of the image.

The set of four primitives T_a1–T_a4 defining the alternate Time graph gives a graph 12-tuple associated with two points (not necessarily distinct) and a connecting line

$$TT_a^*(m+i, m+i+1) = (t_i, q_i^*, x_i'', x_i', z_i', z_i'', s_i, f_{i+1}, d_{i+1}, c_{i+1}, q_{i+1}^*, t_{i+1}) \quad \text{3.3GT*}$$

such that $\Gamma^{*\prime}(t_i, q_i^*, x_i'', x_i', z_i', z_i'', s_i) = (f_{i+1}, d_{i+1}, c_{i+1}, q_{i+1}^*, t_{i+1})$, and the line is directed from t_i of the domain element to t_{i+1} of the image. Note that this tuple does not contain the recorded response b_{i+1}.

The set of four primitives T_p1–T_p4 defining the principal time graph gives a graph 15-tuple associated with two points (not necessarily distinct) and a connecting line

$$TT_p^*(m+i, m+i+1) \\ = (t_i, e_i, q_i^*, x_i'', x_i', z_i', z_i'', s_i, b_{i+1}, f_{i+1}, d_{i+1}, c_{i+1}, q_{i+1}^*, e_{i+1}, t_{i+1}) \qquad 3.4GT^*$$

such that $\Lambda^*(t_i, e_i, q_i^*, x_i'', x_i', z_i', z_i'', s_i) = (b_{i+1}, f_{i+1}, d_{i+1}, c_{i+1}, q^*_{i+1}, e_{i+1}, t_{i+1})$, and the line is directed from t_i of the domain element to t_{i+1} of the image.

The set of four primitives E_a1–E_a4 defining the alternate environment graph gives a graph 15-tuple associated with two points (not necessarily distinct) and a connecting line

$$TE_a^*(m+i, m+i+1) \\ = (e_i, t_i, q_i^*, x_i'', x_i', z_i', z_i'', s_i, b_{i+1}, f_{i+1}, d_{i+1}, c_{i+1}, q_{i+1}^*, t_{i+1}, e_{i+1}) \qquad 3.5GT^*$$

such that $\Lambda^{*\prime}(e_i, t_i, q^*_i, x_i'', x_i', z_i', z_i'', s_i) = (b_{i+1}, f_{i+1}, d_{i+1}, c_{i+1}, q^*_{i+1}, t_{i+1}, e_{i+1})$, and the line is directed from e_i of the domain element to e_{i+1} of the image.

3.3 TRANSFORMATION RESPONSE COMPONENTS

Recall, there are the transformation response, the spatial change response, and the time change response. The transformation response element f_{i+1} produced by an automaton at some moment $m+(i+1)$ has five components that provide distinction of interpreted stimuli, and this detail must still be established (Koenig 1978). The component responses exist for identifying the performing automaton through the five senses of an observing automaton a_0. They are the responses that provide an awareness by an observer a_0 through his sight, smell, hearing, touch, taste. The set of transformation responses F is now defined as $F \subseteq FS \times FM \times FH \times FT \times FA$ and an automaton receiving a stimulus at moment $m+i$ may produce a transformation response $(fs, fm, fh, ft, fa)_{i+1} \in F$. $b_{i+1} \in B$ is the recorded response of a transformation response and is recorded in an environment $e_j \in E$.

The elements of a graph n-tuple and an element of a set of automata A may be considered basic word elements and determine the meaning of sentences (to be discussed in Chapter 4). An example is presented to demonstrate the components of the transformation response element. **John** as an automaton who **strikes** a **table** with a **hammer** and leaves a **dent** also perspires to produce an odor, where **strikes** is a transformation response component fs_{i+1} that is detectable by the observing automaton a_0 through sight and is recorded as a **dent**. The production of an odor is a component fm_{i+1} detectable by a_0 through the sense of smell.

The 15-tuple for either the alternate environment graph 3.5GT* or the principal time graph 3.4GT* for a given automaton with non-null elements is considered to contain complete knowledge for the pair of moments of observation involved. The alternate environment graph is the most informative in that an automaton is observed to move from one environment to another in a direct correspondence to a graph as demonstrated in Section 2.2. The 15- and 10-tuples of the principal time and processor graphs can be determined from that of the alternate environment graph to obtain a complete graph model for representing knowledge structures. To facilitate the identification of the graph, the automaton word for identifying the automaton is included as another element in the graph n-tuple. It is positioned between the elements from the domain and the elements from the co-domain of the graph functions. After considering the transformation response components and the automaton identification, the alternate environment graph tuple of 3.5GT* becomes

$$TE_a(m+i, m+i+1) = (e_i, t_i, q^*_i, x''_i, x'_i, z'_i, z''_i, s_i, a, b_{i+1}, (fs, fm, fh, ft, fa)_{i+1}, d_{i+1}, c_{i+1}, q^*_{i+1}, t_{i+1}, e_{i+1}) \qquad \text{3.5GT}$$

There are graph tuples 3.1GT–3.4GT corresponding to 3.1GT*–3.4GT*.

3.4 NONSHARED ENVIRONMENTS INTERPRETED AS DISTINGUISHABLE

A natural automaton a_0 obtains knowledge about automata and later performs like automata previously observed. a_0 has the knowledge about the automata stored in its nonshared environments (memory). If a_0 is a human, the stored knowledge may be obtained either by direct observation or by communication with other automata in the natural language. The performance of a_0 must involve feedback in order for a_0 to perform satisfactorily like the automaton previously observed. a_0 must also be capable of performing like the automaton previously observed. a_0 can be observed performing like another automaton if some of the environments are shared by the observing automaton $\bar{a}_0$, and $\bar{a}_0$ can draw a graph model of a_0's performance in these shared environments as demonstrated in Section 2.2. But to construct an artificial automaton that will replace a_0, a model must also describe a_0 operating in the nonshared environments during the performance and must also include a description of the feedback performance.

This section is concerned with the construction of the model for a_0 performing in both nonshared and shared environments with feedback when given the observed knowledge of a_0 performing in only the shared environments and the assumption that a_0 is performing like an automaton previously observed (Koenig 1975b). This requires that certain nonshared environments be made distinguishable from the processor of the automaton. Automata theory has been concerned with the realization of state and input-output behaviors of machines using feedback-free networks, that is, using cascades (Krohn, Rhodes 1965). Feedback has often been excluded by simplification in that feedback is included in "black boxes." But feedback has been generally accepted as an important feature of systems when they exhibit complex behavior. For one machine to realize another, feedback from the state of a realizing machine to an input encoder has been introduced and a class of mappings between digraphs established (Geller 1975). In this section, the performance of an automaton a_0 is observed and a_0 is either capable of performing like (realizing) another automaton in recording a response similar to that of another automaton (satisfactory, stable operation) or it is incapable and aborts the performance (unsatisfactory, unstable operation). The model is simplified when infrequent moments of observations obscure feedback. The analysis is based on the work of previous sections.

In Section 2.1, a first set of disciplines enabled an observing automaton to model automata as they are observed operating over a finite time interval and to model histories of automata. The models are in the form of graphs consisting of a processor graph, environment graph, and time graph. In Appendix A, a second set of disciplines enables an automaton to operate on graph models to describe effective operations of the modeled automata, i.e., to determine effective operation paths through the graph models. Appendix B presents corresponding disciplines for general systems of interactive automata.

3.4.1 Model for Performance in Both Shared and Nonshared Environments

An automaton a_0 has knowledge structures stored in its nonshared environments (memory) that are an integral part of the processor p^*. The knowledge is about an observed automaton a_x that recorded responses in shared environments. The goal of a_0 is to record similar responses in its own shared environments. a_0 obtained the knowledge about a_x either by direct observation or by communication in the natural language with another automaton who may have made the observation directly.

Given are knowledge structures stored in the nonshared environments e_c and e_{c+n+1} of a_0. Each of the knowledge structures consist of a graph of two points and an adjacent line represented by a (15+1)-tuple for an alternate environment graph defined by $TE_a(k, k+1)$ and $TE_a(k+1, k+2)$. The first describes operations of the observed automaton a_x at moments $m+k$, $m+k+1$ and the second of moments $m+k+1$, $m+k+2$.

Figure 3.2 shows the nonshared environments of a_0 in e_c and e_{c+n+1}, and the (15+1)-tuples contained in them are $TE_a(k, k+1)$ and $TE_a(k+1, k+2)$. a_0 is assumed to have already selected the principal and auxiliary receptors and effectors for receiving stimuli and recording the responses from knowledge

Figure 3.2 **(1)** s_c interpreted at nonshared environment e_c, q^*_{c+1} entered for recording a response similar to b_{k+1} in e_{c+1}. **(2)** A first transformation response f_{c+1} recorded at the shared environment e_{c+1}. **(3)** s_{c+2} interpreted from a_0's first recorded response at e_{c+1}.

(1)

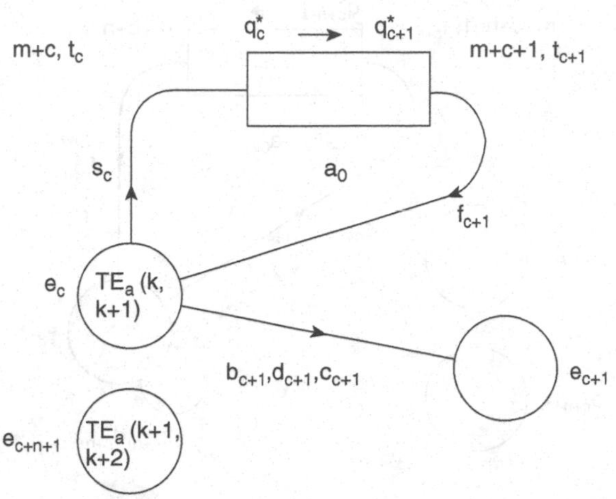
q_c^* → q_{c+1}^*
m+c, t_c
m+c+1, t_{c+1}
s_c
a_0
f_{c+1}
e_c
TE_a (k, k+1)
b_{c+1},d_{c+1},c_{c+1}
e_{c+1}
e_{c+n+1}
TE_a (k+1, k+2)

(2)

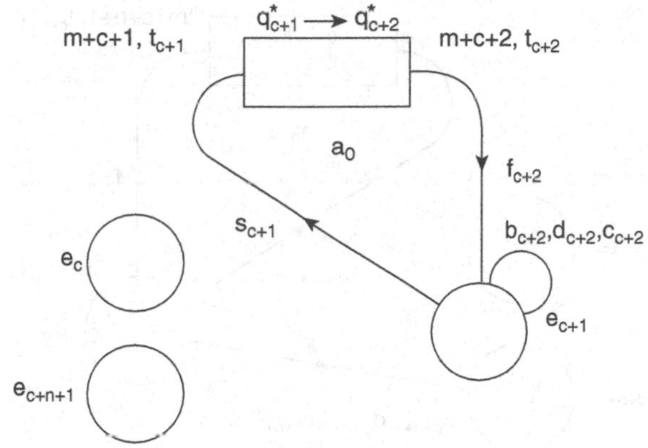
q_{c+1}^* → q_{c+2}^*
m+c+1, t_{c+1}
m+c+2, t_{c+2}
a_0
f_{c+2}
s_{c+1}
b_{c+2},d_{c+2},c_{c+2}
e_c
e_{c+1}
e_{c+n+1}

(3)

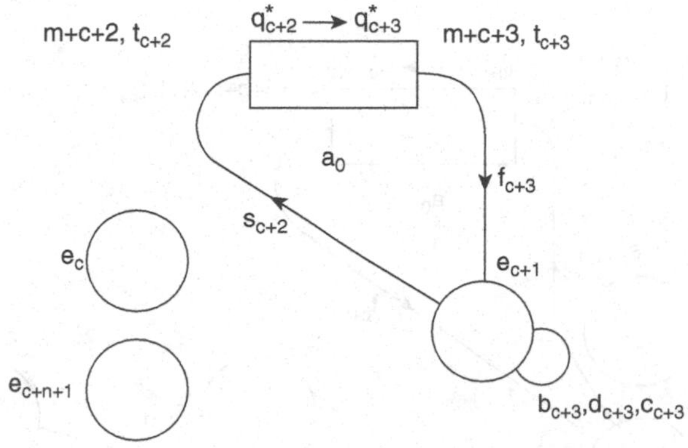
q_{c+2}^* → q_{c+3}^*
m+c+2, t_{c+2}
m+c+3, t_{c+3}
a_0
f_{c+3}
s_{c+2}
e_c
e_{c+1}
e_{c+n+1}
b_{c+3},d_{c+3},c_{c+3}

(*n*)

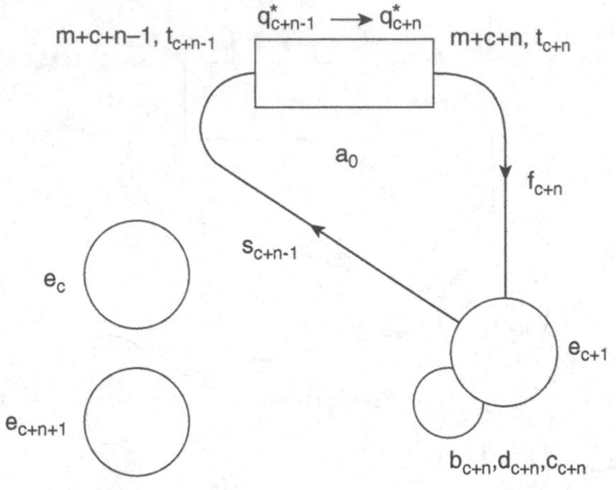
$q^*_{c+n-1} \rightarrow q^*_{c+n}$
$m+c+n-1, t_{c+n-1}$
$m+c+n, t_{c+n}$
a_0
f_{c+n}
s_{c+n-1}
e_c
e_{c+1}
e_{c+n+1}
$b_{c+n}, d_{c+n}, c_{c+n}$

(*n*+1)

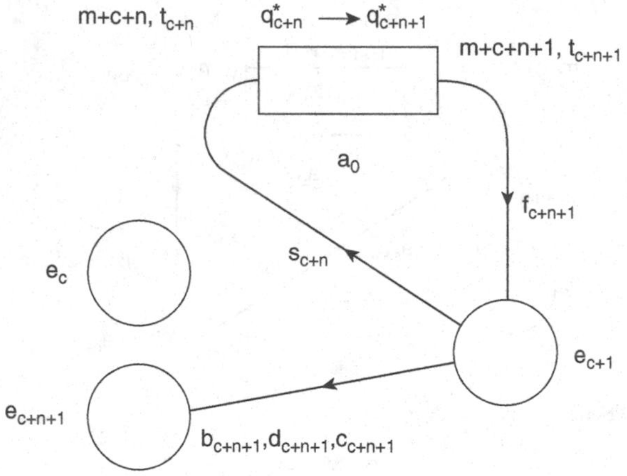
$m+c+n, t_{c+n}$
$q^*_{c+n} \rightarrow q^*_{c+n+1}$
$m+c+n+1, t_{c+n+1}$
a_0
f_{c+n+1}
s_{c+n}
e_c
e_{c+1}
e_{c+n+1}
$b_{c+n+1}, d_{c+n+1}, c_{c+n+1}$

(*n*+2)

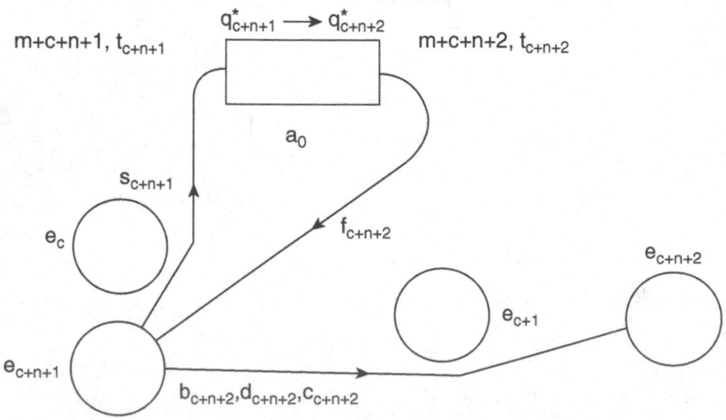
$q^*_{c+n+1} \rightarrow q^*_{c+n+2}$
$m+c+n+1, t_{c+n+1}$
$m+c+n+2, t_{c+n+2}$
a_0
s_{c+n+1}
f_{c+n+2}
e_c
e_{c+n+2}
e_{c+1}
e_{c+n+1}
$b_{c+n+2}, d_{c+n+2}, c_{c+n+2}$

Figure 3.2 (cont.) **(n)** s_{c+n-1} interpreted from a_0's recorded response b_{c+n-1}. **(n+1)** s_{c+n} interpreted from b_{c+n}, which is acceptably like b_{k+1}, and feedback is terminated. **(n+2)** s_{c+n+1} interpreted at nonshared environment e_{c+n+1} and sequence of operations repeat.

contained in the (15+1)-tuple of e_c. These receptors and effectors are not normally observable.

In Figure 3.2(1), a stimulus s_c, a function of the (15+1)-tuple of e_c, is received from e_c at moment $m+c$ by a_0 while in state q_c^* and the component d_{c+1} of the triadic response $(f_{c+1}, d_{c+1}, c_{c+1})$ produced at moment $m+c+1$ determines the environment e_{c+1} in a_0's shared environment corresponding to the originally observed environment e_k. In order that $TE_a(k, k+1)$ stored in e_c is not changed, f_{c+1} is required to be an identity transformation response. A state q_{c+1}^* is entered at moment $m+c+1$ such that when the automaton a_0 next receives a stimulus from e_{c+1}, it will record a response in e_{c+1} similar to b_{k+1} in the (15+1)-tuple of e_c. The (15+1)-tuple that describes the diagram of Figure 3.2(1) is $TE_a(c, c+1)$.

Figure 3.2(2) shows a_0 receiving the stimulus s_{c+1} from e_{c+1} at moment $m+c+1$ while a_0 is in state q_{c+1}^* and producing a response $(f_{c+2}, d_{c+2}, c_{c+2})$ at moment $m+c+2$, where s_{c+1}, q_{c+1}^*, f_{c+2} are similar to s_k, q_k^*, f_{k+1} in the (15+1)-tuple of e_c. The transformation response f_{c+2} is recorded as b_{c+2} in e_{c+1}, which is similar to b_{k+1} originally observed to have been the recorded response in e_k and described in the knowledge structure of e_c. Since e_{c+1} is a shared environment, a different set of principal and auxiliary receptors and effectors x''_{c+1}, x'_{c+1}, z'_{c+1}, z''_{c+1}, similar to x''_{k+1}, x'_{k+1}, z'_{k+1}, z''_{k+1}, are put in operation. These are assumed to have been previously selected by a_0. Note that s_{c+1} is not interpreted by a_0 from its own recorded response.

Before the automaton a_0 leaves the environment e_{c+1}, it will interpret its own recorded response as a stimulus, and if the difference between b_{k+1} and b_{c+2} is not sufficiently small, it will attempt to decrease the difference by recording a second response at e_{c+1}. But if the difference is sufficiently small (as determined by its own personal satisfaction or by command of some other automaton), it will proceed to a next environment location. In comparison with feedback theory for continuous systems, b_{k+1} associates with a reference R, and b_{c+2} associates with an output C (Chestnut, Mayer 1959). The difference between b_{k+1} and b_{c+2} associates with an error E (must compare b_{c+2} by the use of s_{c+2} interpreted from b_{c+2}). For a_0 to interpret its own recorded response as a stimulus at moment $m+c+2$, d_{c+2} is required to be null. The (15+1)-tuple that describes the diagram of Figure 3.2(2) is $TE_a(c+1, c+2)$.

The diagram of Figure 3.2(3) shows a_0 accessing e_{c+1} a second time, at moment $m+c+2$, receiving a stimulus s_{c+2}, which a_0 interprets from its own recorded response b_{c+2}. a_0 receives s_{c+2} while in state q_{c+2}^* at moment $m+c+2$ and enters state q_{c+3}^* at moment $m+c+3$. The (15+1)-tuple that describes the diagram of the figure is $TE_a(c+2, c+3)$.

There are diagrams for Figures 3.2(4), ---, 3.2(n – 1) (not shown), that are similar to the diagram for Figure 3.2(3), where the spatial change responses d_{c+4}, ---, d_{c+n} are null, $n \geq 4$. A triadic response is produced with d_{c+n+1} not null and f_{c+n+1} an identity response; i.e., feedback is terminated when the difference between b_{k+1} and b_{c+n} is satisfactorily small (associated with stability and with satisfactory operation). This feedback terminating operation is described by the diagram of Figure 3.2(n+1). The stimulus s_{c+n} results in a response which contains a spatial change response d_{c+n+1} that causes access of the environment e_{c+n+1} at moment $m+c+n+1$. For b_{c+n} not to be changed, f_{c+n+1} is required to be an identity transformation response, and b_{c+n+1} is identical with b_{c+n}. The (15+1)-tuples that describe the diagrams of Figures 3.2(4), ---, 3.2(n), 3.2(n+1) are $TE_a(c+3, c+4)$, ---, $TE_a(c+n-1, c+n)$, $TE_a(c+n, c+n+1)$. The corresponding (10+1)-tuples for the processor graph are $TP(c+3, c+4)$, ---, $TP(c+n-1, c+n)$, $TP(c+n, c+n+1)$. All stimuli s_{c+2}, ---, s_{c+n} are different from each other and produce corresponding different responses. Therefore, from Theorem 2.18, each of the states of the sequence q^*_{c+1}, q^*_{c+2}, ---, q^*_{c+n} might be identical with q^*_{c+1}.

Figure 3.2(n+2) follows and is similar to Figure 3.2(1). A change in the set of principal and auxiliary receptors and effectors employed in the operations is required. A recorded response b_{k+2} in the (15+1)-tuple of the nonshared environment e_{c+n+1} is to be duplicated in the shared environment e_{c+n+2} just as b_{k+1} in the (15+1)-tuple of e_c was to be duplicated in the shared environment e_{c+1}, and the operating procedure repeats. f_{c+n+2} is an identity response leaving $TE_a(k+1, k+2)$ in e_{c+n+1} unchanged. The (15+1)-tuple for the diagram is $TE_a(c+n+1, c+n+2)$.

There may be a different sequence of operations with d_{c+4}, ---, d_{c+t} null, $t \geq 4$ and with d_{c+t+1} not null for terminating feedback. This condition exists when the difference between b_{k+1} and b_{c+t} is prohibitively large (associated with instability and with unsatisfactory operation that can occur when a_0 is incapable of performing like the other automaton) and the operation of recording of similar sequences of responses in a shared environment is terminated. The stimulus s_{c+t} interpreted from b_{c+t} is received at moment $m+c+t$ while a_0 is in state q^*_{c+t} (identical with q^*_{c+1}) and produces a spatial change response d_{c+t+1} (not null), which causes a_0 to next access a nonshared environment e_{c+t+1}. This environment contains a (15+1)-tuple $TE_a(k'+1, k'+2)$, which initiates through s_{c+t+1}, a different sequence of recorded responses in shared environments. s_{c+t+1} interpreted at this environment at moment $m+c+t+1$ causes a response (f_{c+t+2}, d_{c+t+2}, c_{c+t+2}) where f_{c+t+2} is an identity transformation response and d_{c+t+2} causes access of a next shared environment e_{c+t+2} and the operation procedure repeats in the recording of a new sequence of responses. The complete environment graph for the shared and nonshared environments describing both conditions of feedback termination is shown in Figure 3.3, and the corresponding processor graph is shown in Figure 3.4. In both cases, only the points are identified.

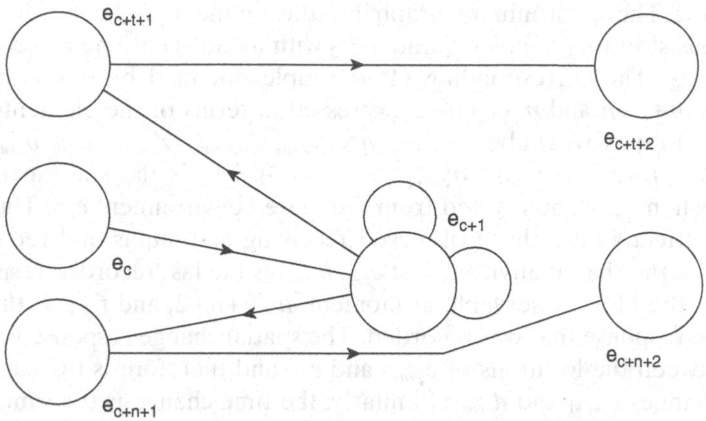

Figure 3.3 Environment graph for shared and nonshared environments for two conditions for feedback termination.

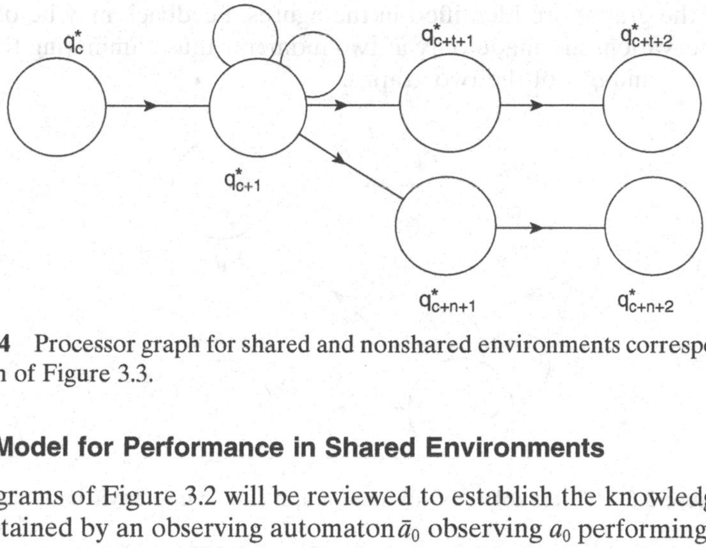

Figure 3.4 Processor graph for shared and nonshared environments corresponding to the graph of Figure 3.3.

3.4.2 Model for Performance in Shared Environments

The diagrams of Figure 3.2 will be reviewed to establish the knowledge structures obtained by an observing automaton $\bar{a}_0$ observing a_0 performing in environments e_{c+1} and e_{c+n+2}, which are the shared environments and which can be the only environments normally observed. The performance by Figure 3.2(2) is the first that can be observed because the stimulus is the first received from a shared environment, and the response is the first recorded in a shared environment. For the performances described by Figure 3.2(3), ---, 3.2(n), the only environment involved is the shared environment e_{c+1} and the performances are observable by $\bar{a}_0$. The combined performances for Figure 3.2(n+1) and 3.2(n+2) for moments $m+c+n$, $m+c+n+1$, $m+c+n+2$ involve a nonshared environment e_{c+n+1} and can be observed only as a single performance with moment $m+c+n+1$ obscured. Suppose no observation was made at moment

$m+c+n+1$. The environment graph for the moments $m+c+n$, $m+c+n+2$ would consist of two points e_{c+1} and e_{c+n+2} with an adjacent line directed from e_{c+1} to e_{c+n+2}. The corresponding (15+1)-tuple obtained by observations at moments $m+c+n$ and $m+c+n+2$ expressed in terms of the elements of the two (15+1)-tuples would be $(e_{c+1}, t_{c+n}, q^*_{c+n}, x''_{c+n}, x'_{c+n}, z'_{c+n}, z''_{c+n}, s_{c+n}, a_0, b_{c+n+1}, f_{c+n+1}, (d_{c+n+1}+d_{c+n+2}), (c_{c+n+1}+c_{c+n+2}), q^*_{c+n+2}, t_{c+n+2}, e_{c+n+2})$. q^*_{c+n} is the state at moment $m+c+n$ when s_{c+n} is interpreted from the shared environment e_{c+1}. The receptors and effectors are those observed receiving a stimulus and recording a response at the shared environment e_{c+1}. b_{c+n+1} is the last recorded response in e_{c+1} and is therefore observable at moment $m+c+n+2$, and f_{c+n+1} is the transformation response that was recorded. The spatial change response is the distance between the locations of e_{c+n+2} and e_{c+1} and therefore is the sum of the spatial changes d_{c+n+1} and d_{c+n+2}. Similarly, the time change is the sum of c_{c+n+1} and c_{c+n+2}. Finally, the state at moment $m+c+n+2$ is q_{c+n+2}.

The environment graph obtainable by an observing automaton $\bar{a}_0$ observing a_0 for the complete performance described in Figure 3.2 is shown in Figure 3.5, and the corresponding processor graph is shown in Figure 3.6. Only the points of the graphs are identified in the figures. Feedback may be obscured when observations are made only at two moments thus eliminating the loops at points e_{c+1} and q^*_{c+1} of the two graphs.

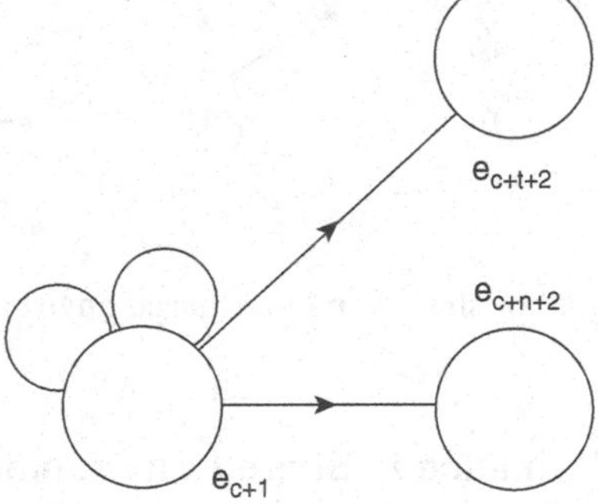

Figure 3.5 Observable environment graph (for shared environments).

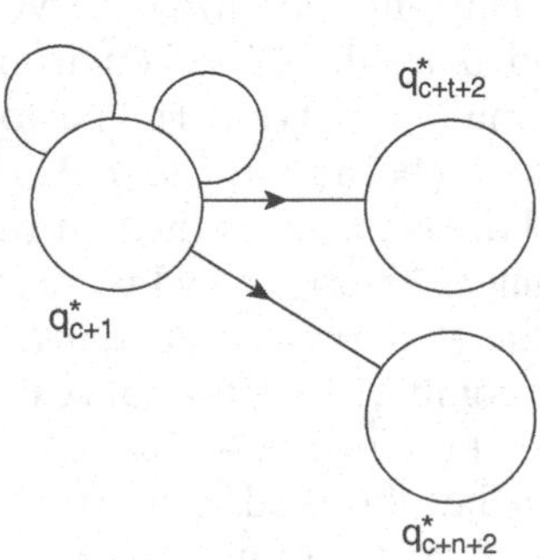

Figure 3.6 Processor graph corresponding to the graph of Figure 3.5.

The preceding remarks pertaining to an automaton performing in shared environments like an automaton previously observed are summarized in the following theorem and corollary:

Theorem 3.1. *Given the following knowledge structure obtained at moments* $m+i, m+i+1, \text{---}, m+i+u+1, u>1$, *by an observing automaton* $\bar{\alpha}_0$ *observing an automaton* a_0 *performing in shared environments* e_i *and* e_{i+u+1}: $TE_a(i, i+1)$. $TE_a(i+1, i+2), \text{---}, TE_a(i+u, i+u+1)$, where $a_i, a_{i+1}, \text{---}, a_{i+u}, (a_k = x''_k, x'_k, z'_k, z''_k, e_k; k=i, i+1, \text{---}, i+u)$, *are identical with* a_i. *If* a_0 *performed like an automaton previously observed in producing a similar recorded response* b_{i+u+1} *in* e_i *as a goal, and if this knowledge about the observed automaton,* $TE_a(k, k+1)$, $TE_a(k+1, k+2)$ *with* b_{k+1} *similar to* b_{i+u+1} *is stored in* a_0*'s nonshared environments* e_c *and* e_{c+n+1}, *the structure that models the performance of* a_0 *operating in both its nonshared and shared environments is represented by either of the following sets of* (15+1)*-tuples*: (i) *For the condition of stable, satisfactory performance of* a_0, $TE_a(c, c+1)$, $TE_a(c+1, c+2)$, ---, $TE_a(c+n+1, c+n+2)$, $n=u+1, n>2$, *obtained at moments* $m+c, m+c+1, \text{---}, m+c+n+2$. *The moments* $m+i+w, w=0, 1, \text{---}, u$, *identify with the moments* $m+c+w+1$, *and for any element* y_{i+w} *of the elements of* $TE_a(i, i+1)$, $TE_a(i+1, i+2)$, ---, $TE_a(i+u-1, i+u)$, y_{i+w} *identifies with the element* y_{c+w+1} *for* $TE_a(c+1, c+2)$, $TE_a(c+2, c+3)$, ---, $TE_a(c+n-1, c+n)$. *Moment* $m+i+u+1$ *identifies with* $m+c+n+2$, *and elements* $b_{i+u+1}, f_{i+u+1}, d_{i+u+1}, c_{i+u+1}$ *of* $TE_a(i+u, i+u+1)$ *identify with* $b_{c+n+1}, f_{c+n+1}, (d_{c+n+1}+d_{c+n+2}), (c_{c+n+1}+c_{c+n+2})$, *of* $TE_a(c+n, c+n+1)$, $TE_a(c+n+1, c+n+2)$. *Any element* y_{i+u}, *and* y_{i+u+1} *of the remaining elements identify with* y_{c+n} *and* y_{c+n+2}. (ii) *For the condition of unstable, unsatisfactory operation,* $TE_a(c, c+1)$, $TE_a(c+1, c+2)$, ---, $TE_a(c+t+1, c+t+2)$, $t=u+1$, $t \geq 2$. *To identify* $TE_a(i, i+1)$, $TE_a(i+1, i+2)$, ---, $TE_a(i+u, i+u+1)$ *with the structure, n is replaced by t in* (i). *The shared environments* e_i *and* e_{i+u+1} *identify with* e_{c+i} *and* e_{c+i+2}, *and* e_c *and* e_{c+i+1} *are the nonshared environments, and* b_{i+u+1} *and* b_{k+1} *are dissimilar.*

Corollary 3.1. $s_{i+1}, s_{i+2}, \text{---}, s_{i+u}, u \geq 1$ *are interpretations of* a_0*'s own recorded responses and* $q_i^*, q_{i+1}^*, \text{---}, q_{i+u}^*$ *may each be identical with* q_i^*.

EXERCISES

3.1 Detail the primitives for the following graphs to include the detailed analyses of Sections 3.1 and 3.2 for distinguishable receptors and effectors and for nonhomogeneous environments: (a) processor graph, (b) alternate environment graph, and (c) principal time graph.

3.2 Using Theorem 3.1(i), construct the alternate environment graph for $u=1$, $n=3$ for (a) observed operation in shared environments only, and (b) operation in both the shared and nonshared environments.

[illegible] summarized in the [illegible]

[illegible]

EXERCISES

3.1 [illegible]

3.2 [illegible]

CHAPTER 4

Processing of Knowledge About Automata

In this book knowledge is considered to be about automata and histories of automata.

Four basic combinations of types of stimuli and responses of an automaton processor can be observed over a finite time period. The combinations depend on whether the stimuli and responses are in the form of natural language. The four combinations are as follows:

1. No natural language is involved as a stimulus and as a response: Knowledge about an automaton can be observed by an automaton a_0, and later a_0 can perform like the automaton it previously observed.
2. Natural language is involved only as a response: Knowledge is observed by the automaton a_0, and later a_0 communicates to another automaton in the natural language the knowledge it observed.
3. Natural language is involved only as a stimulus: Knowledge is communicated to the automaton a_0 in the natural language, and later a_0 performs like the automaton described by the knowledge.
4. Natural language is involved both as a stimulus and as a response: Knowledge is communicated to the automaton a_0 in the natural language, and later a_0 communicates the knowledge to another automaton in the natural language.

Extracting and storing the meanings of sentences is then an important part of knowledge processing. Consistent with the above opening remark about knowledge, the following Language Information Theory (L.I. Theory) is proposed:

Theory 4.1. Much of what humans communicate over any finite time interval is knowledge about automata and histories of automata which they obtain by observing their environments.

Knowledge Structures for Communications in Human-Computer Systems: General Automata-Based, by Eldo C. Koenig

The environments of humans include their nonshared environments (their memory and parts of their automaton structure). The formulation of the L.I. Theory is presented in Section 4.1 and, in the most part, is limited to noninteracting automata extending the disciplines presented in Chapters 2 and 3. Some considerations for formulating the theory for interacting automata are given in Chapter 6 extending the disciplines presented in Chapter 5 (Koenig 1986a).

The disciplines for finding effective operation paths in knowledge structures after knowledge is stored as automaton graphs are presented in Appendix A for a general automaton and in Appendix B for Interacting automata.

Knowledge received at a current time by an automaton is associated with knowledge previously stored, thereby unifying knowledge structures and filling in missing knowledge elements. This knowledge processing for knowledge association is discussed in Section 4.3.

Arguments received by an automaton in the form of natural language may be either checked for validity and the conclusions stored along with the premises, or the conclusions may be ignored. Also, an automaton may establish conclusions as responses from knowledge received as premises. Processing knowledge for the deductive processes is discussed in Section 4.4.

To infer is to conclude or decide from something known or assumed and is considered here to demonstrate the simplest of intelligence in a conversation; for example, "From your smile, I infer that you are pleased." Establishing inferences is discussed in Section 4.5 and requires knowledge association, which is discussed in Section 4.3.

4.1 FORMULATION OF A LANGUAGE INFORMATION THEORY

The formulation of the L.I. Theory presented in this section extends the disciplines presented in Chapters 2 and 3 for a general automaton. The sentences considered are simple sentences, and the knowledge of a sentence is structured into a graph of two points and an adjacent line (or a single point and a loop) corresponding to a minimum knowledge structure. Seven principal classes of sentences are identified with a varying number of types (Koenig 1973, 1974). A relation for each type establishes the ordering of the basic sentence elements, and the relations are on the automaton sets.

In Section 2.1, a formal analysis was made of a general automaton, and graph primitives were established for the processor, environment, and time graphs. Further analysis was made on the automaton in Chapter 3, a) to make the receptors and effectors distinguishable (Section 3.1), b) to include nonhomogeneous environments (Section 3.2), and c) to establish transformation response components (Section 3.3). A summary of the operation of a general automaton and of the results of the analyses are given here.

The principal notations required are as follows:

$a_j \in A$	jth automaton of set A
$q_i^* \in Q^*$	state of the automaton A at some moment $m+i$
$s_i \in S$	stimulus received at moment $m+i$
$f_{i+1} \in F$	transformation response as a first component of a total response produced at moment $m+i$
$d_{i+1} \in D$	spatial change response as a second component of the total response
$c_{i+1} \in C$	time change response as a third component of the total response
$r_{i+1} \in R$	total response, $R \subseteq F \times D \times C$ at moment $m+i+1$
$q_{i+1}^* \in Q^*$	state at moment $m+i+1$
$e_i \in E$	environment accessed at moment $m+i$
$b_{i+1} \in B$	the recorded transformation response f_{i+1} in e_i at moment $m+i+1$
$e_{i+1} \in E$	environment accessed at moment $m+i+1$ whose location is the addition of the vectors for location e_i and d_{i+1}
$t_i \in T$	time of $m+i$ when e_i is accessed
$t_{i+1} \in T$	time of $m+i+1$ when e_{i+1} is accessed and equals t_i+c_{i+1}
$x_i'' \in X''$	auxiliary receptor at moment $m+i$
$x_i' \in X'$	principal receptor at moment $m+i$
$z_i' \in Z'$	principal effector at moment $m+i$
$z_i'' \in Z''$	auxiliary effector at moment $m+i$
$fs_{i+1} \in FS$	component transformation response that provides an awareness for an observer through sight at moment $m+i+1$
$fm_{i+1} \in FM$	component transformation response that provides an awareness for an observer through smell at moment $m+i+1$
$fh_{i+1} \in FH$	component transformation response that provides an awareness for an observer through hearing at moment $m+i+1$
$ft_{i+1} \in FT$	component transformation response that provides an awareness for an observer through touch at moment $m+i+1$
$fa_{i+1} \in FA$	component transformation response that provides an awareness for an observer through taste at moment $m+i+1$
$Tu_k(m+i, m+i+1)$	graph tuple for moments $m+i$ and $m+i+1$; $u=E, T, Q$; $k=a$ (alternate), p (principal); $Tq_a(m+i, m+i+1)$ does not exist

$T_u(m+i, m+i+1)$	graph tuple for moments $m+i$ and $m+i+1$ for either alternate or principal graphs
-,	an unknown or unimportant value for an element in a graph tuple
0,	a nonexistent value or zero-value for an element in a graph tuple
Auv (r-variable)	a function of r-variables, $u=E, T, Q$; $v=1, 2, ---$
Auv (r-values)	r-place proposition, $u=E, T, Q$; $v=1, 2, ---$
Auv (t)	a proposition, $u=E, T, Q$; $v=1, 2, ---$, and t denotes a graph (n+1)-tuple of (n+1)-value places
Tuy	graph tuple for arbitrary moments and for either alter nate or principal graphs, $u=E, T, Q$; $y=1, 2, ---$
$WFFij$	well-formed formulas for the ith class and jth type sentences
$Fr'\ nj$	set of elements of the transformation response compo nent for an nth class and jth type sentence, $r=S, M, H, T, A$; $n=1, 2, 5, 6, 7$

A single general automaton was analyzed over a moment base $M=\{m+i \mid i \in I\}$, $I=\{0, 1, 2, ---, h\}$ in an environment of space R^n and was considered to operate by the following principle (see Figures 2.1 and 3.1):

General Automaton Principle. Given a discrete moment $m+i$, a general automaton is accessing one and only one environment (stimulus location) from a set $E \subseteq R^n$ of the environment space, receiving one and only one stimulus from a set S of stimuli through one and only one auxiliary receptor and one and only one principal receptor from sets X'' and X' of receptors, is in one and only one state from a set Q^* of possible states, and has one and only one time associated with the moment $m+i$ from a set T of times. Multiple accesses of a same environment during a moment base M are made only by the subject automaton. Furthermore, during the next moment $m+(i+1)$, the automaton produces one and only one triadic response from a set R of responses through one and only one principal effector and one and only one auxiliary effector from sets Z' and Z'' of effectors based on the aforementioned conditions. The triadic response consists of an ordered triple from a set F of transformation responses, a set D of spatial change responses, and a set C of time change responses. The pentadic transformation response at moment $m+(i+1)$ consists of the ordered quintuple from sets FS, FM, FH, FT, FA of responses interpreted as stimuli for sight, smell, hearing, touch, and taste and is recorded as a response from the set B of recorded responses at the environment accessed at moment $m+i$.

The diagram of Figure 4.1 is an extension of that of Figure 2.1 and illustrates the general automaton principle. The broken line part of the diagram could

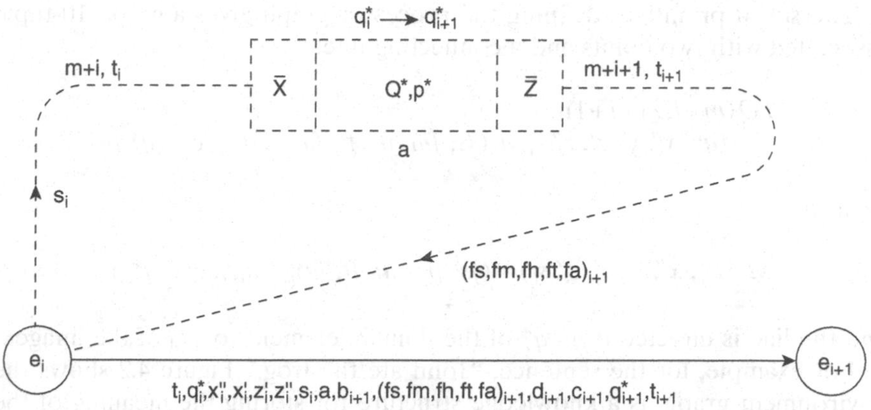

Figure 4.1 A diagram to illustrate the general automaton principle.

be drawn for each environment to show the operation of the automaton in its environments. Instead, all the information pertaining to the automaton is brought together in two points and an adjacent line of the alternate environment graph shown by the solid line part of the diagram. That is, the graph (15+1)-tuple is considered to contain complete knowledge of the automaton related to the consecutive moments $m+i$, $m+i+1$.

The alternate environment graph, principal time graph, and processor graph are chosen for the graph model of a general automaton. Recall, both the alternate environment graph and the principal time graph contain maximum knowledge, i.e., their graph tuples contain a maximum number of elements. In summary, the set of four primitives defining the alternate environment graph gives a graph (15+1)-tuple associated with two points (not necessarily distinct) and a connecting line

$$TE_a(m+i, m+i+1) = (e_i, t_i, q_i^*, x_i'', x_i', z_i', z_i'', s_i, a, \\ b_{i+1}, (fs, fm, fh, ft, fa)_{i+1}, d_{i+1}, c_{i+1}, q_{i+1}^*, t_{i+1}, e_{i+1})$$

such that

$$\Lambda^{*\prime}(e_i, t_i, q_i^*, x_i'', x_i', z_i', z_i'', s_i) \\ = (b_{i+1}, (fs, fm, fh, ft, fa)_{i+1}, d_{i+1}, c_{i+1}, q_{i+1}^*, t_{i+1}, e_{i+1})$$

and the line is directed from e_i of the domain element to e_{i+1} of the image.

The set of primitives defining the principal time graph also gives a graph (15+1)-tuple associated with two points and a connecting line. It is the same as that of the alternate environment graph except the elements e_i and t_i, and the elements e_{i+1} and t_{i+1} are interchanged, and the function $\Lambda^{*\prime}$ is replaced with the graph function Λ^*.

The set of primitives defining the processor graph gives a graph 10-tuple associated with two points and a connecting line

$$TQ(m+i, m+i+1)$$
$$= (q_i^*, x_i'', x_i', z_i', z_i'', s_i, a, (fs, fm, fh, ft, fa)_{i+1}, d_{i+1}, c_{i+1}, q_{i+1}^*)$$

such that

$$\Omega^*(q_i^*, x_i'', x_i', z_i', z_i'', s_i) = ((fs, fm, fh, ft, fa)_{i+1}, d_{i+1}, c_{i+1}, q_{i+1}^*)$$

and the line is directed from q_i^* of the domain element to q_{i+1}^* of the image.

For example, for the sentence, "John ate the frog," Figure 4.2 shows the environment graph as a knowledge structure for storing the meaning of the sentence using the general automaton concept. John$\in A$, (frog, -, -, -)$\in E$ at some moment $m+i$, ate$\in FS$ at moment $m+i+1$, -frog$\in B$ at $m+i+1$. This example will be used again later to demonstrate how past knowledge is used to fill in missing knowledge elements.

Seven classes of sentences are analyzed here for their meanings with a varying number of types per class. Let the ordering of the basic sentence elements for the ith class and jth type sentence that contribute meaning be given by the set of well-formed formulas $WFFij$, which is a relation on automaton sets; $i=1, 2, ---, 7; j=1, 2, ---, m_i$. What is observed about automata and histories of automata at two consecutive moments of observation refers to words, defined symbols, or numbers for elements of the 16 automaton sets. Let the set of elements of the transformation response component Fr for an nth class and jth type sentence be the set $Fr'nj$, $r=S, M, H, T, A$; $n=1, 2, 5, 6, 7$. Let set $Frnj$ be a dyadic relation on the sets $Fr'nj$, FAD; $Frnj \subseteq Fr'nj \times FAD$, where FAD is a set of modifier words that modify the words of the set $Fr'nj$. The sets $Fr'nj$, FAD, and all other automaton sets each contain a null element, and these null elements may appear in the $(n+1)$-tuples of the orderings of the basic sentence elements when other elements of the sets do not appear in sentences, or the knowledge that these other elements represent are not observed. There is a very limited number of word symbols in the English language as elements of some of the sets of component transformation responses, e.g., the set $FMij$.

4.1.1 Class 1 Sentence

There is a class of sentence that conveys knowledge related to an automaton producing a transformation response on an environment with the use of a

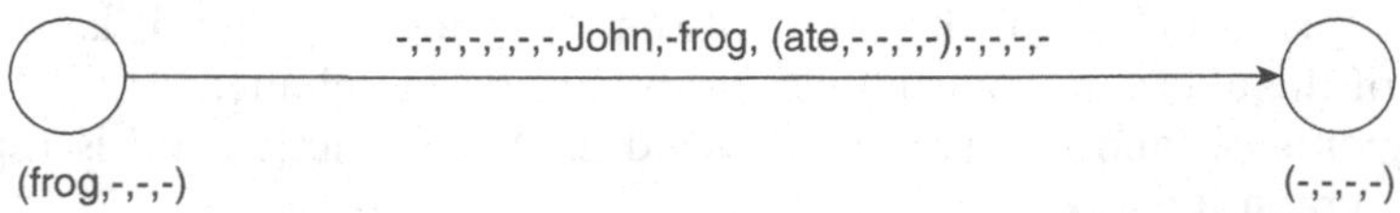

Figure 4.2 Storing the meaning of the sentence "John ate a frog."

principal effector and with or without the use of an auxiliary effector. This most often involves the transformation response component for a sight stimulus *FS*. Three principal types for this class are presented. The orderings of the basic sentence elements of the sets *WFF*11, *WFF*12, *WFF*13 and all have second acceptable orderings. The sets *Fr*11, *Fr*12, *Fr*13 are the sets of words that refer to the transformation responses for the three principal types of sentences.

Type 1. A Type 1, Class 1 sentence is illustrated by the sentences

"Today, John struck the table very hard with a hammer."
"Today, John struck the table with his fist."

For the first example, today$\in T$, John$\in A$, struck$\in FS'11$, table$\in E$, (very hard)$\in FAD$, hammer$\in Z''$, (struck, very hard)$\in FS11$. *WFF*11 is a hexadic relation

$$WFF11 \subseteq T \times A \times Fr'11 \times E \times FAD \times (Z' \cup Z'')$$

or

$$WFF'11 \subseteq A \times Fr'11 \times E \times FAD \times (Z' \cup Z'') \times T$$
$$(\text{today, John, struck, table, very hard, hammer}) \in WFF11$$

The sextuple for the sentence

"John struck the table."

contains null elements for T, FAD, and Z''.

Consider the alternate environment graph for structuring the knowledge of the first example sentence. The minimum graph consists of two points and an adjacent line (or a point and a loop) described by the (15+1)-tuple

$$TE_a(m+i, m+i+1) = (e_i, t_i, q_i^*, x_i'', x_i', z_i', z_i'', s_i, a, b_{i+1}, (fs, fm, fh, ft, fa)_{i+1}, d_{i+1}, c_{i+1}, q_{i+1}^*, t_{i+1}, e_{i+1})$$

The tuple for the first example is

(table, today, -, -, -, -, hammer, -, John,
-, ((struck, very hard), -, -, -, -), -, -, -, today, -)

All but six elements of the (15+1)-tuple are assumed null. All elements may not be null if knowledge about that same automaton and environment had been previously received and the knowledge given by the example sentence had been added to that same knowledge structure. For example, if the automaton that received this knowledge had previous knowledge about a hammer, he would have associated the current knowledge with that previous knowledge to establish that the hammer was held in the hand. This is described as knowledge association and is discussed in Section 4.3.

A Type 1A sentence conveys the same knowledge as Type 1 and is demonstrated by the example sentence

"Today, a hammer was used by John to strike the table very hard."

The relation is

$$WFF1A \subseteq T \times (Z' \cup Z'') \times Fr'11A \times A \times (Fr'11 \times E \times FAD)$$

Here, the set $Fr'\,11A$ may be necessary to classify a sentence for extracting its meaning, but the element of $Fr'\,11A$ need not be stored in a graph, and the graph for this Type 1A and Type 1 will be the same.

Type 2. A second type sentence is similar to the first type for this first class as demonstrated by the sentences

"Today, Jim's hammer struck the table very hard."
"Yesterday, John's fist struck the table."

*WFF*12 is a pentadic relation

$$WFF12 \subseteq T \times (Z' \cup Z'') \times Fr'12 \times E \times FAD$$

or

$$WFF'12 \subseteq (Z' \cup Z'') \times Fr'12 \times E \times FAD \times T$$

The quintuple for the first example sentence is

$$(\text{today, Jim's hammer, struck, table, very hard}) \in WFF12$$

A Type 2A sentence conveys the same knowledge as Type 2. It is illustrated by a sentence consistent with the above example sentence

"Today, the table was struck very hard with Jim's hammer."

The relation is

$$WFF12A \subseteq T \times E \times Fr'12A \times FAD \times (Z' \cup Z'')$$

Type 3. The elements of the recorded response set B appear in a Type 3, Class 1, sentence. This type is illustrated by the sentences

"Today, Jim's hammer quickly made a dent in the table."
"Mary unknowingly made a scratch on the car today."
"Today, Mary's fingernail made a scratch on John's arm."

*WFF*13 is a hexadic relation

$$WFF13 \subseteq T \times (A \cup Z' \cup Z'') \times FAD \times Fr'13 \times B \times E$$

or

$$WFF'13 \subseteq (A \cup Z' \cup Z'') \times FAD \times Fr'13 \times B \times E \times T$$

The tuple for the first example is

(table, today, -, -, -, -, Jim's hammer, -, -,
dent, ((made, quickly), -, -, -, -), -, -, -, today, -)

where

$$\text{table} \in E,\ \text{today} \in T,\ (\text{Jim's hammer}) \in Z'',\ \text{dent} \in B,\ (\text{made, quickly}) \in FS13$$

The knowledge of Type 3 sentence is also conveyed by a Type 3A which is demonstrated by the sentence

"Today, a dent was made quickly in the table by Jim's hammer."

$$WFF13A \subseteq T \times B \times Fr'13 \times FAD \times E \times (A \cup Z' \cup Z'')$$

4.1.2 Class 2 Sentence

There is a class of sentences that conveys knowledge related to the movement of an automaton structure from one environment location to another with or without an auxiliary effector and with or without the relocation of an environment. Eight types of this class are presented. The orderings of the basic sentence elements are the elements of the sets *WFF*21, *WFF*22, ---, *WFF*28, and all types have alternate orderings.

Type 1. A first type of this class is illustrated by the sentence

"Today, John drove the car carefully three blocks from the house to the store in five minutes."

The relation *WFF*21 is

$$WFF21 \subseteq T \times A \times Fr'21 \times Z'' \times FAD \times D \times E \times E \times C$$

Three alternates are

$$WFF'21 \subseteq A \times Fr'21 \times Z'' \times FAD \times D \times E \times E \times T \times C$$

$$WFF''22 \subseteq T \times A \times FAD \times Fr'21 \times Z'' \times D \times E \times E \times C$$

$$WFF'''21 \subseteq A \times FAD \times Fr'21 \times Z'' \times D \times E \times E \times T \times C$$

The 9-tuple for the example sentence and for the condition of a maximum of null elements are the elements of the set

{(today, John, drove, car, carefully, three blocks, house, store, five minutes), (-, John, drove, -, -, -, -, -, -)} $\subseteq WFF21$

The corresponding (15+1)-tuple of the sentence for the alternate environment graph is

(house, today, -, -, -, -, car, -, John, -, ((drove, carefully), -, -, -, -), 3 blocks, 5 minutes, -, today, store)

Type 2. A second type is similar to the first type for this second class as demonstrated by the sentence

"Today, the car traveled slowly three blocks from the house to the store in five minutes."

*WFF*22 is an octadic relation

$$WFF22 \subseteq T \times Z'' \times Fr'22 \times FAD \times D \times E \times E \times C$$

or

$$WFF'22 \subseteq Z'' \times Fr'22 \times FAD \times D \times E \times E \times T \times C$$

Type 3. A third type is demonstrated by the sentence

"Today, John promptly left the house by car at 10 a.m."

*WFF*23 is a heptadic relation

$$WFF23 \subseteq T' \times A \times FAD \times Fr'23 \times E \times Z'' \times T', \ T \subseteq T' \times T'$$

Three alternates are

$$WFF'23 \subseteq A \times FAD \times Fr'23 \times E \times Z'' \times (T' \times T')$$

$$WFF''23 \subseteq T' \times A \times Fr'23 \times E \times FAD \times Z'' \times T', T \subseteq T' \times T'$$

$$WFF'''23 \subseteq A \times Fr'23 \times E \times FAD \times Z'' \times (T' \times T')$$

The graph (15+1)-tuple for the example sentence is

(house, (today, 10 a.m.), -, -, -, -, car, -, John,
-, ((left, promptly), -, -, -, -), -, -, -, -, -)

Type 4. A fourth type is similar to the third type of this second class as illustrated by the sentence

"Today, the car slowly left the house at 10 a.m."

*WFF*24 is a hexadic relation

$$WFF24 \subseteq T' \times Z'' \times FAD \times Fr'24 \times E \times T', T \subseteq T' \times T'$$

Three alternates are

$$WFF'24 \subseteq Z'' \times FAD \times Fr'24 \times E \times (T' \times T')$$

$$WFF''24 \subseteq T' \times Z'' \times Fr'24 \times E \times FAD \times T', T \subseteq T' \times T'$$

$$WFF'''24 \subseteq Z'' \times Fr'24 \times E \times FAD \times (T' \times T')$$

Type 5. A fifth type of sentence parallels the third type and is illustrated by the sentence

"Today, John arrived late at the store by car at 10:05 a.m."

*WFF*25 is a heptadic relation

$$WFF25 \subseteq T' \times A \times Fr'25 \times FAD \times E \times Z'' \times T', T \subseteq T' \times T'$$

Three alternates are

$$WFF'25 \subseteq A \times Fr'25 \times FAD \times E \times Z'' \times (T' \times T')$$

$$WFF''25 \subseteq T' \times A \times Fr'25 \times E \times FAD \times Z'' \times T', T \subseteq T' \times T'$$

$$WFF'''25 \subseteq A \times Fr'25 \times E \times FAD \times Z'' \times (T' \times T')$$

The graph (15+1)-tuple for the example sentence is

(-, -, -, -, -, -, car, -, John, -, ((arrived, late), -, -, -, -), -, -, -, (today, 10:05 a.m), store)

Type 6. A Type 6 sentence is similar to the Type 5 sentence and is demonstrated by the sentence

"Today, the car arrived abruptly at the store at 10:05 a.m."

*WFF*26 is a hexadic relation

$$WFF26 \subseteq T' \times Z'' \times Fr'26 \times FAD \times E \times T', T \subseteq T' \times T'$$

Three alternates are

$$WFF'26 \subseteq Z'' \times Fr'26 \times FAD \times E \times (T' \times T')$$

$$WFF''26 \subseteq T' \times Z'' \times Fr'26 \times E \times FAD \times T', T \subseteq T' \times T'$$

$$WFF'''26 \subseteq Z'' \times Fr'26 \times E \times FAD \times (T' \times T')$$

Type 7. A Type 7 sentence and its companion, Type 8, to be presented, refers to the relocation of an environment through the movement of an automaton structure from one environment to another. It is illustrated by the sentence

"Today, John cautiously took a package three blocks by car from the house to the store in five minutes."

The relation is

$$WFF27 \subseteq T \times A \times FAD \times Fr'27 \times B' \times D \times Z'' \times E' \times C, E \subseteq E' \times B'$$

where the elements of B have the negatives of the elements of B', $B = -B'$. Three alternates are

$$WFF'27 \subseteq A \times FAD \times Fr'27 \times B' \times D \times Z'' \times E' \times E' \times T \times C$$

$$WFF''27 \subseteq T \times A \times Fr'27 \times B' \times FAD \times D \times Z'' \times E' \times E' \times C$$

$$WFF'''27 \subseteq A \times Fr'27 \times B' \times FAD \times D \times Z'' \times E' \times E' \times T \times C$$

The graph (15+1)-tuple for the example sentence is

((package, -, -, house), today, -, -, -, -, car, -, John, -package, ((took, cautiously), -, -, -, -), 3 blocks, 5 minutes, -, today, (package, -, -, store))

The identifications for the environments in the (15+1)-tuple will be defined in the discussion of Class 4 sentences.

Type 8. A Type 8 sentence is similar to Type 7 and is demonstrated by a similar sentence

"Today, the car slowly carried the package three blocks from the house to the store in five minutes."

$$WFF28 \subseteq T \times Z'' \times FAD \times Fr'28 \times B' \times D \times E' \times E' \times C, E \subseteq E' \times B', B = -B'$$

There are the alternates

$$WFF'28 \subseteq Z'' \times FAD \times Fr'28 \times B' \times T \times D \times E' \times E' \times C$$

$$WFF''28 \subseteq T \times Z'' \times Fr'28 \times B' \times FAD \times D \times E' \times E' \times C$$

$$WFF'''28 \subseteq Z'' \times Fr'28 \times B' \times FAD \times D \times E' \times T \times C$$

4.1.3 Class 3 Sentence

A class of sentence is related primarily to the states of an automaton. Given an automaton that receives a same stimulus at two different moments, by the principles of operation, if the states of the automaton are the same for the two moments, the responses are the same. If the states are different, the responses are different. Some of the states of an automaton can be observed and the knowledge conveyed in the word language. This knowledge helps the receiver to understand the performance of the automaton. If the states and the moments that the states change cannot be observed, the knowledge can sometimes be determined by deduction as demonstrated in Section 2.2.

Type 1. There is a single type for this Class 3 sentence. It is illustrated by the following sentences:

a. "John is happy."
b. "John became angry."
c. "John went into a state of shock."
d. "John remained intoxicated."

*WFF*31 is a dyadic relation

$$WFF31 \subseteq A \times Q^*$$

The elements of the automaton sets other than those of A and Q^* are not involved. The words **became**, **remain**, **is**, and similar words are not elements of the transformation response but determine the structuring of the knowledge of sentences.

The processor graph can serve as the knowledge structure for this class sentence. Recall, the graph (10+1)-tuple, $TQ\ (m+i,\ m+i+1)$, is

$$(q_i^*, x_i'', x_i', z_i', z_i'', s_i, a, (fs, fm, fh, ft, fa)_{i+1}, d_{i+1}, c_{i+1}, q_{i+1}^*)$$

and the following (10+1)-tuples are for the example sentences (a), (b), (d):

(happy, -, -, -, -, -, John, (-, -, -, -, -), -, -, -)

(-, -, -, -, -, -, John, (-, -, -, -, -), -, -, angry)

(intoxicated, -, -, -, -, -, John, (-, -, -, -, -), -, -, intoxicated)

4.1.4 Class 4 Sentence

Thus far, attention has been given to knowledge pertaining to performing automata. Humans also wish to communicate what they observe of histories of automata and of the structural features of automata. Identification is accomplished through functions performed and physical descriptions. Types of sentences of this class relate to identification for environments and for automata as special environments. Effectors and receptors are also considered as special cases.

Identification of Environments. Two types of sentences relate to the identification of environments. Identifiers are expressed in the form of n-tuples. Let E' be a family of sets of environments, $E'=\{E_i'\}$, $i=1, 2, ---, m$, where E_i' is a set of subenvironments. Let $EN \subseteq ENN^n$ and let $EA \subseteq EA1 \times EA2 \times --- \times EAk$, where ENN is a set of identifiers of special functions and classifications of the subenvironments and EAj, $j=1, 2, ---, k$, are sets of identifiers that convey special knowledge obtained through the observer's receptors directly. The elements of EN affect the operation of the automaton whereas those of EA do not. Each of the sets include a null element. Let E_i' serve as the domain set for the functions f_i and g_i. For $f_i(E_i')$ and $g_i(E_i')$, the range sets, let $f_i(E_i') \subseteq EN$ and $g_i(E_i') \subseteq EA$. Then E as a set of identifiers of environments is given by

$$E \subseteq \{(e_i, f_i(e_i), g_i(e_i), E_i') | e_i \in E_i' \in E', f_i(e_i) \in EN, g_i(e_i) \in EA\}$$

To identify composite environments, two or more of these elements of E may be associated. The association of the two elements (**1**, **2**, **3**, **4**) and (**1***, **2***, **3***, **4***) may be written in either of two ways depending on the composition of the environments.

$$((\mathbf{1}^*, \mathbf{2}^*, \mathbf{3}^*), (1, \mathbf{2}, \mathbf{3}), 4) \in E$$

$$((\mathbf{1}^*, \mathbf{2}^*, \mathbf{3}^*), \mathbf{1}, \mathbf{2}, \mathbf{3}, \mathbf{4}) \in E$$

The first identifies, for example, an environment consisting of a book and two of its chapters, and the second identifies, for example, an environment consisting of a book, a chapter of the book, and a section of the chapter.

In Section 4.1.1, there was the example sentence

"Today, John struck the table very hard with a hammer."

Now, with (table, -, -, library) $\in E$, the sentence can be modified to become

"Today, John struck the table very hard with a hammer in the library."

Examples are presented later in this section and in subsequent sections using this definition of E. For the example sentence for *WFF*27, (package, -, -, house) $\in E$; **package** is e_i and **house** is E_i'. Without the package, house $\in E$.

Type 1. A first type, Class 4 sentence is illustrated by the following sentences:

"Gold is beautiful."
"Gold is a beautiful metal."
"Copper is a hardening metal."
"Gold is a metal of beauty."
"The alloy of gold is a good hardener."
"Gold's alloy is a good hardener."
"Gold's alloy is a beauty."
"Gold has a copper alloy."
"The yellow metal is gold."
"The hardening alloy is copper."

*WFF*41 is a tetradic relation

$$WFF41 \subseteq E_i' \times EA \times EN \times E'$$

Alternates are

$$WFF'41 \subseteq E_i' \times E' \times EA \times EN$$

$$WFF''41 \subseteq E' \times E_i' \times EA \times EN$$

$$WFF'''41 \subseteq EA \times EN \times E' \times E_i'$$

The word **is** and similar words do not identify with any of the elements of the automaton sets. When generating sentences, the words are obtained from separate lists for the different orderings.

The ordered quadruple for the second example sentence is

$$(\text{gold, beautiful, -, metal}) \in WFF41$$

and the environment is identified as

$$(\text{gold}, \text{-}, \text{beautiful}, \text{metal}) \in E$$

(from first definition of E).

Type 2. A second type sentence requires two elements of the set E for identification, and these two may then be associated to form a composite environment. An example sentence is

"The beautiful metal with the copper alloy is gold."

$WFF42$ is a pentadic relation

$$WFF42 \subseteq EA \times EN \times E' \times E^{*\prime}_i \times E'_i$$

And the composite $e \in E$ is identified as

$$((e^*_i \in E^{*\prime}_i, \text{-} \in EN, \text{-} \in EA), e_i \in E'_i, f_i(e_i) \in EN, g_i(e_i) \in EA, E'_i \in E') \in E$$

where "-" is the null element. For the example sentence, the composite e is

$$((\text{copper alloy}, \text{-}, \text{-}), \text{gold}, \text{-}, \text{beautiful}, \text{metal}) \in E$$

Identification of Automata. When an automaton is observed in an environmental role, it is considered to have identifier elements in the set E, and knowledge for the identifier elements come from sentences of Types 1, 2. When an automaton is observed in a performance role, it is considered to have identifier elements in the set A, and knowledge for the elements will be considered as coming from sentences of Types 3, 4 as special cases of Types 1, 2.

Let A' be a family of sets of automata, $A' = \{A'_i\}$, $i = 1, 2, ---, m$, where A'_i is a set of parts of automata. Let $AN \subseteq ANN^n$ and let $AA \subseteq AA1 \times AA2 \times --- \times AAk$, where ANN is a set of identifiers of special functions and classifications of the automaton parts and AAj, $j = 1, 2, ---, k$, are sets of identifiers that convey special knowledge obtained directly through the observer's receptors. The elements of AN affect the operation of the automaton whereas those of AA do not. Each of the sets include a null element. Let A'_i serve as the domain set for the functions f'_i and g'_i. For $f'_i(A'_i)$ and $g'_i(A'_i)$, the range sets, let $f'_i(A'_i) \subseteq AN$ and $g'_i(A'_i) \subseteq AA$. Then A as a set of identifiers of automata is given by

$$A \subseteq \left\{ \left(a_i, f'_i(a_i), g'_i(a_i), A'_i\right) \middle| a_i \in A'_i \in A', f'_i(a_i) \in AN, g'_i(a_i) \in AA \right\}$$

To identify automata in detail, two or more of these elements of A may be associated. The association may be written in the same manner as that for environments

$$((\mathbf{1}^*, \mathbf{2}^*, \mathbf{3}^*), (1, \mathbf{2}, \mathbf{3}), 4) \in A$$

$$((\mathbf{1}^*, \mathbf{2}^*, \mathbf{3}^*), \mathbf{1}, \mathbf{2}, \mathbf{3}, \mathbf{4}) \in A$$

The first identifies, for example, an automaton and its two hands, and the second identifies, for example, an automaton, its arm, and a hand as part of the arm.

Type 3. A Type 3, Class 4 sentence is illustrated by the following example sentences:

"Mary is beautiful."
"Mary is a beautiful girl."
"Mary is a secretary."
"Mary is a girl of beauty."
"The hair of Mary is yellow."
"Mary's hair is a good sun-shade."
"Mary has strong arms."
"The tall girl is Mary."

*WFF*43 is a tetradic relation

$$WFF43 \subseteq A_i' \times AA \times AN \times A'$$

Alternates are as follows:

$$WFF'43 \subseteq A_i' \times A' \times AA \times AN$$

$$WFF''43 \subseteq A' \times A_i' \times AA \times AN$$

$$WFF'''43 \subseteq AA \times AN \times A' \times A_i'$$

The ordered quadruple for the second example sentence is

$$(\text{Mary, beautiful, -, girl}) \in WFF43$$

and the automaton is identified as

$$(\text{Mary, -, beautiful, girl}) \in A$$

Type 4. A fourth type sentence requires two elements of the set A for identification, and these two may then be associated. An example sentence is

"The index finger of Joe's hand is long."

*WFF*44 is a pentadic relation

$$WFF44 \subseteq A_i^{*\prime} \times A' \times A_i' \times AA^* \times AN^*$$

And the composite automaton $a \in A$ is identified as

$$((a_i^*, \in A_i^{*\prime}, f_i^{*\prime}(a_i^*) \in AN^*, g_i^{*\prime}(a_i^*) \in AA), a_i \in A_i', - \in AN,$$
$$- \in AA, A_i' \in A') \in A$$

For the example sentence, the composite a is

$$((\text{index finger}, -, \text{long}), \text{hand}, -, -, \text{Joe}) \in A$$

Identification of Receptors and Effectors. Just as an automaton is a special case of an environment when the automaton is observed in a performance role, the receptors may be considered special cases of environments and automaton structure when they are observed in a performance role. That is, x'' and x' appear in graph $(n+1)$-tuples.

There are the types of sentences. Types 5, 6, Types 7, 8, Types 9, 10, Types 11, 12 of Class 4 that provide knowledge for each of the sets X', Z', X'', Z''. By replacing the letter A with X', Z' and E with X'', Z'' in the definitions and relations established in the discussions above, there are obtained corresponding sets and relations for receptor and effector identification.

4.1.5 Class 5 Sentence

Thus far, the stimulus set has not been referenced. There is a 5th class of sentence that names the type of stimulus received by an observing automaton and conveys what was observed. The type of stimulus implies the type of receptor used. An observing automaton interprets recorded responses of automata in environments (observes histories of automata) and the responses of observed automata. An observing automaton does not produce responses that change the environments, i.e., the responses are null responses.

A set of words that refers to the stimuli for a Class 5 sentence is the set $ST5$. Let $ST5$ be a dyadic relation on the sets $ST'5$ and SAD; $ST5 \subseteq ST'5 \times SAD$, where the elements of SAD are the modifier elements and the elements of $ST'5$ are the modified elements. The set SAD contains a null element.

Type 1. Responses of automata provide all types of stimuli. This suggests a first type of sentence, Class 5. Performing automata that produce responses may also be referenced in this type sentence. Examples are as follows:

"Today, John heard the record that Mary cut last week."
"Today, Jim saw the dent in the table that John made yesterday."
"John quickly tasted the pudding today that Mary made yesterday."
"John smelled the pudding cautiously that Mary made in the bowl yesterday."
"Today, Jim felt the dent that John made in the table yesterday."

*WFF*51 is a hexadic relation

$$WFF51 \subseteq T \times A \times ST'5 \times (E \cup B) \times SAD \times WFF'131$$

Alternates are

$$WFF'51 \subseteq A \times SAD \times ST'5 \times (E \cup B) \times T \times WFF'131$$

$$WFF''51 \subseteq T \times A \times ST'5 \times (E \cup B) \times E \times SAD \times WFF'132$$

$$WFF'''51 \subseteq A \times SAD \times ST'5 \times (E \cup B) \times E \times T \times WFF'132$$

where *WFF'* 131 and *WFF'* 132 are modified relations of *WFF'* 13 and may have null elements

$$WFF'131 \subseteq A \cup Z' \cup Z'' \times FAD \times Fr'13 \times E \times T$$

$$WFF'132 \subseteq A \cup Z' \cup Z'' \times FAD \times Fr'13 \times T$$

Consider the graph tuple for the second example sentence. First determine the 7-tuple for *WFF''* 51

(today $\in T$, Jim $\in A$, saw $\in ST'5$, dent $\in (E \cup B)$, table $\in E$, - $\in SAD$,
(John $\in A \cup Z' \cup Z''$, - $\in FAD$, made $\in Fr'13$, yesterday $\in T) \in WFF'132$)
$\in WFF''51$

The (15+1)-tuple for the alternate environment graph, $TE_a(m+i, m+i+1)$, is

(table $\in E$, today $\in T$, - $\in Q^*$, - $\in X''$, eye $\in X'$, - $\in Z'$, - $\in Z''$, (dent, $TE_a(m+k, m+k+1)) \in S$, Jim $\in A$, - $\in B$, ((saw, -), -, -, -, -) $\in F$, - $\in D$, - $\in C$, - $\in Q^*$, today $\in T$, - $\in E$)

where (from *WFF'*13), $TE_a(m+k, m+k+1)$ is

(table, yesterday, -, -, -, -, -, -, John, dent, ((made, -), -, -, -, -), -, -, -, yesterday, -)

which is referred to by dent $\in S$.

Types k. Knowledge can be acquired by an observing automaton directly from a performing automaton or an environment, or through another automaton by interpreting the responses that convey knowledge about the performing automaton or the environment. Types *k* of Class 5 sentences describe the former case. (The latter case will be discussed later as Class 6 sentence.). Example sentences are as follows:

"Today, Jim heard John strike the table very hard with a hammer."
"Jim vaguely saw John strike the table with a hammer today."

The relations are

$$WFF5k \subseteq T1 \times A \times SAD \times ST'5 \times WFFij1$$

$$WFF'5k \subseteq A \times SAD \times ST'5 \times WFF'ij$$

$$WFF''5k \subseteq T1 \times A \times SAD \times ST'5 \times WFF''ij1$$

$$WFF'''5k \subseteq A \times SAD \times ST'5 \times WFF'''ij$$

where *WFF' ij*, *WFF''' ij* are the alternate sets for the Type *j*, Class *i* sentences, and the sets *WFFij*1 and *WFF'' ij*1 are modified relations of *WFFij*, *WFF'' ij* involving *T*. $T1 = T$, where *T*1 indicates that *WFFij* and *WFF'' ij* have been modified to exclude *T* from each of the relations when it exists. The relations above, *WFF5k*, *WFF' 5k*, *WFF'' 5k*, *WFF''' 5k*, $k=2, 3, ---, 25$, may be generated by the following formulation:

$$k(i, j) = f(i) + j$$

where

$$i \in \{1, 2, 3, 4\},\ j \in \{1, 2, ---, f(i+1) - f(i)\},\ f(1)=1,\ f(2)=4,\ f(3)=12,\ f(4)=13,\ f(5)=25$$

Those generated that are nonexistent are

$$WFF'5n_1,\ n_1 = 13, 15, 17, 19, 21, 23, 25$$

$$WFF''5n_2,\ n_2 = 2, 3, 4, 6, 13, 15, 17, 19, 21, 23, 25$$

$$WFF'''5n_3,\ n_3 = n_2$$

For example, with $i=2$, $j=4$

$$k(2, 4) = f(2) + 4 = 4 + 4 = 8$$

$$WFF58 \subseteq T1 \times A \times SAD \times ST'5 \times WFF241$$

Note that *WFF*11*A*, *WFF*12*A*, *WFF*13*A* do not appear as sets in the product sets above.

4.1.6 Class 6 Sentence

The Class 6 sentence describes the case when knowledge is acquired by an observing automaton through *another* automaton by interpreting his responses, which convey the knowledge about a performing automaton or an environment. Example sentences are as follows:

> "Today, Jim faintly heard that yesterday John struck the table very hard with a hammer."
>
> "Jim read today that John struck the table with a hammer yesterday."
>
> "Jim heard that Mary is a beautiful girl."

A set of words that refers to the stimuli for a Class 6 sentence is the set *ST*6. Let *ST*6 be a dyadic relation on the sets *ST′* 6 and *SAD*

$$ST6 \subseteq ST'6 \times SAD$$

The relations are as follows:

$$WFF6u \subseteq T \times A \times SAD \times ST'6 \times WFFij$$
$$WFF_1 6u \subseteq A \times SAD \times ST'6 \times T \times WFFij$$
$$WFF'6u \subseteq T \times A \times SAD \times ST'6 \times WFF'ij$$
$$WFF_1'6u \subseteq A \times SAD \times ST'6 \times T \times WFF'ij$$
$$WFF''6u \subseteq T \times A \times SAD \times ST'6 \times WFF''ij$$
$$WFF_1''6u \subseteq A \times SAD \times ST'6 \times T \times WFF''ij$$
$$WFF'''6u \subseteq T \times A \times SAD \times ST'6 \times WFF'''ij$$
$$WFF_1'''6u \subseteq A \times SAD \times ST'6 \times T \times WFF'''ij$$

The above relations for $u=1, 2, ---, 24$ may be generated by the following formulation:

$$u(i, j) = g(i) + j$$

where

$i \in \{1, 2, 3, 4\}$, $j \in \{1, 2, ---, g(i+1) - g(i)\}$, $g(1)=0$, $g(2)=3$, $g(3)=11$, $g(4)=12$, $g(5)=24$

Those generated that are nonexistent are

$$WFF'6t_1,\ WFF_1'6t_1,\ t_1=12, 14, 16, 18, 20, 22, 24$$

$$WFF''6t_2,\ WFF_1''6t_2,\ t_2=1, 2, 3, 5, 12, 14, 16, 18, 20, 22, 24$$

$$WFF'''6t_3,\ WFF_1'''6t_3,\ t_3=t_2$$

The following relations are included but are not generated:

$$WFF61A \subseteq T \times A \times SAD \times ST'6 \times WFF11A$$

$$WFF_1 61A \subseteq A \times SAD \times ST'6 \times T \times WFF11A$$

$$WFF62A \subseteq T \times A \times SAD \times ST'6 \times WFF12A$$

$$WFF_1 62A \subseteq A \times SAD \times ST'6 \times T \times WFF12A$$

$$WFF63A \subseteq T \times A \times SAD \times ST'6 \times WFF13A$$

$$WFF_1 63A \subseteq A \times SAD \times ST'6 \times T \times WFF13A$$

4.1.7 Class 7 Sentence

The responses of an automaton related to his communication by the natural language are conveyed through a class of sentences that is called a Class 7 sentence. The types of responses conveyed by the sentences imply the types of effectors used. Class 7 sentences are illustrated by the example sentences

"Today, Jim wrote that John struck the table with a hammer yesterday."
"Jim casually said today that the hammer made a dent in the table yesterday."

The relations are as follows:

$$WFF7v \subseteq T \times A \times FAD \times Fr'7 \times WFFij$$

$$WFF_1 7v \subseteq A \times FAD \times Fr'7 \times T \times WFFij$$

$$WFF'7v \subseteq T \times A \times FAD \times Fr'7 \times WFF'ij$$

$$WFF_1'7v \subseteq A \times FAD \times Fr'7 \times T \times WFF'ij$$

$$WFF''7v \subseteq T \times A \times FAD \times Fr'7 \times WFF''ij$$

$$WFF_1''7v \subseteq A \times FAD \times Fr'7 \times T \times WFF''ij$$

$$WFF'''7v \subseteq T \times A \times FAD \times Fr'7 \times WFF'''ij$$

$$WFF_1'''7v \subseteq A \times FAD \times Fr'7 \times T \times WFF'''ij$$

Relations for types $u=1$, 2, ---, 24 may be generated from the above by the formulation

$$v(i, j) = h(i) + j$$

where

$$i \in \{1, 2, 3, 4\},\ j \in \{1, 2, ---, h(i+1) - h(i)\},\ h(1) = 0,\ h(2) = 3,\ h(3) = 11,\ h(4) = 12,\ h(5) = 24$$

Those generated that are nonexistent are

$$WFF'6s_1,\ WFF_1'6s_1,\ s_1 = 12, 14, 16, 18, 20, 22, 24$$
$$WFF''6s_2,\ WFF_1''6s_2,\ s_2 = 1, 2, 3, 5, 12, 14, 16, 18, 20, 22, 24$$
$$WFF'''6s_3,\ WFF_1'''6s_3,\ s_3 = s_2$$

The following relations are included but are not generated:

$$WFF71A \subseteq T \times A \times FAD \times Fr'7 \times WFF11A$$
$$WFF_1 71A \subseteq A \times FAD \times Fr'7 \times T \times WFF11A$$

$$WFF72A \subseteq T \times A \times FAD \times Fr'7 \times WFF12A$$
$$WFF_1 72A \subseteq A \times FAD \times Fr'7 \times T \times WFF12A$$

$$WFF73A \subseteq T \times A \times FAD \times Fr'7 \times WFF13A$$
$$WFF_1 73A \subseteq A \times FAD \times Fr'7 \times T \times WFF13A$$

4.2 EXTRACTING AND STORING THE MEANING OF SENTENCES BY COMPUTER

Extracting and storing the meaning of natural language by computer is very important for effective human-computer interaction. Programs for understanding the Class 1 sentences have been written (Koenig, Mason, Jakubowski 1978). The discussion is limited to simple sentences that were discussed in the previous section, excluding quantifiers or logical connectives. However, as would be expected from the previous discussion, the method of extracting the meaning is *not* limited to sentences describing a specific field. This implies that a system using the method is *not* text based. That is, there is a database to which data can be added, and a system can do more than store a mass of information and simply retrieve it.

All words that contribute to the meaning of sentences are elements of the 16 automaton sets and those that are not, such as **the**, **to** are called *filler words*. They are not elements of the 16 automaton sets. Recall, the alternate environment and principal time graphs for a given automaton are (15+1)-tuple graphs and each stores complete knowledge. The (15+1)-tuple of the alternate environment graph is used to store the meaning of sentences in the application to be presented here. The $(n+1)$-tuples for the processor and time graphs can be obtained from the (15+1)-tuple of the environment graph.

Template matching is the method used here for extracting the meaning by computer in application to the Class 1 sentence. The parsing method might be used in application to some of the classes, or used in combination with template matching.

4.2.1 Description of an Algorithm

The unique features consisted of the following:

1. A dictionary structure that contains both vocabulary entries with associated grammatical definitions and class-type lists representative of the various types of sentences.
2. A method for matching complete and partial grammatical skeletons with stored class-type templates.

The procedure is to break down the input sentence word by word. This involves the following:

1. A dictionary lookup is made yielding an associated characteristic (e.g., John $\rightarrow A$).
2. The growing tuple is searched for an occurrence of this characteristic. If the characteristic is not found, it is appended to the tuple.
3. The actual source word is included in the class designated by its characteristic.

Upon reaching the end of the input sentence, the growing tuple is terminated. This tuple (a symbolic representation of the input sentence) is now compared with the store of class-type templates. If a match occurs, the operation succeeds, and the class-type template is revealed. Otherwise, one of two cases might have been present: 1) the sentence is not of Class 1, or 2) the sentence is of Class 1 but contains null elements. In the latter case, a second set of templates representing the minimal number of characteristics per given type needed to maintain a valid English sentence is employed. The comparison proceeds by breaking off elements from the minimal class-type template. For each term that is broken off, a search is made within the input generated list. As this input list is searched, it too decreases in size. No term is compared twice. In this manner, one is assured that the order of characteristics, both in the template and in the input sentence, remains preserved.

While executing the minimal template comparison, one of two states is reached. Either 1) the input tuple reduces to a null list, or 2) the minimal template reduces to a null list. In the case of the former, the input sentence is unable to satisfy that particular template's conditions. The next minimal template is fetched and comparison restarts. In the latter case, the match is successful and any remaining items from the input-generated tuple are placed on a null list called RESIDUE. RESIDUE may then be searched to distinguish a regular sentence from its corresponding prime.

In either case, once a match occurs between template and input tuple, a message is printed declaring the type. In addition, both the standard list of tuple characteristics (E, T, Q^*, X'', X', ---) as well as the list containing and indicating relative position of the actual English words is provided.

4.3 KNOWLEDGE ASSOCIATION

Knowledge association is accomplished by bringing together separate knowledge structures into wholes, i.e., by establishing connected graphs (when expressed in automaton terms) (Koenig 1972b, 1978, 1982). This eliminates duplication of separate structures or parts and facilitates operating on structures. The various aspects of knowledge association can best be discussed with the aid of an example. Recall the example sentence, "John ate a frog," from the opening paragraph of Section 4.1. The storing of the meaning using the automaton concepts was shown in Figure 4.2. A human automaton receiving this sentence as a stimulus may associate the knowledge with knowledge previously stored from eating experiences and from generally accepted eating knowledge called world knowledge obtainable from a dictionary for natural language. The automaton, upon receiving the sentence, may then be able to obtain full meaning for the performance of John eating a frog. The analysis for a more complete meaning will now be made and the knowledge structure drawn.

From an English dictionary (giving world knowledge), there is the definition of **eat**.

eat—to put in the mouth, chew and swallow

Again from the dictionary, there are the definitions

put—to thrust; push; drive; impel

chew—to bite and grind or crush with the teeth

swallow—to pass food, etc. from the mouth through the esophagus into the stomach

In automaton terms, there are three distinct transformation responses **put**, **chew**, **swallow** representing greater detail for the response for **eat**. That is, there are three distinct pairs of moments of observations required for obtaining detailed knowledge related to eating while that for **eat** requires only one pair (see Figure 4.2). There are graph (15+1)-tuples for representing the structure for knowledge from the dictionary for three pair of moments.

$TE_{a1}(m+i, m+i+1)$=(-, -, -, -, -, -, -, -, -,
-, ((put, -), -, -, -, -), -, -, -, -, (mouth, -, -, -))

$TE_{a1}(m+i+1, m+i+2)$=((mouth, -, -, -), -, -, -, -, (teeth, -, -, -), -, -, -,
chewed, ((chew, -), -, -, -, -), -, -, -, -, (mouth, -, -, -))

$TE_{a1}(m+i+2, m+i+3)$=((mouth, -, -, -), -, -, -, -, esophagus, -, -, -,
-, ((swallow, -), -, -, -, -), -, -, -, -, (stomach, -, -, -))

The graph (15+1)-tuple for representing the structure for knowledge related directly to the response **eat** is

$TE_{a1}(m+i, m+i+3)$=(-, -, -, -, -, -, -, -, -, -,
((eat, -), -, -, -, -), -, -, -, -, (stomach, -, -, -))

The complete graph structure for the four definitions that appeared in the dictionary is shown in Figure 4.3 and was obtained from the above four tuples (j=1).

There is still more knowledge that the receiver of the sentence is likely to associate with the meaning of the sentence, and this knowledge comes from past personal experience associated with eating. For example, by experience it would be known that the hand is likely to be used to put nutrients into the mouth for chewing, that the stimulus to do this is received through the eyes, and that the person eating is hungry. This knowledge that is gained through personal experience is combined with the general knowledge that can be obtained through the dictionary to give the following graph (15+1)-tuples:

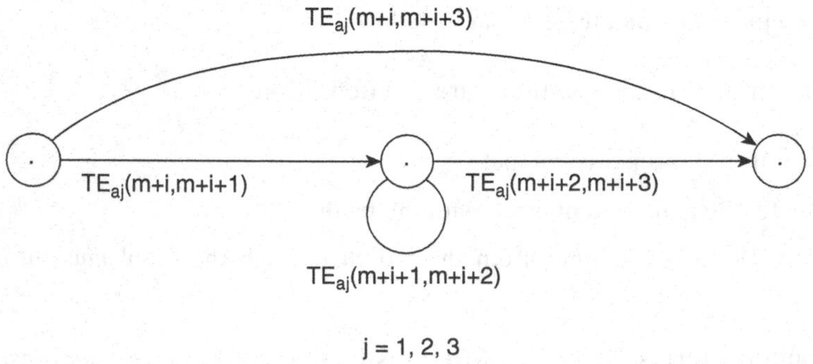

Figure 4.3 Graph structure describing the meaning of the definition of "eat" from the English dictionary.

$TE_{a2}(m+i, m+i+1)$=((-, nutrient, -, -), -, hungry, -, eyes, (hand, -, -, -), -, -, -, -, ((put, -), -, -, -, -), -, -, hungry, -, (mouth, -, -, -))

$TE_{a2}(m+i+1, m+i+2)$= ((mouth, -, -, -), -, hungry, -, taste buds, (teeth, -, -, -), -, -, -, chewed, ((chew, -), -, -, -, -), -, -, hungry, -, (mouth, -, -, -))

$TE_{a2}(m+i+2, m+i+3)$=((mouth, -, -, -), -, hungry, -, -, esophagus, -, -, -, -, ((swallow, -), -, -, -, -), -, -, hungry, -, (stomach, -, -, -))

The graph (15+1)-tuple for representing the structure for knowledge related directly to the response **eat** is

$TE_{a2}(m+i, m+i+3)$=((-, nutrient, -, -), -, hungry, -, eyes, (hand, -, -, -), -, -, -, -, ((eat, -), -, -, -, -), -, -, hungry, -, (stomach, -, -, -))

The form of the complete graph structure is the same as that of Figure 4.3 with $j=2$.

Now associate the knowledge of the structure of Figure 4.2 for the specific sentence "John ate a frog," with the knowledge of the structure of Figure 4.3 obtained from the dictionary and from experience. The graph (15+1)-tuples are the following:

$TE_{a3}(m+i, m+i+1)$= ((frog, nutrient, -, -), -, hungry, -, eyes, (hand, -, -, John), -, -, John,-frog, ((put, -), -, -, -, -), -, -, hungry, -, (frog, nutrient, -, (mouth, -, -, John)))

$TE_{a3}(m+i+1, m+i+2)$=((frog, nutrient, -, (mouth, -, -, John)), -, hungry, -, taste buds, (teeth, -, -, John), -, -, John, chewed, ((chew, -), -, -, -, -), -, -, hungry, -, (frog, nutrient, -, (mouth, -, -, John)))

$TE_{a3}(m+i+2, m+i+3)$= ((frog, nutrient, -, (mouth, -, -, John)), (-, hungry, -, -, esophagus, -, -, John,-frog, ((swallow, -), -, -, -, -), -, -, hungry, -, (frog, nutrient, -, (stomach, -, - John)))

$TE_{a3}(m+i, m+i+3)$= ((frog, nutrient, -, -), -, hungry, -, eyes, ((hand, -, -, John), -, -, John, -frog, ((eat, -), -, -, -, -), -, -, hungry, -, (frog, nutrient, -, (stomach, -, -, John)))

The form of the complete graph structure is again the same as that of Figure 4.3 with $j=3$.

Now, if the frog was uncooked, John may become sick (general knowledge), and if one more observation was made at $m+i+4$, John would show signs of sickness. The graph (15+1)-tuple resulting from this additional observation (or anticipated observation) is

$TE_{a3}(m+i+3, m+i+4)$ = ((frog, nutrient, -, (stomach, -, -, John)), -, hungry, -, -, -, -, -, John, -, (-, -, -, -, -), -, -, sick, -, (frog, nutrient, -, (stomach, -, -, John)))

The knowledge structure for the above five graph (15+1)-tuples using the automaton concepts is shown in Figure 4.4.

Suppose the human automaton who received the sentence "John ate a frog" had limited knowledge about eating and did not, or could not, use a dictionary. His internal knowledge structure might be represented by that shown in Figure 4.2. To enable him to build up to the knowledge structure shown in Figure 4.4 (i.e., to obtain additional meaning of the sentence), additional sentences giving the detail must follow. For example, the following sequence of sentences may be presented:

> "John ate a frog. He was hungry. He saw the frog and put it in his mouth with his hand. John chewed the frog with his teeth. He swallowed the frog with his esophagus. The frog went into John's stomach. John became sick."

Note, that given the knowledge structure of Figure 4.4 stored in a computer, the above sequence of sentences might be required to be generated by the computer during interaction with a human automaton in adapting to that automaton's background and experience.

It should be clear that there is required an association of knowledge currently received with knowledge previously stored and with knowledge of previous sentences received in a sequence. All associations of knowledge involving knowledge currently being received may be considered as occurring in temporary memory. Basic disciplines for knowledge association for the automaton concept follow. The associations are categorized as 1) associations by combining graphs through common points, and 2) associations by combining graph $(n+1)$-tuples.

4.3.1 Association by Combining Graphs Through Common Points

Knowledge received during different time periods may be combined through common points. For the environment graph, this means combining points

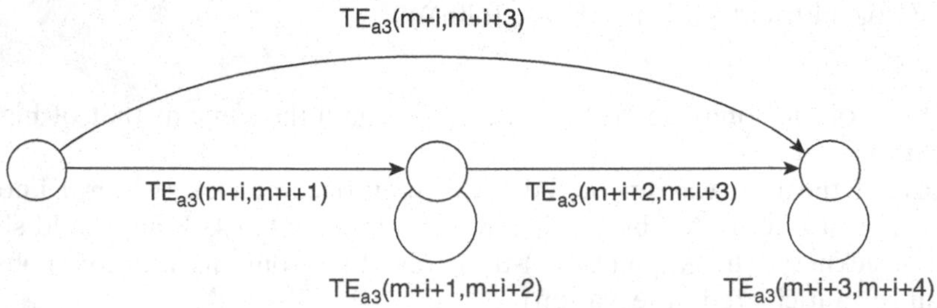

Figure 4.4 Detailed knowledge structure related to sentence "John ate a frog."

representing environments; for the time graph, it means combining points representing time; and for the processor graph, it means combining points representing states. Consider for the moment two environment graphs to be associated by combining graphs through common points. Let E_i and E_j be the sets of points for the two graphs. Association is accomplished when $|E_i|+|E_j|>|E_i \cup E_j|$ and similarly for the processor and time graphs. When knowledge pertaining to a particular automaton is to be communicated, only knowledge structures are combined that have that automaton name as an element in the graph $(n+1)$-tuples.

Consider an example for combining graphs through common points. Given are the following (15+1)-tuples for the environment graph:

1. (Supermarket Dee, today, pleasant, -, -, -, -, -, Mrs. Bee, -, ((shops, -), -, -, -, -), -, -, angry, today, Supermarket Dee)
2. (Supermarket Dee, today, pleasant, -, -, -, -, -, Mrs. Cee, -, ((shops, -), -, -, -, -), -, -, pleasant, today, Supermarket Dee)

The knowledge is associated through a common point denoted by **Supermarket Dee**. The (15+1)-tuples could have resulted either from the observer a_0 being told the information, or from a_0 observing the performances of Mrs. Bee and Mrs. Cee during two different pairs of moments. Assume the information was communicated in the natural language. The first (15+1)-tuple might have been established in temporary memory by the following sentences before being transferred to permanent memory:

1a. "Today, Mrs. Bee shopped at Supermarket Dee."
1b. "She was pleasant."
1c. "Then, she became angry."

Later the same day, a_0 was told the following to establish the second (15+1)-tuple (other intervening information might have been received):

2a. "Mrs. Cee shopped at Supermarket Dee today."
2b. "She remained pleasant."

The association of the knowledge through the common point denoted by **Supermarket Dee** facilitates operating on the Knowledge Structure to generate the following response of a_0:

"Mrs. Cee and Mrs. Bee shop at Supermarket Dee."

The environment and time elements are interchanged directly in the environment (15+1)-tuples to obtain the (15+1)-tuples for the Time graphs. The two time (15+1)-tuples for the previous example are

1. (today, Supermarket Dee, pleasant, -, -, -, -, -, Mrs. Bee, -, ((shops, -), -, -, -, -), -, -, angry, Supermarket Dee, today)
2. (today, Supermarket Dee, pleasant, -, -, -, -, -, Mrs. Cee, -, ((shops, -), -, -, -, -), -, -, pleasant, Supermarket Dee, today)

After knowledge association takes place by combining graphs through common points, there is a single point identified by **today** and two loops. The association facilitates generating the following response of a_0:

"Mrs. Cee and Mrs. Bee shopped today."

An important historical example of time association is evidenced by the comment

"During the Renaissance period, the people of Venice built the Library of St. Mark, and the Romans built the Church of St. Peter."

Lastly, knowledge association for combining processor graphs through common points will be considered. The (10+1)-tuples for the processor graph can be obtained directly from the environment or time (15+1)-tuples. The two (10+1)-tuples for the previous example become

1. (pleasant, -, -, -, -, -, Mrs. Bee, ((shops, -), -, -, -, -), -, -, angry)
2. (pleasant, -, -, -, -, -, Mrs. Cee, ((shops, -), -, -, -, -), -, -, pleasant)

The first describes two distinct points representing states **pleasant** and **angry** and an adjacent line, and the second (10+1)-tuple describes a single point and a loop. The point of origin for the line of the first tuple is common with the point for the second in accomplishing knowledge association, and the result is a single connected graph. The association facilitates a_0 generating the following response:

"Mrs. Cee and Mrs. Bee were pleasant."

Clearly, the associations can take place through a common point for more than two structures, and a generated response can reference elements from more than two of the structures. Deductive processes associated with the generation of this type of response will be discussed in Section 4.4.

4.3.2 Associations by Combining Graph (n+1)-Tuples

The associations by combining graph tuples begin with the processor graph following the hierarchical development of the graph functions. Although two processor graph tuples may be combined, it may not be possible to combine the corresponding time and environment (15+1)-tuples. The number of elements with known values for a combined $(n+1)$-tuple is greater or equal to the number of elements with known values for either of the two noncombined $(n+1)$-tuples.

Recall, the processor (10+1)-tuple of a graph is

$$TQ(m+i, m+i+1) = (q_i^*, x_i'', x_i', z_i', z_i'', s_i, a, (fs, fm, fh, ft, fa)_{i+1}, d_{i+1}, c_{i+1}, q_{i+1}^*)$$

such that

$$\Omega^*(q_i^*, x_i'', x_i', z_i', z_i'', s_i) = ((fs, fm, fh, ft, fa)_{i+1}, d_{i+1}, c_{i+1}, q_{i+1}^*)$$

Let $TQ1$ be a (10+1)-tuple from a first processor graph and $TQ2$ a (10+1)-tuple from a second processor graph for the same automaton. Let both (10+1)-tuples have known values for the domain elements, and let corresponding elements have the same values. Any of the image elements of each of the two (10+1)-tuples may have known values. By the single-valueness of the function Ω^*, the values of corresponding image elements of $TQ1$ and $TQ2$ are the same, and the values can be combined into a single tuple. The set of elements that have known values for the combined (10+1)-tuple is the union of the sets of elements that have known values for $TQ1$ and $TQ2$. The same reasoning applies in the association of knowledge for the time and environment graphs involving the graph functions $\Lambda^{*\prime}$ and Λ^*.

An example is presented to demonstrate the combining of tuples under the above mentioned conditions. Suppose a child a_0 as an observer sees a dog called Rover jabbed with a stick at moment $m1$ while the dog is in a pleasant mood. A moment later, $m1+1$, a_0 observes that Rover becomes angry and sees his ears lay back. The observations are made through a window. The (10+1)-tuple of the processor graph that contains this knowledge is

(pleasant, 0, *rt*, *em*, 0, jabbed, Rover, ((ears lay back) pattern,
(-, -, -, -), -, -, angry)

where 0 indicates no value and *rt* and *em* indicate a natural receptor associated with touch and a muscular effector. Sometime later in the summer at moment $m2$, a_0 observes the same conditions as observed at moment $m1$, and a moment later, $m2+1$, a_0 hears Rover bark. The (10+1)-tuple representing the knowledge structure is

(pleasant, 0, *rt*, *em*, 0, jabbed, Rover, (-, -, (bark) sound, -, -,), -, -, -)

The values of the elements on the left of Rover $\in A$ for the one (10+1)-tuple are the same as the values of the corresponding elements of the other tuple, as required of the aforementioned conditions, and the two tuples can be combined, i.e., the knowledge they represent can be associated, and if the information for each tuple is true, the information of the combined tuple is true. The resulting combined tuple is

(pleasant, 0, *rt*, *em*, 0, jabbed, Rover, ((ears lay back) pattern, -, (bark) sound, -, -), -, -, angry)

The above discussion establishes the proof of the following theorem:

Theorem 4.1. *Given two graph* $(n+1)$*-tuples,* Tuy, $u=E, T, Q$; $y=1, 2$ *stored in the nonshared environments (memory) of an observing automaton* a_0. *If the values of the elements on the left of the element* $\boldsymbol{a} \in A$ *including* $\boldsymbol{a}$, *are known for both* $(n+1)$*-tuples and the values of corresponding elements are the same, and if values that are known for corresponding elements of the* $(n+1)$*-tuples on the right of the element* $\boldsymbol{a}$ *are the same, knowledge association can take place by combining the two* $(n+1)$*-tuples into one, and if the information represented by the values for each of the noncombined tuples is true, the information represented by the values of the combined* $(n+1)$*-tuple is true. The set of elements that have known values for the combined* $(n+1)$*-tuple is the union of the sets of elements that have known values for* $Tu1$ *and* $Tu2$.

4.3.3 Computer Methods for Association of Knowledge

In Section 4.2, algorithms were discussed that extracted the meaning of sentences and established the alternate environment graph (15+1)-tuples as knowledge structures. Here, principal features are discussed for an algorithm that illustrates association of knowledge structures by combining graphs through common points. (A program was written by a student in computer sciences, John Clatanoff, as a problem assignment.) The algorithm assumes the alternate environment graph (15+1)-tuples have been established and proceeds to form the principal time graph (15+1)-tuples and the processor graph (10+1)-tuples from these given (15+1)-tuples. Next, the environment, time, and processor graph points are identified and mapped in matrices. Within the knowledge association routine of the program for combining graphs through common points, a check is made to determine if two parallel lines are the same, and if so, deletes one of the lines.

Temporary and permanent memory are simulated. Alternate environment graph (15+1)-tuples are considered to be in temporary memory. A number of sequenced sentences may have been required to establish the assumed (15+1)-tuples, and the association occurs in temporary memory. Processing follows and produces the other $(n+1)$-tuples of the time and processor graphs. Permanent memory contains the matrices, which, once in that memory, remain there and are only altered by the knowledge association routine.

A simple method for processing the data contained in an environment graph (15+1)-tuple is to separate its elements. Each environment graph (15+1)-tuple that is received by the knowledge association routine is given a next integer number in ascending order. Points of the (15+1)-tuples are placed in a list as they become available and are numbered in a similar manner. Corresponding lists of time graph (15+1)-tuples and processor graph (10+1)-tuples are obtained from the list of environment graph (15+1)-tuples, and corresponding lists of points arc obtained from these lists of graph $(n+1)$-tuples.

The matrices communicate how the $(n+1)$-tuples are connected in graphs. The rows and columns of the matrices give the points of the graphs identified by their numbers in the point lists. Points from which lines are directed are identified by the numbered rows, and points to which lines are directed are identified by the numbered columns. The entries in the matrices are the identification numbers from the $(n+1)$-tuple lists for corresponding points. For each nonidentical parallel line, an additional matrix is required (identical parallel lines are deleted). In combining graphs by points for the processor graphs, a state represented by a point is for a single automaton. This requires that there be a matrix for each automaton in the list of processor graph (10+1)-tuples. A state represented by a point may be common to more than one automata when comparisons between automata are made, and there is a single matrix for all the automata.

4.4 DEDUCTIVE PROCESSES

Both association types, by combining graphs through common points (Section 4.3.1) and by combining graph $(n+1)$-tuples (Secton 4.3.2), facilitate operations on the structures in accomplishing basic deductive processes by computer (Koenig 1989, 1990, 1994a). The associations provide access to premises received as knowledge at different periods.

4.4.1 Deductive Processes Related to Association Through Common Points

Consider how knowledge association by combining graphs through common points relates to basic deductive processes (Koenig 1978). A stimulus to a_0 for initiating the process can take various forms, and there are correspondingly different forms of output. Recall the example of Section 4.3.1.

1. (Supermarket Dee, today, pleasant, -, -, -, -, -, Mrs. Bee, -, ((shops, -), -, -, -,-), -, -, angry, today, Supermarket Dee)
2. (Supermarket Dee, today, pleasant, -, -, -, -, -, Mrs. Cee, -, ((shops, -), -, -, -, -), -, -, pleasant, today, Supermarket Dee)

A first question as stimulus to the observer a_0

"Who shops at Supermarket Dee?"

results in the answer (response)

"Mrs. Cee and Mrs. Bee shop at Supermarket Dee."

The question may be presented in a different form requiring only a *true* or *yes* for a response

"Do Mrs. Cee and Mrs. Bee shop at Supermarket Dee?"

Also, when a statement

"Mrs. Cee and Mrs. Bee shop at Supermarket Dee."

is received, a_0 establishes as an internal response whether any part of the information is already stored and takes appropriate action.

Let *Auv*(*r*-variables) denote a function of *r*-variables corresponding to the number of element places in the graph tuples

$$u = E, T, Q$$

$$v = 1, 2, ---, w$$

Let the order of the elements of the domain of a function be the same as the order of the elements of a graph tuple. For a set of values of the variables, the function becomes an *r*-place proposition, *Auv*(*r*-values). For $u=E$, the vth premise for a deductive process related to knowledge association for the environment graph is $AEv(t)$, where t denotes a graph $(n+1)$-tuple of $(n+1)$-value places. For the example, there are two premises, $v=1, 2$

1. *AE*1(Supermarket Dee, -, -, -, -, -, -, -, Mrs. Bee, -, ((shops, -), -, -, -, -), -, -, -, -, -)

That is, "Mrs. Bee shops at Supermarket Dee.".

2. *AE*2(Supermarket Dee, -, -, -, -, -, -, -, Mrs. Cee, -, ((shops, -), -, -, -, -), -, -, -, -, -)

That is, "Mrs. Cee shops at Supermarket Dee." The premises were determined from a single graph point identified by **Supermarket Dee**. It remains to show that $AE1(t), AE2(t) \vdash BE2$; i.e., to show that the argument is valid, where $BE2$ is the conclusion, and is true if the premises are true and the argument is valid. $BE2$ is a_0's response to the first form of a stimulus. For the example, $BE2$ is the answer to the question, "Who shops at Supermarket Dee?" What is the sequence of formulas in the simulated deductive process that will give $BE2$—"Mrs. Cee and Mrs. Bee shop at Supermarket Dee"? The following sequence results when the deductive process is described by the procedures for mathematical logic (⊃-implication, &-conjunction (Kleene 1967)) (Koenig, Schultz 1972–1973):

1a. $AE1(t)$—premise (all premises obtained upon finding of single point of graph)
b. $AE2(t)$—premise
c. $AE2(t) \supset (AE1(t) \supset AE2(t) \,\&\, AE1(t))$—axiom schema $(A \supset (B \supset A \,\&\, B))$, and substitution rule
d. $AE1(t) \supset AE2(t) \,\&\, AE1(t)$—implication rule using 1b, c
2a. $AE2(t) \,\&\, AE1(t)$—implication rule, 1a, d
b. $AE3(t)$—premise
c. $AE3(t) \supset ((AE2(t) \,\&\, AE1(t)) \supset AE3(t) \,\&\, (AE2(t) \,\&\, AE1(t)))$—axiom schema $(A \supset (B \supset A \,\&\, B))$ and substitution rule
d. $(AE2(t) \,\&\, AE1(t)) \supset AE3(t) \,\&\, (AE2(t) \,\&\, AE1(t))$—implication rule 2b, c
3a. $AE3(t) \,\&\, AE2(t) \,\&\, AE1(t)$—implication rule 2a, d

•
•
•
•

$(k-1)$a. $AE(k-1)(t) \,\&\, AE(k-2)(t) \,\&\, \text{---} \,\&\, AE1(t)$—implication rule $(k-2)$a, d
b. $AEk(t)$—premise
c. $AEk(t) \supset (AE(k-1)(t) \,\&\, AE(k-2)(t) \,\&\, \text{---} \,\&\, AE1(t) \supset AEk(t) \,\&\, (AE(k-1)\,(t) \,\&\, AE(k-2)(t) \,\&\, \text{---} \,\&\, AE1(t)))$—axiom schema and substitution rule
d. $AE(k-1)(t) \,\&\, AE(k-2)(t) \,\&\, \text{---} \,\&\, AE1(t) \supset AEk(t) \,\&\, (AE(k-1)(t) \,\&\, AE(k-2)(t) \,\&\, \text{---} \,\&\, AE1(t))$—implication rule $(k-1)$b, c
ka. $AEk(t) \,\&\, AE(k-1)(t) \,\&\, \text{---} \,\&\, AE1(t)$—implication rule $(k-1)$a, d

Where, for $AEv(t)$, $v=1, 2, \text{---}, k$; $2<k<w$, w is the number of (15+1)-tuples combined through a common point. The conclusion BEk is formula (ka), $AEk(t) \,\&\, AE(k-1)(t) \,\&\, \text{---} \,\&\, AE1(t)$. Therefore

$$AE1(t), AE2(t), \text{---}, AEk(t) \vdash AEk(t) \,\&\, AE(k-1)(t) \,\&\, \text{---} \,\&\, AE1(t)$$

and the deductive process by formal logic correctly models the natural deductive process. For the example with $k=2$

$$AE1(t), AE2(t) \vdash AE2(t) \,\&\, AE1(t)$$

Rules of introduction and elimination of the propositional calculus established by Gentzen (1934) can be used effectively to reduce the number of formulas in many deductions (see also Kleene 1967). There are a total of 14 rules involving all the propositional connectives. For the above example, using the &-introduction rule, the list of formulas reduces to

$AE1(t), AE2(t) \vdash AE2(t)$ & $AE1(t)$—the &-introduction rule ($A, B \vdash A$ & B) and substitution rule

Consider how knowledge association combining principal time graphs through common points relate to basic deductive processes. The νth premise, $\nu=1, 2$, ---, k for a deductive process related to the association for the time graph is $AT\nu(t)$, and the two premises for the example follow:

1. $AT1$(today, -, -, -, -, -, -, -, Mrs. Bee, -, ((shops, -), -, -, -, -), -, -, -, -, -)

That is

"Today, Mrs. Bee shopped."

2. $AT2$(today, -, -, -, -, -, -, -, Mrs. Cee, -, ((shops, -), -, -, -, -), -, -, -, -, -)

That is

"Today, Mrs. Cee shopped."

For $AT1(t), AT2(t) \vdash BT2$, $BT2$ is the direct answer to the question

"Who shopped today?"

a_0's answer is

$BT2$—"Mrs. Cee and Mrs. Bee shopped today."

Note that if **clerks** was substituted for **shops** in $AT2(t)$, the answer would be

$BT2$—"Today, Mrs. Cee clerked and Mrs. Bee shopped."

The other forms of stimuli to a_0 for initiating the deductive process are similar to those for the environment graph, and the sequence of formulas as established by formal logic is the same when E is replaced by T for the propositional symbols.

The νth premise $\nu=1, 2$, ---, k for a deductive process related to association for the Processor graphs is $AQ\nu(t)$. For the example, the two premises selected through the common point follow:

1. $AQ1$(pleasant, -, -, -, -, -, Mrs. Bee, (-, -, -, -, -), -, -, -)

That is

"Mrs. Bee was pleasant."

2. $AQ2$(pleasant, -, -, -, -, -, Mrs. Cee, (-, -, -, -, -), -, -, -)

That is

"Mrs. Cee was pleasant."

And $AQ1(t)$, $AQ2(t) \vdash BQ2$, where $BQ2$ is the direct answer to the question

"Who was pleasant?"

a_0's answer is

$BQ2$—"Mrs. Cee and Mrs. Bee were pleasant."

The other forms of stimuli to a_0 for initiating the deductive process are similar to those for the environment graph or time graph, and the sequence of formulas established by formal logic is the same when E or T are replaced by Q for the proposition symbols.

The above discussion provides proof of the following theorem:

Theorem 4.2. *Given w graph tuples Tuy, u=E, T, Q; y=1, 2, ---, w stored in the nonshared environments (memory) of an observing automaton a_0 and associated through a common point identified by identical values for the leftmost element of the tuples. Let $Auv(t)$, v=1, 2, ---, k; $2 \leq k \leq w$, be internal premises established from Tuy and containing values for corresponding elements with values of the leftmost element and the automaton element as a minimum set of values. Then $Au1(t)$, $Au2(t)$, ---, $Auk(t) \vdash Auk(t)$ & --- & $Au2(t)$ & $Au1(t)$ and $AEk(t)$ & --- & $AE2(t)$ & $AE1(t)$ gives information about automata that accessed a same environment; $ATk(t)$ & --- & $AT2(t)$ & $AT1(t)$ gives information about automata of a same time period; and $AQk(t)$ & --- & $AQ2(t)$ & $AQ1(t)$ gives information about an automaton in a particular state or about automata in a same state.*

4.4.2 Deductive Processes Related to Association by Combining Graph Tuples

The associations by combining graph tuples relate to basic deductive processes and begin with the processor graph following the hierarchal development of the graph functions of Chapter 2 (Koenig 1978). The example of Section 4.3.2 will be used to further explore the deductive processes.

When a_0 of the example sees **Rover** sometime later (moment $m3$) in a pleasant mood about to be jabbed with a stick, or if the question is asked of a_0

"What will happen if Rover, in a pleasant mood, is jabbed with a stick?"

the verbal response in either case may be either

"Rover will bark."
"Rover will lay back his ears."

or

"Rover will lay back his ears and will bark."

In preparation for the formalizing of the deductive process, the premises must be established, and this can now be done from the single combined tuple. The knowledge is expressed in sentence form, then put in logical symbolism.

"Rover is pleasant and is jabbed."

$$AQ0(\text{pleasant}, 0, rt, em, 0, \text{jabbed}, \text{Rover}, (\text{-}, \text{-}, \text{-}, \text{-}, \text{-}), \text{-}, \text{-}, \text{-})$$

$AQ0(t)$ is an external premise.

"When Rover is pleasant and is jabbed, he lays back his ears."

$$AQ1(\text{-}, \text{-}, \text{-}, \text{-}, \text{-}, \text{-}, \text{Rover}, ((\text{lay back ears})\text{pattern-}, \text{-}, \text{-}, \text{-},), \text{-}, \text{-}, \text{-})$$

$AQ0(t) \supset AQ1(t)$ is an internal premise.

"When Rover is pleasant and is jabbed, he barks."

$$AQ2(\text{-}, \text{-}, \text{-}, \text{-}, \text{-}, \text{-}, \text{Rover}, (\text{-}, \text{-}, (bark)\text{Sound}, \text{-}, \text{-},), \text{-}, \text{-}, \text{-})$$

$AQ0(t) \supset AQ2(t)$ is an internal premise. The premise $Au0(t)$, $u = Q, T, E$ can come from the values of the elements to the left of the element $\boldsymbol{a}$ for the $(n+1)$-tuples of the processor, time, and environment graphs (by Theorem 4.1). Also, as many as all five component elements of the transformation response can be involved in the premises of an argument. It is required to show by the formal deductive procedure that the generalized argument is valid. The generalized argument is

$$Au0(t), Au0(t) \supset Au1(t), Au0(t) \supset Au2(t), ---, Au0(t) \supset Auk(t)$$
$$\vdash Au1(t) \,\&\, Au2(t) \,\&\, --- \,\&\, Auk(t), v=1, 2, ---, \text{k}; 1 \leq k \leq w;$$
$$1 \leq w \leq 5$$

w is the number of transformation response components with known values.

1a. $Au0(t)$—external premise
 b. $Au0(t) \supset Au1(t)$—internal premise
 c. $Au1(t)$—implication rule 1a, b

2a. $Au0(t) \supset Au2(t)$—internal premise
 b. $Au2(t)$—implication rule 1a, 2a
 c. $Au1(t) \supset (Au2(t) \supset Au1(t) \,\&\, Au2(t))$—axiom schema $(A \supset (B \supset A \,\&\, B))$ and substitution rule
 d. $Au2(t) \supset Au1(t) \,\&\, Au2(t)$—implication rule 1c, 2c
 e. $Au1(t) \,\&\, Au2(t)$—implication rule 2b, d

•

•

•

•

*k*a. $Au0(t) \supset Auk(t)$—internal premise
 b. $Auk(t)$—implication rule 1a, *k*a
 c. $(Au1(t) \,\&\, Au2(t) \,\&\, --- \,\&\, Au(k-1)(t)) \supset (Auk(t) \supset (Au1(t) \,\&\, Au2(t) \,\&\, --- \,\&\, Auk(t)))$—axiom schema $(A \supset (B \supset A))$ and substitution rule
 d. $Auk(t) \supset (Au1(t) \,\&\, Au2(t) \,\&\, --- \,\&\, Auk(t))$—implication rule $(k-1)$e, *k*c
 e. $Au1(t) \,\&\, Au2(t) \,\&\, --- \,\&\, Auk(t)$—implication rule *k*b, d

The generalized argument is valid, and the deductive process by formal logic correctly models the natural deductive process. For the example, $w=2$, $u=Q$, then $k=2$, and

$$AQ0(t), AQ0(t) \supset AQ1(t), AQ0(t) \supset AQ2(t) \vdash AQ1(t) \,\&\, AQ2(t)$$

The proof of the following theorem is given by the above discussion:

Theorem 4.3. *Given a graph tuple Tuy, $u=Q$, T, E; $y=1$, 2, --- stored in the nonshared environments (memory) of an observing automaton a_0 resulting from the association of knowledge by combining graph tuples under the conditions of Theorem 4.1. Let Au0(t) be an external premise acting as a stimulus to a_0 with the elements on the left of element **a**, including **a**, having the same values as the corresponding elements of Tuy. Let $Au0(t) \supset Auv(t)$,*

*$v=1, 2, ---, k$; $1 \leq k \leq w \leq 5$ be an internal premise where w is the number of transformation response components with known values. $Auv(t)$ is established by giving two of its elements, element **a** and an element for one of the transformation response components, values of the corresponding elements of Tuy. Then $Au0(t), Au0(t) \supset Au1(t), Au0(t) \supset Au2(t), ---, Au0(t) \supset Auk(t) \vdash Au1(t)$ & $Au2(t)$ & --- & $Auk(t)$ and the response of a_0 to the stimulus $Au0(t)$ is $Au1(t)$ & $Au2(t)$ & --- & $Auk(t)$.*

4.4.3 Deductive Processes with Aristotelian Form A as a Premise

Some deductive processes that relate to the predicate calculus will now be considered (Koenig 1979a). There are four categorical forms of the Aristotelian logic *A*, *E*, *I*, *O*. The Form *A* involving the universal quantifier *V* will be discussed here, and the most complex premise contained in an argument will be of this form. A complete argument consisting of a statement of Form *A*, a second premise, and the conclusion falls within the Class 4 sentence capability for storing the meanings (Section 4.1.4).

The Form *A* is

$$\forall x (S(x) \supset P(x))$$

symbolizing "All *S* are *P*." (Kleene 1967, p. 138). The letters *S* and *P* correspond to the two parts of the sentence considered as *subject* and *predicate*. The meanings of *S*(*x*) and *P*(*x*) are as follows:

$S(x)$—*x* possesses the property *S*

$P(x)$—*x* possesses the property *P*

"All *S* are *P*." can be written as

"If *x* possesses the property *S*, then *x* possesses the property *P*."

for all *x*. *S* and *P* may relate to the identification of an environment, automaton, principal or auxiliary receptor or effector. In the case of an environment, the meaning of "All *S* are *P*." is stored in the (15+1)-tuple for the Environment graph in the following manner:

$$((S(x), P(x), \text{-}, \text{-}), \text{-}, \text{-}, \text{-}, \text{-}, \text{-}, \text{-}, \text{-}, \text{-}, \text{-}, (\text{-}, \text{-}, \text{-}, \text{-}, \text{-}), \text{-}, \text{-}, \text{-}, \text{-}, \text{-})$$

It should be clear the position (*S*(*x*), *P*(*x*), -, -) takes in the (15+1)-tuple when defining ***a***, *x′*, *x″*, *z″*, or *z′*. For the example sentence

"All metals are malleable."

the (15+1)-tuple containing the meaning is

((metals, malleable, -, -), -, -, -, -, -, -, -, -, -, (-, -, -, -, -), -, -, -, -, -)

Suppose there was the statement (proposition)

$S(e)$—e possesses the property S

The (15+1)-tuple containing the meaning is

((e, -, -, $S(e)$), -, -, -, -, -, -, -, -, -, (-, -, -, -, -), -, -, -, -, -)

From the discussion of Section 4.1.4 for the Class 4 sentence, the association of ($S(x)$, $P(x)$, -, -) and (e, -, -, $S(e)$) for identifying the environment yields

((e, -, -), $S(x)$, $P(x)$, -, -)

and the (15+1)-tuple that contains the total meaning is

(((e, -, -), $S(x)$, $P(x)$, -, -), -, -, -, -, -, -, -, -, -, (-, -, -, -, -), -, -, -, -, -)

For an example sentence (proposition)

"Copper is a metal."

the corresponding (15+1)-tuple is

((copper, -, -, metal), -, -, -, -, -, -, -, -, -, (-, -, -, -, -), -, -, -, -, -)

The association of (metals, malleable, -, -) and (copper, -, -, metals) for identifying the environment yields

((copper, -, -), metals, malleable, -, -)

and the (15+1)-tuple that contains the total meaning is

(((copper, -, -), metals, malleable, -, -), -, -, -, -, -, -, -, -, -, (-, -, -, -, -), -, -, -, -, -)

Now, if $\forall x(S(x) \supset P(x))$, $S(e)$ are two premises, is there a conclusion such that the corresponding argument is valid? From proof theory for the predicate calculus, it can be shown that

$P(e)$—e possesses the property P

is deducible from the two premises; that is

$$\forall x(S(x) \supset P(x)), S(e) \vdash P(e)$$

For the example, the corresponding argument that can be shown to be valid is the following:

"All metals are malleable.
Copper is a metal.
Therefore, copper is malleable."

Validity will be established with the use of the following theorems taken from mathematical logic (Kleene 1967): Γ is a list of zero or more formulas so that $\Gamma, A \vdash B$ is written instead of A_1, ---, $A_{m-1}, A_m \vdash B$, which states that formula B is deducible from the m assumption formulas A_1, ---, A_m.

Theorem 4.4. *For any finite list of (zero or more) formulas Γ, and any formulas A, B, C*

	Introduction	*Elimination*
$\supset$	*If* $\Gamma, A \vdash B$ *then* $\Gamma \vdash A \supset B$	$A, A \supset B \vdash B$
&	$A, B \vdash A \& B$	$A \& B \vdash A$ $A \& B \vdash B$
V	$A \vdash AVB$ $B \vdash AVB$	*If* $\Gamma, A \vdash C$ *and* $\Gamma, B \vdash C$, *then* $\Gamma, AVB \vdash C$
$\neg$	*If* $\Gamma, A \vdash B$ *and* $\Gamma, A \vdash \neg B$ *then* $\Gamma \vdash \neg A$	$\neg\neg A \vdash A$ (*$\neg\neg$-elim.*) $A, \neg A \vdash B$ (*$\neg$-elim.*)
~	$A \supset B, B \supset A \vdash A \sim B$	$A \sim B \vdash A \supset B$ $A \sim B \vdash B \supset A$

Theorem 4.5.

i) *For* $m \geq 1$:
A_1, ---, $A_m \vdash A_1$

A_1, ---, $A_m \vdash A_m$

ii) *For* $m, p \geq 0$:
If A_1, ---, $A_m \vdash B_1$

A_1, ---, $A_m \vdash B_p$ *and* B_1, ---, $B_p \vdash C$ *then* A_1, ---, $A_m \vdash C$

Theorem 4.6. *Let x be any variable, $A(x)$ be any formula, r be any variable not necessarily distinct from x, and $A(r)$ be the result of substituting r for the free occurrences of x in $A(x)$. Also, let Γ be any list of (zero or more) formulas, and C be any formula. Then the following rules hold, provided:*

A. For $\forall$-elimination and $\exists$-introduction, r is free for x in $A(x)$.
B. For $\forall$-introduction and $\exists$-elimination, Γ does not contain x free.
C. For $\exists$-elimination, C does not contain x free.

	Introduction	*Elimination*
$\forall$	*If* $\Gamma \vdash A(x)$, *then* $\Gamma \vdash \forall x A(x)$	$\forall x A(x) \vdash A(r)$
$\exists$	$A(r) \vdash \exists x A(x)$	*If* $\Gamma, A(x) \vdash C$, *then* $\Gamma, \exists x A(x) \vdash C$

With the use of the above three theorems from mathematical logic, the following list of formulas establish that $P(e)$ is deducible from $\forall x(S(x) \supset P(x))$, $S(e)$:

1. $\forall x(S(x) \supset P(x)), S(e) \vdash \forall x(S(x) \supset P(x))$—Theorem 4.5i
2. $\forall x(S(x) \supset P(x)) \vdash S(e) \supset P(e)$—Theorem 4.6, $\forall$—elimination, substitution of e for x free
3. $\forall x(S(x) \supset P(x)), S(e) \vdash S(e) \supset P(e)$—Theorem 4.5ii, 1, 2
4. $\forall x(S(x) \supset P(x)), S(e) \vdash S(e)$—Theorem 4.5i
5. $S(e), S(e) \supset P(e) \vdash P(e)$—Theorem 4.4, $\supset$—elimination 3, 4
6. $\forall x(S(x) \supset P(x)), S(e) \vdash P(e)$—Theorem 4.5ii, 3, 4, 5

The following (15+1)-tuple contains the meaning of the valid argument:

$$(((e, P(e), \text{-}), S(x), P(x), \text{-}, \text{-}), \text{-}, \text{-}, \text{-}, \text{-}, \text{-}, \text{-}, \text{-}, \text{-}, \text{-}, (\text{-}, \text{-}, \text{-}, \text{-}, \text{-}), \text{-}, \text{-}, \text{-}, \text{-}, \text{-})$$

The (15+1)-tuple for the example argument is

(((copper, malleable, -), metal, malleable, -, -), -, -, -, -, -, -, -, -, -, (-, -, -, -, -), -, -, -, -, -)

Consider the following argument defining an automaton ***a***:

"All men are mortal.
Socrates is a man.
Therefore, Socrates is mortal."

The following (15+1)-tuple contains the meaning of this valid argument (the form of the logical symbolism for the argument is the same as above):

(-, -, -, -, -, -, -, -, ((Socrates, mortal, -), men, mortal, -, -), -, (-, -, -, -, -), -, -, -, -)

There are various ways an automaton can receive the statements of an argument. If the three statements making up the argument are received in a definite order, one following the other, the argument can be checked for validity and the meaning immediately stored, if valid, in a single (15+1)-tuple. This has been demonstrated by the sequence the material was presented in the discussion. Certain conflicts might also be resolved.

If the statements of an argument are received at various times and other elements of the (15+1)-tuples also have values, a (15+1)-tuple can be created for the purpose of storing only the meaning of an argument. For making distinctions between the statements, it should be remembered that $S \subseteq P$ and $a \in S$.

The premise "All S are P" might be established by the computer in that many findings from a search through its knowledge structures establishes the statement to be true.

4.5 INFERENCES

To infer suggests the arriving at a decision or opinion by reasoning from known facts or evidence; for example

"From your smile, I infer that you are pleased."

A computer might infer its output from its input; for example

Input: John is a student.
Output: John probably writes on a chalkboard.

Because the general automaton concept for representing knowledge is built upon order and organization of the world, it defines complete knowledge and produces a well defined set of inferences.

Inferences are based on knowledge previously acquired, and knowledge association becomes paramount in establishing inferences. Recall (Section 4.3), knowledge association is accomplished by bringing together knowledge structures into wholes, i.e., by establishing connected graphs (Koenig 1982). Associations based on knowledge about automata were categorized as 1) associations by combining graphs through common points, and 2) associations by combining graph tuples (Koenig 1978). Inferences are either related to a single graph tuple of associated knowledge or to more than one tuple of associated knowledge (Koenig 1983).

4.5.1 Inferences Related to a Single Graph Tuple of Associated Knowledge

Given, a single graph tuple of associated knowledge that was established by knowledge association discussed in Section 4.3: (The knowledge elements are assigned values from an example.)

$$TE_{aj}(m+i, m+i+1)=(e_i, t_i, q_i, x_i'', x_i', z_i', z_i'', s_i, a, b_{i+1}, (fs, fm, fh, ft, fa)_{i+1}, d_{i+1}, c_{i+1}, q_{i+1}, t_{i+1}, e_{i+1})$$

$e_i=(e_i, f_i(e_i), g_i(e_i), E_i')$=(chalkboard, communication, rectangular, classroom)

t_i=daytime

q_i=tense

$x_i''=(x_i'', fr_i''(x_i''), gr_i''(x_i''), X_i'')$=(no hearing aids, -, -, -)

$x_i'=(x_i', fr_i'(x_i'), gr_i'(x_i'), X_i')$=(ears, hearing, oval shaped, head)

$z_i'=(z_i', fe_i'(z_i'), ge_i'(z_i'), Z_i')$=(right hand, writing, size 8, right arm)

$z_i''=(z_i'', fe_i''(z_i''), ge_i''(z_i''), Z_i')$=(chalk, writing, cylindrical, mineral)

$s_i=b_{n+1}$=" Write on chalkboard.", from $Te_{ak}(m+n, m+n+1)$= (students, -, -, -, -, -, -, -, (-, teachers, -, -), (write on chalkboards), (-, -, tell, -, -), -, -, -, -, -)

$a=(Y, f'(a), g'(a), A')=(Y$, student, 5′6″ tall, young person) (No specific automaton is named; hence the space for the name carries the symbol Y in the identification of a.)

b_{i+1} =marks

$f_{i+1}=(fs, fm, fh, ft, fa)_{i+1}$=(write, -, -, -, -)

d_{i+1} =near

c_{i+1} =less than an hour

q_{i+1} =relieved

t_{i+1} =daytime

$e_{i+1}=(e_{i+1}, f_{i+1}(e_{i+1}), g_{i+1}(e_{i+1}), E_{i+1}')$ =(chalkboard, communication, rectangular, classroom)

The representations of e_i, x_i'', x_i', z_i', z_i'', a, e_{i+1} all contain declarative knowledge for identification purposes and are discussed in Section 4.1.4. The stimulus s_i for a comes from the recorded response b_{n+1} of the tuple $Te_{ak}(m+n, m+n+1)$, which is the knowledge structure for the sentence

"Teachers tell students to write on chalkboards."

A graph tuple of associated knowledge, $TE_{aj}(m+i, m+i+1)$, is established over an extended period of time from discourses *and* visual observations of various individuals, Joe, Mary, Sue, ---, writing on chalkboards. If the graph tuple is established by a computer, the assumption can be made that it is established only from a series of discourses. The knowledge from any given discourse is associated with the knowledge of related discourses that were previously stored. Consider, for example, the knowledge element

$$a=(a, f'(a), g'(a), A')$$

for the above graph tuple. a could come from the following groups of sentences of discourses:
Sentences of a first discourse

Joe is a student.
Joe wrote on a chalkboard.
etc.

Sentences of a more recent discourse

Mary, who is a student, wrote on a chalkboard.
etc.

There are similar discourses related to writing on chalkboards occurring at different times for a number of other people, mostly students, and the value of a as a knowledge element of the tuple becomes

$$a=(Y, f'(a), g'(a), A')=(Y, \text{student}, 5'\ 6''\ \text{tall}, \text{young person})$$

No specific student (automaton) is named; hence, the space for the name carries the symbol Y in the identification of a. Also, most of the students writing on chalkboards are 5′ 6″ tall and are young persons. Similarly, there are probabilities associated with other knowledge elements of the tuple.

We have defined a graph tuple of associated knowledge and determined how it may be established. Consider now the different types of sentences that are required for communication of the knowledge elements pertaining to the tuple of associated knowledge. They may be input or output sentences and may be factual or inferred. A factual sentence is defined as a sentence whose knowledge elements have been obtained by observation of an automaton a. The example introduced above will be carried along in parallel. The general forms of the sentences hold for many cases, but in other cases, the sentences may sound awkward when specific values are used.

1. $a=(a, f'(a), g'(a), A')=(Y, \text{student}, 5'\ 6''\ \text{tall}, \text{young person})$
   a. a is/are $f'(a)$.
      John is a student.
   b. a (probably) is/are $g'(a)$.
      John (probably) is 5′ 6″ tall.
   c. a (probably) is/are A'.
      John (probably) is a young person.

2. $e_i=(e_i,\ f_i(e_i),\ g_i(e_i),\ E_i')$ = (chalkboard, communication, rectangular, classroom)
   $f_{i+1}=(fs, fm, fh, ft, fa)_{i+1}$ = (write, -, -, -, -)
   a (probably) started $f_{i+1}e_i$ in the E_i'.
   John (probably) started writing on a chalkboard in the classroom.
   a. e_i is/are $f_i(e_i)$.
      The chalkboard is for communication.
   b. e_i (probably) is/are $g_i(e_i)$.
      The chalkboard (probably) is rectangular.
   c. e_i (probably) is/are in E_i'.
      The chalkboard (probably) is in a classroom.
3. t_i=daytime
   a (probably) started $f_{i+1}e_it_i$.
   John (probably) started writing on a chalkboard in the daytime.
4. q_i=tense
   a (probably) was/were q_i when he/she/it/they started $f_{i+1}e_i$.
   John (probably) was tense when he started writing on the chalkboard.
5. $s_i=b_{n+1}$=Write on a chalkboard
   a (probably) received information to s_i.
   John (probably) received information to write on a chalkboard.
6. $x_i=(x_i', fr_i'(x_i'), gr_i'(x_i'), X_i')$ = (ears, hearing, oval shaped, head)
   a (probably) saw/smelled/heard/felt/tasted the information to s_i.
   John (probably) heard the information to write on a chalkboard.
   (x_i'=ears, implies *heard*)
   a. *a*'s x_i' (probably) is/are for $fr_i'(x_i')$.
      John's ears (probably) are for hearing.
   b. *a*'s x_i' (probably) is/are $gr_i'(x_i')$.
      John's ears (probably) are oval shaped.
   c. *a*'s x_i' (probably) is/are part of X_i'.
      John's ears (probably) are part of the head.
7. $x_i''=(x_i'', fr_i''(x_i''), gr_i''(x_i''), X_i'')$ = (no hearing aids, -, -, -)
   a probably) required x_i'' to receive information to s_i.
   John (probably) required no hearing aids to receive information to write on a chalkboard.
   a. x_i'' is/are used for $fr_i''(x_i'')$.
   b. x_i'' (probably) is/are $gr_i''(x_i'')$.
   c. x_i'' (probably) is/are a part of X_i''.
8. $z_i'=(z_i', fe_i'(z_i'), ge_i'(z_i'), Z_i')$ = (right hand, writing, size 8, right arm)
   a (probably) used (use) his/her/its/their z_i' to f_{i+1} e_i.
   John (probably) used his right hand to write on the chalkboard.

a. a's z_i' (probably) is/are used for $fe_i'(z_i')$.
John's right hand (probably) is used for writing.
b. a's z_i' (probably) is/are $ge_i'(z_i')$.
John's right hand (probably) is size 8.
c. a's z_i' (probably) is/are part of Z_i'.
John's right hand (probably) is part of the right arm.

9. $z_i''=(z_i'', fe_i''(z_i''), ge_i''(z_i''), Z_i'')$=(chalk, writing, cylindrical, mineral)
a (probably) used z_i'' to $f_{i+1}e_i$.
John (probably) used chalk to write on the chalkboard.
a. z_i'' is/are used for $fe_i''(z_i'')$.
The chalk is used for writing.
b. z_i'' (probably) is/are $ge_i''(z_i'')$.
The chalk (probably) is cylindrical.
c. z_i'' (probably) is/are Z_i''.
The chalk (probably) is a mineral.

10. b_{i+1}=marks
a (probably) made b_{i+1} when $f_{i+1}e_i$.
John (probably) made marks when writing on the chalkboard.

11. $f_{i+1}=(fs, fm, fh, ft, fa)_{i+1}$ implies was seen/smelled/heard/felt/tasted f_{i+1}.
a (probably) was/were seen/smelled/heard/felt/tasted $f_{i+1}e_i$.
John (probably) was seen writing on the chalkboard.

12. d_{i+1}=near
a (probably) was/were $d_{i+1}e_i$ after $f_{i+1}e_i$.
John (probably) was near the chalkboard after writing on the chalkboard.

13. c_{i+1}=less than an hour
a (probably) $f_{i+1}c_{i+1}e_i$.
John (probably) wrote less than an hour on the chalkboard.

14. q_{i+1}=relieved
a (probably) was/were q_{i+1} after $f_{i+1}e_i$.
John (probably) was relieved after writing on the chalkboard.

15. t_{i+1}=daytime
a (probably) finished $f_{i+1}e_it_{i+1}$.
John (probably) finished writing on the chalkboard in the daytime.

16. $e_{i+1}=(e_{i+1}, f_{i+1}(e_{i+1}), g_{i+1}(e_{i+1}), E_{i+1})$=(chalkboard, communication, rectangular, classroom).
a (probably) was/were (is/are) at e_{i+1} after he/she/it/they finished $f_{i+1}e_i$.

John (probably) was at the chalkboard after he finished writing on the chalkboard.

a. e_{i+1} is/are $f_{i+1}(e_{i+1})$.
The chalkboard is for communication.

b. e_{i+1} (probably) is/are $g_{i+1}(e_{i+1})$.
The chalkboard (probably) is rectangular.

c. e_{i+1} (probably) is/are in E'_i.
The chalkboard (probably) is in a classroom.

When knowledge elements of a tuple $TE_{aj}(m+i, m+i+1)$ of associated knowledge are not known, the sentences requiring the unknown values are assumed not to exist for that tuple.

Consider the operation of an automaton when its input names the automaton a of the tuple of associated knowledge and names the function of a.

Algorithm 4.5i. Given a factual sentence 1a as input to automaton a_2 from automaton a_1

1a. Factual: a is/are $f'(a)$.
e.g., John Doe is a student.

Also, given none or any one or more than one of the sentences, 1b, 1c, 2 through 16, each either factual or inferred.

If a_2 has in its memory a tuple $TE_{aj}(m+i, m+i+1)$ of associated knowledge that contains $f'(a)$, and/or if a_2 has in its memory a tuple that contains corresponding factual knowledge about a, the output of a_2 is the factual and inferred sentences 1b, 1c, 2 through 16, each of which contains at least one knowledge element not contained in the input sentences.

Otherwise, there is no output, and a_2 associates the knowledge contained in the input sentences with knowledge previously stored.

In the *conversation* mode, the number of output sentences of a_2 of an uninterrupted sequence, which follows a set of input sentences of a_1 of an uninterrupted sequence, is not to exceed the number of input sentences of a_1 in the latest set.

In the *examination* mode of a_2 by a_1, the output of a_2, which follows the first set of input sentences of a_1 of an uninterrupted sequence, is one uninterrupted sequence of all the sentences, 1b, 1c, 2 through 16, each of which contains at least one knowledge element not contained in the input sentences.

When a factual and inferred sentence have identical knowledge elements, the factual sentence should dominate and be presented.

The above Algorithm 4.5i operating in the conversation mode is applied to the example: a_2 has in its memory a tuple $TE_{aj}(m+i, m+i+1)$ of associated knowledge that contains $f'(a)$ and has a tuple in its memory that contains factual knowledge about a.

Input

1a. Factual: John is a student.
1b. Factual: He is 5′ 6″ tall.
2. Inferred: He probably started writing on a chalkboard in the classroom.

Output

1c. Inferred: John probably is a young person.
3. Inferred: He probably started writing on a chalkboard in the daytime.
4. Factual: He was tense when he started writing on the chalkboard.

Input

5. Factual: John received information to write on a chalkboard.

Output

6. Inferred: John probably heard the information to write on a chalkboard.

Input

7. Inferred: John probably required no hearing aids to receive information to write on a chalkboard.
8. Factual: He used his right hand to write on the chalkboard.

Output

9. Inferred: John probably used chalk to write on the chalkboard.
10. Inferred: He probably made marks when writing on the chalkboard.

Input

11. Inferred: John probably was seen writing on the chalkboard.
12. Factual: He was near the chalkboard after writing on the chalkboard.

Output

13. Inferred: John probably wrote less than an hour on the chalkboard.
14. Inferred: He probably was relieved after writing on the chalkboard.

Input

15. Factual: John finished writing on the chalkboard in the daytime.

Output

16. Factual: John was at the chalkboard after he finished writing on the chalkboard.

Consider the operation of an automaton when its input names the environment of the tuple of associated knowledge and names the function of environment.

Algorithm 4.5ii. Given a factual sentence 2a as input to automaton a_2 from automaton a_1

2a. Factual: e_i is/are $f_i(e_i)$.
    e.g., The chalkboard is for communication.

Also, given none or any one or more than one of the sentences 2b, 2c, 3 through 16, each either factual or inferred with f' (a) substituted for a.

If a_2 has in its memory a tuple $TE_{aj}(m+i, m+i+1)$ of associated knowledge that contains e_i and $f_i(e_i)$, and/or if a_2 has in its memory a tuple that contains corresponding factual knowledge about e_i, the output of a_2 is the factual and inferred sentences 2b, 2c, 3 through 16, each of which contains at least one knowledge element not contained in the input sentences.

Otherwise, there is no output, and a_2 associates the knowledge contained in the input sentences with knowledge previously stored.

The operation can be either in the conversation mode or in the examination mode.

The above Algorithm 4.5ii operating in the examination mode is applied to the example. a_2 has in its memory a tuple $TE_{aj}(m+i, m+i+1)$ of associated knowledge that contains e_i and $f_i(e_i)$ and has a tuple in its memory that contains factual knowledge about e_i.

Input

2a. Factual: The chalkboard is for communication.
9. Inferred: Students probably use chalk to write on the chalkboard.

Output

2b. Inferred: The chalkboard probably is rectangular.
2c. Factual: The chalkboard is in a classroom.

3. Inferred: Students probably start writing on the chalkboard in the daytime.
4. Inferred: They probably are tense when they start writing on the chalkboard.
5. Inferred: Students probably receive information to write on the chalkboard.
6. Factual: They hear the information to write on the chalkboard.
7. Inferred: Students probably require no hearing aids to receive information to write on the chalkboard.
8. Inferred: They probably use their right hand to write on the chalkboard.
9. Factual: Students use chalk to write on the chalkboard.
10. Inferred: They probably make marks when writing on the chalkboard.
11. Inferred: Students probably are seen writing on the chalkboard.
12. Inferred: They probably are near the chalkboard after writing on the chalkboard.
13. Factual: Students write less than an hour on the chalkboard.
14. Factual: They are relieved after writing on the chalkboard.
15. Inferred: Students probably finish writing on the chalkboard in the daytime.
16. Factual: They are at the chalkboard after they finish writing on the chalkboard.

Consider the operation of an automaton when its input names the auxiliary receptor and the function of the auxiliary receptor. The sentences that are directly relevant to the auxiliary receptor are 7a, 7b, 7c.

Algorithm 4.5iii. Given a factual sentence 7a

7a. Factual: x_i'' is/are used for $fr_i''(x_i'')$.

Also, given none or any one or more than one of the sentences 7b, 7c, each either factual or inferred.

If a_2 has in its memory a tuple $TE_{aj}(m+i, m+i+1)$ of associated knowledge that contains x_i'' and $fr_i''(x_i'')$, and/or if a_2 has in its memory a tuple that contains corresponding factual knowledge about x_i'', the output of a_2 is the factual and inferred sentences 7b, 7c, each of which contains at least one knowledge element not contained in the input sentences.

Otherwise, there is no output, and a_2 associates the knowledge contained in the input sentences with knowledge previously stored.

The operation can be either in the conversational mode or in the examination mode.

Consider the operation of an automaton when its input names the auxiliary effector and the function of the auxiliary effector. The sentences that are directly relevant to the auxiliary effector are 9a, 9b, 9c.

Algorithm 4.5iv. Sentences 9a, 9b, 9c and z_i'' may be substituted for 7a, 7b, 7c and x_i'' in the above Algorithm 4.5iii.

The above Algorithm 4.5iv, operating in the examination mode, is applied to the example. a_2 has in its memory a tuple $TE_{aj}(m+i, m+i+1)$ of associated knowledge that contains z_i'' and $fe_i''(z_i'')$ and has a tuple in its memory that contains factual knowledge about z_i''

Input

9a. Factual: The chalk is used for writing.

Output

9c. Inferred: The chalk probably is a mineral.
9b. Factual: It is cylindrical.

Consider the operation of an automaton when its input names the automaton a of the tuple of associated knowledge and names a's principal receptor. Here, a first factual input sentence is implied, namely

a has/have x_i'.

Algorithm 4.5v. Given, any one or more than one of the sentences 6a, 6b, 6c, factual or inferred, as input to automaton a_2 from automaton a_1.

If a_2 has in its memory a tuple $TE_{aj}(m+i, m+i+1)$ of associated knowledge that contains x_i' and/or if a_2 has in its memory a tuple that contains corresponding factual knowledge about a, the output of a_2 is the factual and inferred sentences, 6a, 6b, 6c, each of which contains at least one knowledge element not contained in the input sentences.

Otherwise, there is no output, and a_2 associates the knowledge contained in the input sentences with knowledge previously stored.

The operation can be either in the conversational mode or in the examination mode.

Consider the operation of an automaton when its input names the automaton a of the tuple of associated knowledge and names the principal effector of a. A first factual input sentence is implied, namely

a has/have z_i'.

Algorithm 4.5vi. Sentences 8a, 8b, 8c, and z_i' may be substituted for 6a, 6b, 6c, and x_i' in the Algorithm 4.5v.

The above Algorithm 4.5vi, operating in the conversation mode, is applied to the example. a_2 has in its memory a tuple $TE_{aj}(m+i, m+i+1)$ of associated knowledge that contains a and z_i'.

Input

8a. Factual: John's right hand is used for writing.

Output

8b. Inferred: John's right hand probably is size 8.

4.5.2 Inferences Related to More than One Graph Tuple of Associated Knowledge

We will now consider knowledge elements of one tuple inferred from knowledge elements of another tuple. Two cases will be considered.

For the two cases, there is a sequence of two or more tuples of associated knowledge $TE_{aj}(m+i, m+i+1)$, ---, $TE_{aj}(m+i+k, m+i+k+1)$. For example, a sequence of three tuples of associated knowledge is $TE_{aj}(m+i, m+i+1)$, $TE_{aj}(m+i+1, m+i+2)$, $TE_{aj}(m+i+2, m+i+3)$, $k=2$. The knowledge of the following sequence of three sentences may have been associated with the knowledge of the tuples:

Mary put the radish in her mouth.
(Knowledge is associated with knowledge of $TE_{aj}(m+i, m+i+1)$)

She chewed the radish.
(Knowledge is associated with knowledge of $TE_{aj}(m+i+1, m+i+2)$)

She swallowed the radish.
(Knowledge is associated with knowledge of $TE_{aj}(m+i+2, m+i+3)$)

Consider now the types of sentences that are required to communicate the knowledge elements of the sequence of tuples of associated knowledge. As before, an automaton a_1 provides input to an automaton a_2 which has the tuples of associated knowledge stored in its memory and which produces the output. The input sentences for communication are factual, and the output sentences may be factual or inferred. The general forms of the sentences are for simple sentences containing only the essential knowledge elements

$$a(\text{probably})\, f_{i+1}e_i$$

•

•

•

$$a(\text{probably})\, f_{i+k+1}e_{i+k}$$

Consider the operation of an automaton when its input is one of the above sentences.

Algorithm 4.5vii. Given a factual sentence as input to automaton a_2 from automaton a_1

$$af_{i+n+1}e_{i+n}$$

If a_2 has in its memory a sequence of tuples of associated knowledge $TE_{aj}(m+i, m+i+1)$, ---, $TE_{aj}(m+i+k, m+i+k+1)$. $0 \leq n \leq k$, and/or if a_2 has in its memory tuples containing corresponding factual knowledge about a, the output of a_2 is a factual or inferred sentence

$$a(\text{probably})\, f_{i+p+1}e_{i+p},\, 0 \leq p \leq k,\, p \neq n$$

Otherwise, there is no output and a_2 associates the knowledge contained in the input sentence with knowledge previously stored. Each of the tuples in the memory of a_2 must contain a and one must contain f_{i+n+1}.

$$\text{Case 1}: n > p$$

$$\text{Case 2}: n < p$$

Example of Case 1, $n>p$. Tuples of associated knowledge in memory of a_2

$TE_{aj}(m+i, m+i+1)$
$TE_{aj}(m+i+1, m+i+2)$
$TE_{aj}(m+i+2, m+i+3)$

Input

$a\ f_{i+3}e_{i+2}$. $(n=2)$
John swallowed the nut.

Output

a (probably) $f_{i+2}e_{i+1}$. $(p=1)$
Inferred: John probably chewed the nut.

Example of Case 2, $n<p$. Tuples of associated knowledge in memory of a_2 are the same as the example above for Case 1.
Input

$a\ f_{i+2}e_{i+1}$. $(n=1)$
John chewed the nut.

Output

a (probably) $f_{i+3}e_{i+2}$. $(p=2)$
Inferred: John probably swallowed the nut.

EXERCISES

4.1 Establish the element of the appropriate relation and the corresponding $TE_{aj}(m+i, m+i+1)$ for the following sentences:

a. for the second example, Class 1, Type 1
b. for the second example, Class 1, Type 2
c. for the second example, Class 1, Type 3
d. "Mary pushed the cart one block." Class 2
e. "Today, the train traveled slowly from Redbud to Baldwin." Class 2
f. "Mary left the airport by taxi." Class 2
g. "The boat left Amsterdam today." Class 2
h. "The plane arrived at the airport today at 2:00 p.m." Class 2
i. "Today, Jim arrived late for class." Class 2
j. "Today, Jane took her lunch to school." Class 2
k. "The train carried the space module to the launch site." Class 2
l. for the third example, Class 3
m. "Copper is a metal." Class 4
n. "Copper is malleable."
o. "The thumb of Jim's right hand is injured."
p. "The handle of the hammer is broken."
q. for the third sentence, Class 5, Type 1
r. "Jim saw John strike the table with a hammer." Class 5
s. for the second example, Class 6
t. for the first example, Class 7

4.2 Generate the relations by the given formulations for the following classes of sentences: a) Class 5, b) Class 6, c) Class 7.

4.3 Write a program in a language of your choice for performing a meaningful analysis of sentences of the following classes: a) Class 2, b) Class 3, c) Class 4, d) Class 5, e) Class 6, f) Class 7.

4.4 In a manner similar to the analysis for the sentence, "John ate a frog," analyze the sentence, "Mary breathed the fragrant air." And establish the alternate environment graph.

4.5 In Section 4.4.1, give the sequence of formulas in the simulated deductive process for a set of three premises.

4.6 In Section 4.4.2, give the sequence of formulas for three transformation response components with known values.

CHAPTER 5

A General System of Interactive Automata

5.1 FORMAL ANALYSIS FOR A GENERAL SYSTEM OF INTERACTIVE AUTOMATA

A set of graphs for modeling a general system of interactive automata is the principal result of the analysis paralleling the set of graphs of the single general automaton of Section 2.1 (Koenig, Frederick 1969, 1970b, 1971). However, here there are two classes of graph models instead of one, the microsystem model and the macrosystem model. Each model consists of three types of graphs, the processor, environment, and time graphs describing the interfacing of the interactive automata in space and time. Some of the details of Chapter 3 are left out of the analysis in order to simplify the initial work. Later, the work is projected to include all the detail represented in Chapter 3.

In Section 5.2, the environment graph is illustrated by a simple human-computer system, and in Appendix B, the graph models are used in the development of recursive methods and graph-theoretic principles for determining the effective operation of a general system of Interactive automata.

5.1.1 General Analysis

Let $A=\{A_j \mid j\in J, J=\{1, 2, ---, k\}\}$ denote a system of interactive general automata (k-automata) analyzed over a moment base M in an environment space R^n, and each component automaton A_j is analyzed over its own moment base $M_j=\{m_j+i \mid i\in I\}$, $I=\{0, 1, 2, ---, h\}$. The system moment base M is a set of time slices $M=\{m+i \mid i\in I\}$ and an arbitrary element $m+i\in M$ corresponds to the time slice necessary to analyze the k-automata at their respective moments m_j+i. That is, $M=\{m+i \mid i\in I, m+i=(m_j+i \mid j\in J)\}$.

Furthermore, in any given system moment, the absolute times associated with the moments corresponding to each component automaton are distinct.

Knowledge Structures for Communications in Human-Computer Systems: General Automata-Based, by Eldo C. Koenig

That is, if $T_n(m_n+i)$ denotes the time associated with an arbitrary component automaton A_n at moment m_n+i, then for all n, $k \in J$, $T_n(m_n+i) \neq T_k(m_k+i)$ where m_n+i, $m_k+i \in m+i=(m_j+i \mid j \in J)$. This specifies nonsimultaneous reading and responding of two or more component automata at a same environment.

Finally, the intervals of time associated with consecutive system moments as time slices are disjoint. That is, if $[\min_{j\in J} T_j(m_j+i), \max_{j\in J} T_j(m_j+i)]$ define the time slice associated with an arbitrary system moment $m+i$, then $\{[\min_{j\in J} T_j(m_j+i), \max_{j\in J} T_j(m_j+i)]\} \cap \{[\min_{j\in J} T_j(m_j+(i+1)), \max_{j\in J} T_j(m_j+(i+1))]\}=\phi$ and $\max_{j\in J} T_j(m_j+i)<\min_{j\in J} T_j(m_j+(i+1))$. This specifies that the operation of the system occurs in an orderly fashion.

There is a fundamental system principle associated with the above defined system moment base. The principle establishes an operating framework for the system.

System Principle. Given a discrete system moment $m+i=(m_j+i \mid j \in J)$, $m+i \in M$, each component automaton A_j is accessing one and only one environment from a set of nonhomogeneous environments $E_j \subseteq R^n$ of the environment space, receiving one and only one stimulus from a set S_j of potential stimuli, is in one and only one state from a set of possible states Q_j, and has one and only one absolute time associated with m_j+i from a set T of times. Multiple accesses of a same environment during M can be made by different component automata of the system. Furthermore, during the next system moment $m+(i+1)=(m_j+(i+1) \mid j \in J)$, each component automaton A_j produces one and only one triadic response based on the aforementioned conditions from a set R_j of possible responses. The triadic response consists of an ordered triple from a set F_j of transformation responses, a set D_j of spatial change responses, and a set C_j of time change responses. A transformation response at $(m_j+(i+1) \mid j \in J)$ is produced at the environment accessed at moment $m_j+i \mid j \in J)$.

An underlying assumption in the concept of interaction is the existence of common or shared environments between the automata. That is, if automaton A_n and A_k are interacting over some period of time, then $E_n \cap E_k \neq \phi$, where E_n and E_k denote their respective environment sets. The notion of shared environments enables a general system of interacting automata to be more precisely defined. Suppose $A=\{A_j \mid j \in J, J=\{1, 2, ---, k\}\}$ denotes a system of general automata in an environment space R^n analyzed over a moment base $M=\{m+i \mid i \in I\}$, where $I=\{0, 1, 2, ---, h\}$. Then, A is said to be a *general system of interactive automata* over a system moment base if the environment sets of the automaton components are not disjoint under any pairwise partition, i.e., there does not exist a proper subset $N \subset 2^J$ in the power set J such that $\cup_{n\in N} E_n \cap \cup_{j\in J/N} E_j=\phi$. It is clear that each component automaton accesses all the locations in its environment set during the moment base. Any such system is said to be *completely interactive* if one or more environments are common to all sets, i.e., $\cap_{j\in J} E_j \neq \phi$. Any interactive system that is not completely interactive is said to be a *partially interactive* system.

The general automaton analyzed in Section 2.1 becomes the jth automaton of a system of interactive automata (k-automata) denoted by $A=\{A_j \mid j\in J, J=\{1, 2, ---, k\}\}$, and the system principle is expressed by Equations 2.1E, 2.2E, ---, 2.7E and Equations 3.4E and 3.5E. The latter two result from the condition of nonhomogeneous environments discussed in Section 3.2. These equations are rewritten in the same sequence incorporating the subscript j.

$$Q_j(m_j+(i+1))=\omega_j(Q_j(m_j+i), S_j(m_j+i)) \quad 5.1E$$

$$R_j(m_j+(i+1))=\beta_j(Q_j(m_j+i), S_j(m_j+i)) \quad 5.2E$$

$$F_j(m_j+(i+1))=\sigma_j(Q_j(m_j+i), S_j(m_j+i)) \quad 5.3E$$

$$D_j(m_j+(i+1))=\delta_j(Q_j(m_j+i), S_j(m_j+i)) \quad 5.4E$$

$$C_j(m_j+(i+1))=\tau_j(Q_j(m_j+i), S_j(m_j+i)) \quad 5.5E$$

$$E_j(m_j+(i+1))=\gamma_j'(E_j(m_j+i), D_j(m_j+(i+1))) \quad 5.6E$$

$$T_j(m_j+(i+1))=\lambda_j'(T_j(m_j+i), C_j(m_j+(i+1))) \quad 5.7E$$

$$B_j(m_j+(i+1))=\eta_j'(E_j(m_j+i), F_j(m_j+(i+1))) \quad 5.8E$$

$$S_n(m_n+(i+p))=\alpha_{nj}(B_j(m_j+i)) \quad 5.9E$$

Since the system is an Interactive system, 3.5E is modified to obtain the last Equation 5.9E for the following reason. An arbitrarily recorded response $B_j(m_j+i)$ of automaton A_j has potential of being interpreted as a stimulus by an arbitrary automaton A_n at any time during the period beginning but not including $T_j(m_j+i)$ and ending at time $T_n(m_n+(h-1))$ assuming common environments. The potential stimulus for the A_n automaton is described by 5.9E restricted in time to $((T_j(m_j+i) < T_n(m_n+(i+p)) \mid i\in I-\{0, h\}) \mid p\in P)$, where $I=\{0, 1, 2, ---, h\}$, $P=\{0, 1, 2, ---, h-(i+1)\}$, $n\in J$.

Recall, the system moment base M is a set of time slices $M=\{m+i \mid i\in I\}$, and an arbitrary element $m+i\in M$ corresponds to the time slice necessary to analyze the k-automata at their respective moments m_j+i; i.e., $M=\{m+i \mid i\in I, m+i=(m_j+i \mid j\in J)\}$, $J=\{1, 2, ---, k\}$, $I=\{0, 1, 2, ---, h\}$. Let S, Q, R, F, D, C, E, T, B denote the system sets associated with M which correspond to the jth automaton sets S_j, Q_j, R_j, F_j, D_j, C_j, E_j, T_j, B_j associated with M_j. Then $S\subseteq\Pi_j S_j$, $Q\subseteq\Pi_j Q_j$, $R\subseteq\Pi_j R_j$, $F\subseteq\Pi_j F_j$, $D\subseteq\Pi_j D_j$, $C\subseteq\Pi_j C_j$, $E\subseteq\Pi_j E_j$, $T\subseteq\Pi_j T_j$, $B\subseteq\Pi_j B_j$.

Let the set of time slices used in describing the operation of the automaton system throughout the *life of the system* be denoted by L. For each time slice, let a moment exist for the jth automaton, and denote the set of moments describing the operation *throughout the life of the jth automaton* by L_j. Let the system sets associated with L be denoted by $\bar{S}$, $\bar{Q}$, $\bar{R}$, $\bar{F}$, $\bar{D}$, $\bar{C}$, $\bar{E}$, $\bar{T}$, $\bar{B}$, and let the jth automaton sets associated with L_j be denoted by $\bar{S}_j$, $\bar{Q}_j$, $\bar{R}_j$, $\bar{F}_j$, $\bar{D}_j$, $\bar{C}_j$, $\bar{E}_j$, $\bar{T}_j$, $\bar{B}_j$. These are maximum sets for the system and the jth automaton. Any subset $M \subseteq L$ and any corresponding subset $M_j \subseteq L_j$ may serve as moment base for the automaton system and the jth automaton. Denote the sets associated with M as *active system sets* $S, Q, R, F, D, C, E, T, B$ and those associated with M_j as *active jth automaton sets* $S_j, Q_j, R_j, D_j, C_j, E_j, T_j, B_j$.

From the operational framework described by the system principle, there is a set of moment functions $\{\theta_{Q_j}, \theta_{S_j}, \theta_{R_j}, \theta_{F_j}, \theta_{D_j}, \theta_{C_j}, \theta_{E_j}, \theta_{T_j}, \theta_{B_j}\}$ mapping a subset of M_j into the jth automaton sets. The moment functions 2.1M, 2.2M, ---, 2.8M of Section 2.1 for the general automaton become those for the jth automaton when they are rewritten with the subscript j.

$$\theta_{Q_j}: M_j \to \theta_{Q_j}(M_j) = Q_j \subseteq \bar{Q}_j \qquad 5.1\text{M}_\text{j}$$

$$\theta_{S_j}: M_j - \{m_j + h\} \to \theta_{S_j}(M_j - \{m_j + h\}) = S_j \subseteq \bar{S}_j \qquad 5.2\text{M}_\text{j}$$

$$\theta_{R_j}: M_j - \{m_j\} \to \theta_{R_j}(M_j - \{m_j\}) = R_j \subseteq \bar{R}_j \qquad 5.3\text{M}_\text{j}$$

$$\theta_{F_j}: M_j - \{m_j\} \to \theta_{F_j}(M_j - \{m_j\}) = F_j \subseteq \bar{F}_j \qquad 5.4\text{M}_\text{j}$$

$$\theta_{D_j}: M_j - \{m_j\} \to \theta_{D_j}(M_j - \{m_j\}) = D_j \subseteq \bar{D}_j \qquad 5.5\text{M}_\text{j}$$

$$\theta_{C_j}: M_j - \{m_j\} \to \theta_{C_j}(M_j - \{m_j\}) = C_j \subseteq \bar{C}_j \qquad 5.6\text{M}_\text{j}$$

$$\theta_{E_j}: M_j \to \theta_{E_j}(M_j) = E_j \subseteq \bar{E}_j \qquad 5.7\text{M}_\text{j}$$

$$\theta_{T_j}: M_j \to \theta_{T_j}(M_j) = T_j \subseteq \bar{T}_j \qquad 5.8\text{M}_\text{j}$$

$$\theta_{B_j}: M_j - \{m_j\} \to \theta_{B_j}(M_j - \{m_j\}) = B_j \subseteq \bar{B}_j \qquad 5.9\text{M}_\text{j}$$

The last, 5.9M_j, relates to the recorded response B_j.

The system sets associated with the system moment base $M = \{m+i \mid i \in I\}$ may now be more specifically defined as

$$Q = \{(Q_j(m_j + i) \mid j \in J) \mid i \in I\}$$

$$S = \{(S_j(m_j + i) \mid j \in J) \mid i \in I - \{h\}\}$$

$$R = \{(R_j(m_j + i) \mid j \in J) \mid i \in I - \{0\}\}$$

$$F = \{(F_j(m_j + i) \mid j \in J) \mid i \in I - \{0\}\}$$

$$D = \{(D_j(m_j + i) \mid j \in J) \mid i \in I - \{0\}\}$$

$$C = \{(C_j(m_j + i) \mid j \in J) \mid i \in I - \{0\}\}$$

$$E = \{(E_j(m_j + i) \mid j \in J) \mid i \in I\}$$

$$T = \{(T_j(m_j + i) \mid j \in J) \mid i \in I\}$$

$$B = \{(B_j(m_j + i) \mid j \in J) \mid i \in I - \{0\}\}$$

And a set of moment functions $\{\theta_Q, \theta_S, \theta_R, \theta_F, \theta_D, \theta_C, \theta_E, \theta_T, \theta_B\}$ describes mappings of a subset of M into the system sets $S, Q, R, F, D, C, E, T, B$ similar to those of 5.1M_j, 5.2M_j, ---, 5.9M_j.

Consider the following notation for the finite sets of active elements of a jth automaton operating over a moment base $M_j \subseteq L_j$: $S_j=\{s_{j0}, s_{j1}, ---, s_{jn_j}\}$, $Q_j=\{q_{j0}, q_{j1}, ---, q_{jm_j}\}$, $R_j=\{r_{j0}, r_{j1}, ---, r_{jp_j}\}$, $F_j=\{f_{j0}, f_{j1}, ---, f_{ju_j}\}$, $D_j=\{d_{j0}, d_{j1}, ---, d_{jv_j}\}$, $C_j=\{c_{j0}, c_{j1}, ---, c_{jw_j}\}$, $E_j=\{e_{j0}, e_{j1}, ---, e_{jx_j}\}$, $T_j=\{t_{j0}, t_{j1}, ---, t_{jy_j}\}$, $B_j=\{b_{j0}, b_{j1}, ---, b_{jz_j}\}$. Then, the following functions relate to the jth automaton of a system:

$$\omega_j: V_j \subseteq Q_j \times S_j \rightarrow \omega_j(V_j) \subseteq Q_j \subseteq \bar{Q}_j \qquad 5.1\text{F}_j$$

$$\beta_j: V_j \subseteq Q_j \times S_j \rightarrow \beta_j(V_j) \subseteq R_j \subseteq \bar{R}_j \qquad 5.2\text{F}_j$$

$$\sigma_j: V_j \subseteq Q_j \times S_j \rightarrow \sigma_j(V_j) \subseteq F_j \subseteq \bar{F}_j \qquad 5.3\text{F}_j$$

$$\delta_j: V_j \subseteq Q_j \times S_j \rightarrow \delta_j(V_j) \subseteq D_j \subseteq \bar{D}_j \qquad 5.4\text{F}_j$$

$$\tau_j: V_j \subseteq Q_j \times S_j \rightarrow \tau_j(V_j) = C_j \subseteq \bar{C}_j \qquad 5.5\text{F}_j$$

$$\gamma'_j: W'_j \subseteq E_j \times D_j \rightarrow \gamma'_j(W'_j) \subseteq E_j \subseteq \bar{E}_j \qquad 5.6\text{F}_j$$

$$\lambda'_j: N'_j \subseteq T_j \times C_j \rightarrow \lambda'_j(N'_j) \subseteq T_j \subseteq \bar{T}_j \qquad 5.7\text{F}_j$$

$$\gamma_j: W_j \subseteq E_j \times V_j \rightarrow \gamma_j(W_j) \subseteq E_j \subseteq \bar{E}_j \qquad 5.8\text{F}_j$$

$$\lambda_j: N_j \subseteq T_j \times V_j \rightarrow \lambda_j(N_j) \subseteq T_j \subseteq \bar{T}_j \qquad 5.9\text{F}_j$$

$$\pi_j: Y_j \subseteq T_j \times W_j \subseteq T_j \times E_j \times V_j \rightarrow \pi_j(Y_j) \subseteq (\gamma_j(W_j) \times \lambda_j(N_j)) \subseteq E_j \times T_j \qquad 5.10\text{F}_j$$

$$\mu_j: Y'_j \subseteq E_j \times N_j \subseteq E_j \times T_j \times V_j \rightarrow \mu_j(Y'_j) \subseteq (\lambda_j(N_j) \times \gamma_j(W_j)) \subseteq T_j \times E_j \qquad 5.11\text{F}_j$$

$$\phi_j: E'_j \subseteq E'_j \rightarrow \phi_j(E'_j) = S_j \qquad 5.12\text{F}_j$$

$$\eta_j: W_j \subseteq E_j \times V_j \rightarrow \eta_j(W_j) = B_j \subseteq \bar{B}_j \qquad 5.13\text{F}_j$$

$$\alpha_{nj}: B_j \subseteq \alpha_{nj}(B_j) = S_n \qquad 5.14\text{F}_j$$

The functions 5.1F$_j$, 5.2F$_j$, ---, 5.7F$_j$ parallel 2.1F, 2.2F, ---, 2.7F, and 5.8F$_j$, 5.9F$_j$, ---, 5.12F$_j$ parallel 2.9F, 2.10F, ---, 2.13F. Functions 5.13F$_j$ and 5.14F$_j$ come from 3.13F and 5.9E.

To understand the meaning of Function 5.14F$_j$, recall that an arbitrarily recorded response $B_j(m_j+i)$ by automaton A_j has the potential of being interpreted as a stimulus by an arbitrary automaton A_n of the system of k-automata. This potential is described by a set of functions $\{\alpha_{nj}\}$, $\forall n, j \in J$ defined by $\alpha_{nj}: B_j \rightarrow \alpha_{nj}(B_j) = S_n$, which details how each component automaton A_n interprets the recorded response B_j as a stimulus. If $n=j$, then α_{nj} is the same as function α introduced in Section 2.1. Given that an arbitrary component automaton A_u has produced a transformation response recorded as b_{um} at some environment $e_{up} \in E_u$, the set of possible stimuli that may be received by the component automata in the system at the next accessing of e_{up} is the set $\{\alpha_{ju}(b_{um}) \mid j \in J\} \subseteq \Pi_j S_j$. In particular, if automaton A_v accesses e_{up} next, the stimulus received is $\alpha_{vu}(b_{um}) \in S_v$. Thus, associated with each environment is a set of k potential stimuli, and a stimulus obtained from an environment by an automaton is an interpreted stimulus from the set associated with that environment. Of course, not all of a set is realized unless the environment is shared by all component automata.

For those Functions 5.1F$_j$, 5.2F$_j$, ---, 5.14F$_j$ for the jth automaton of a system, there are corresponding functions for a general system of interactive automata. Recall the system sets $S \subseteq \Pi_j S_j$, $Q_j \subseteq \Pi_j Q_j$, $R \subseteq \Pi_j R_j$, $F \subseteq \Pi_j F_j$, $D \subseteq \Pi_j D_j$, $C \subseteq \Pi_j C_j$, $E \subseteq \Pi_j E_j$, $T \subseteq \Pi_j T_j$, $B \subseteq \Pi_j B_j$ defined over a system moment base $M \subseteq L$. Then the system functions parallel those of the jth automaton and are

$$\omega : V \subseteq Q \times S \to \omega(V) \subseteq Q \subseteq \bar{Q} \qquad 5.1\text{F}_S$$

$$\beta : V \subseteq Q \times S \to \beta(V) \subseteq R \subseteq \bar{R} \qquad 5.2\text{F}_S$$

$$\sigma : V \subseteq Q \times S \to \sigma(V) \subseteq F \subseteq \bar{F} \qquad 5.3\text{F}_S$$

$$\delta : V \subseteq Q \times S \to \delta(V) \subseteq D \subseteq \bar{D} \qquad 5.4\text{F}_S$$

$$\tau : V \subseteq Q \times S \to \tau(V) \subseteq C \subseteq \bar{C} \qquad 5.5\text{F}_S$$

$$\gamma' : W' \subseteq E \times D \to \gamma'(W') \subseteq E \subseteq \bar{E} \qquad 5.6\text{F}_S$$

$$\lambda' : N' \subseteq T \times C \to \lambda'(N') \subseteq T \subseteq \bar{T} \qquad 5.7\text{F}_S$$

$$\gamma : W \subseteq E \times V \to \gamma(W) \subseteq E \subseteq \bar{E} \qquad 5.8\text{F}_S$$

$$\lambda : N \subseteq T \times V \to \lambda(N) \subseteq T \subseteq \bar{T} \qquad 5.9\text{F}_S$$

$$\pi : Y \subseteq T \times W \subseteq T \times E \times V \to \pi(Y) \subseteq (\gamma(W) \times \lambda(N)) \subseteq E \times T \qquad 5.10\text{F}_S$$

$$\mu : Y' \subseteq E \times N \subseteq E \times T \times V \to \mu(Y') \subseteq (\lambda(N) \times \gamma(W)) \subseteq T \times E \qquad 5.11\text{F}_S$$

$$\phi_i : E' \subseteq \bar{E} \to \phi_i(E') = S \qquad 5.12\text{F}_S$$

$$\eta : W \subseteq E \times V \to \eta(W) = B \subseteq \bar{B} \qquad 5.13\text{F}_S$$

$$\alpha' : B \to \alpha'(B) = S^k \subseteq \bar{S}^k \qquad 5.14\text{F}_S$$

The system function ω is defined by $\omega(q, s) = (\omega_j(p_j(q), p_j(s)) \mid j \in J)$, where p_j is the jth projection function, and ω_j is the jth automaton state function. All other system functions are similarly defined.

The following definitions pertain to the initial and final moments of a given finite moment base as they relate to the associated sets for the jth automaton and system of k-automata and to special response conditions and stimuli.

Definition 5.1. By *initial stimulus, initial environment, initial state*, and *initial time* of a *jth automaton* is meant the stimulus, environment, state, and time for that automaton at the initial moment m_j. The k-tuples of stimuli, environments, states, and times at the initial system moment $m = (m_j \mid j \in J)$, $J = \{1, 2, ---, k\}$, are the *initial stimulus, initial environment, initial state*, and *initial time of the system of k-automata*.

Definition 5.2. For a given moment base $M_j = \{m_j + i \mid i \in I\}$, $I = \{0, 1, 2, ---, h\}$, an *absolute initial state of a jth automaton* is an initial state that occurs only at the moments $m_j + g$, $\forall g = 0, 1, 2, ---, i$, where $i \in \{0, 1, 2, ---, h\}$; if every jth automaton of a system of k-automata has an absolute initial state at the system moments $m+g=(m_j+g \mid j \in J)$, $J=\{1, 2, \text{---}, k\}$, $\forall g=0, 1, 2, \text{---}, i$; $i \in \{0, 1, 2, \text{---}, h\}$, the k-tuple of initial states occurring at these moments is an *absolute initial state of a system of k-automata.* If an initial stimulus of a jth automaton is received only at moment m_j in the absolute initial state, that stimulus is an *absolute initial stimulus of the jth automaton*; if every jth automaton of a system of k-automata has an absolute initial stimulus at the system moment $m=(m_j \mid j \in J)$, the k-tuple of initial stimuli at m is an *absolute initial stimulus of the system of k-automata.* An *absolute initial environment of a jth automaton* is an initial environment that is accessed only at the moments m_j+g, $\forall g=0, 1, 2, \text{---}, i$, where $i \in \{0, 1, 2, \text{---}, h\}$; if every *jth* automaton of a system of k-automata has an absolute initial environment accessed at the system moments $m+g=(m_j+g \mid j \in J)$, $\forall g=0, 1, 2, \text{---}, i$, where $i \in \{0, 1, 2, \text{---}, h\}$, the k-tuple of initial environments accessed only at these moments is an *absolute initial environment for a system of k-automata.*

Definition 5.3. By *final state, final response, final environment*, and *final time* of a *jth automaton* is meant the state, response, environment, and time at the final moment m_j+h of a given moment base, and the *final stimulus of the jth automaton* is the stimulus at the moment $m_j+(h-1)$ preceding the final moment; the k-tuples of states, responses, environments, and times at the final system moment $m+h=(m_j+h \mid j \in J)$, $J=\{1, 2, \text{---}, k\}$, are the *final state, final response, final environment*, and *final time of the system of k-automata*, and the k-tuple of stimuli at the system moment $m+(h-1)=(m_j+(h-1) \mid j \in J)$ is the *final stimulus of the system of k-automata.*

Definition 5.4. For a given moment base $M_j=\{m_j+i \mid i \in I\}$, $I=\{0, 1, 2, \text{---}, h\}$ an *absolute final state of a jth automaton* is a final state that occurs only at the moments m_j+g, $\forall g=h, h-1, \text{---}, h-i$, where $i \in \{0, 1, 2, \text{---}, h\}$; if every jth automaton of a system of k-automata has an absolute final state at the system moments $m+g=(m_j+g \mid j \in J)$, $J=\{1, 2, \text{---}, k\}$, $\forall g=h, h-1, \text{---}, h-i$, where $i \in \{0, 1, 2, \text{---}, h\}$, the k-tuple of final states occurring only at these moments is an *absolute final state of a system of k-automata.* If a final stimulus is received only at the moment $m+(h-1)$ by a jth automaton in the absolute final state, or takes the automaton into that state, that stimulus is an *absolute final stimulus of the jth automaton*; if every jth automaton of a system of k-automata has an absolute final stimulus at the system moment base $m+(h-1)=(m_j+(h-1) \mid j \in J)$, the k-tuple of final stimuli at $m+(h-1)$ is an *absolute final stimulus of the system of k-automata.* An *absolute final environment of a jth automaton* is a final environment that is accessed only at the moment m_j+g, $\forall g=h, h-1, \text{---}, h-i$; $i \in \{0, 1, 2, \text{---}, h\}$; if every jth automaton of a system of k-automata has an absolute final environment accessed at the system moments $m+g=(m_j+g \mid j \in J)$, $\forall g=h, h-1, \text{---}, h-i$; $i \in \{0, 1, 2, \text{---}, h\}$, the k-tuple of final environments accessed only at these moments is an *absolute final environment for a system of k-automata.*

Definition 5.5. A *null spatial change response of a jth automaton* is a spatial change response of zero vectors; if a transformation response of a jth automaton is not recorded at an environment and the existing recorded response is not removed or altered, that transformation response is a *null transformation response of the jth automaton* and its associated recorded response is a *null recorded response of the jth automaton*; if either a null transformation response or a null spatial change response occurs in a triadic response of a jth automaton, the triadic response is a *partial null response of the jth automaton*, and if both occur, the triadic response is a *null response of the jth automaton.* If every jth automaton of a system of k-automata has a null spatial change response at some moment $m+i=(m_j+i \mid j\in J)$, $J=\{1, 2, \text{---}, k\}$, the k-tuple of spatial change responses at that moment is a *null spatial change response of the system of k-automata*; if every jth automaton of a system of k-automata has a null transformation response at a system moment $m+i=(m_j+i \mid j\in J)$, the k-tuple of transformation responses at that moment $m+i$ is a *null transformation response of the system of k-automata* and the associated k-tuple of recorded responses is a *null recorded response* of the system of k-automata; if either a k-tuple of null spatial change responses or a k-tuple of null transformation responses occurs in a triadic response of the system, the triadic response is a *partial null response of the system of k-automata*, and if both occur, the triadic response is a *null response of the system of k-automata.*

The total null response defined in Section 2.1 cannot exist for a jth automaton or for a system of k-automata. By definition, to have a total null response, the condition of a null time change response must be present in addition to the two conditions given above for the transformation and spatial change component responses of the triadic response. In the definition of the system moment base, time slices are disjoint precluding the occurrence of a null time change response.

Definition 5.6. The *null stimulus of a jth automaton* is interpreted as the absence of a stinulus but is considered a member of the set S_j of active stimuli; if every jth automaton of a system of k-automata has a null stimulus at some system moment $m+i=(m_j+i \mid j\in J)$, $J=\{1, 2, \text{---}, k\}$, the k-tuple of stimuli at that moment is a null stimulus of the system of k-automata and is considered a member of the system set of active stimuli $S\subseteq\Pi_j S_j$.

Let the jth automaton sets Q_j and S_j be finite sets of active states and stimuli associated with a given moment base M_j, and let the system sets $Q\subseteq\Pi_j Q_j$ and $S=\Pi_j S_j$, $J=\{1, 2, \text{---}, k\}$ be finite sets of active states and stimuli associated with a given moment base M. Recall that the dyadic relations V_j on $Q_j\times S_j$ and V on $Q\times S$ serve as the domain sets for the jth automaton functions σ_j, δ_j, τ_j and system functions σ, δ, τ. Other pertinent automaton sets result from these mappings. For example, $\sigma_j(V_j)=F_j$ and $\sigma(V)=F$ are the sets of active transformation responses; $\delta_j(V_j)=D_j$ and $\delta(V)=D$ are sets of active spatial change responses; and $\tau_j(V_j)=C_j$ and $\tau(V)=C$ are the sets of active time change responses. The bounds for $|F_j|$, $|D_j|$, $|C_j|$, and for $|F|$, $|D|$, $|C|$ are given by the following theorems. Their proofs are similar to those for the parallel theorems of Section 2.1 for a general automaton and are omitted here.

Theorem 5.1. *Let* $|Q_j|=m_j+1$, $|S_j|=n_j+1$, $F_j=\{f_{j0}, f_{j1}, ---, f_{ju_j}\}$, $D_j=\{d_{j0}, d_{j1}, ---, d_{jv_j}\}$, $C_j=\{c_{j0}, c_{j1}, ---, c_{jw_j}\}$ *for a jth automaton; then* $|F_j|=u_j+1\leq|V_j|\leq|Q_j||S_j|=(m_j+1)(n_j+1)$; $|D_j|=v_j+1\leq(m_j+1)\ (n_j+1)$; $|C_j|=w_j+1\leq|V_j|\leq(m_j+1)\ (n_j+1)$. *Let* $|Q|=m+1$, $|S|=n+1$, $F\subseteq\Pi_jF_j$, $D\subseteq\Pi_jD_j$, $C\subseteq\Pi_jC_j$ *for a system of k-automata,* $A=\{A_j \mid j\in J, J=\{1, 2, ---, k\}\}$; *then* $|F|=(u+1)\leq|V|\leq|Q|\ |S|=(m+1)\ (n+1)$; $|D|=(v+1)\leq|V|\leq(m+1)\ (n+1)$, $|C|=(w+1)\leq|V|\leq(m+1)\ (n+1)$.

Assuming the jth automaton sets Q_j and S_j finite and system sets Q and S finite, R_j and R are finite sets. The following results are observed for $|R_j|$ and $|R|$:

Theorem 5.2. *Let* $R_j=\{r_{j0}, r_{j1}, ---, r_{jP_j}\}$, $|F_j|=u_j+1$, $|D_j|=v_j+1$, $|C_j|=w_j+1$ *for a jth automaton; then* $max\ (u_j+1, v_j+1, w_j+1)\leq p_j+1\leq|V_j|$. *Let* $R\subseteq\Pi_j R_j$, $|F|=u+1$, $|D|=v+1$, $|C|=w+1$ *for a system of k-automata,* $A=\{A_j \mid j\in J, J=\{1, 2, ---, k\}\}$; *then* $max\ (u+1, v+1, w+1)\leq p+1\leq|V|$.

The bounds placed on the number of elements in the finite sets are supplemented by the following relationship between the number of active states and active stimuli:

Theorem 5.3. *Let* $V_j\subseteq Q_j\times S_j$, $|Q_j|=m_j+1$, $|S_j|=n_j+1$, *and put max* $(|F_j|, |D_j|, |C_j|)=v_j$ *for a jth automaton; then* $|Q_j|\geq(v_j+x_j)/(n_j+1)$, *where* $v_j+x_j\equiv 0$ *modulo* n_j+1, *and* x_j *is the minimum nonnegative integer with this property. Let* $V=Q\times S$, $|Q|=m+1$, $|S|=n+1$, *and put max* $(|F|, |D|, |C|)=v$ *for a system of k-automata,* $A=\{A_j \mid j\in J, J=\{1, 2, ---, k\}\}$; *then* $|Q|\geq(v+x)/(n+1)$, *where* $v+x\equiv 0$ *modulo* $n+1$, *and x is the minimum nonnegative integer with this property.*

Corollary 5.3. *Given the hypothesis of Theorem* 5.3, $|Q_j|$ *can be replaced by* $|S_j|$ *and* n_j+1 *by* m_j+1, *and* $|Q|$ *can be replaced by* $|S|$ *and* $n+1$ *by* $m+1$. *Thus,* $|S_j|\geq(v_j+x_j)/(m_j+1)$ *and* $|S|\geq(v+x)/(m+1)$, *where* $v_j+x_j\equiv 0$ *modulo* m_j+1 *and* $v+x\equiv 0$ *modulo* $m+1$, *and* x_j *and x are the minimum nonnegative integers with this property.*

General classifications of stimuli of a general automaton of Section 2.1 can be extended to a jth automaton of a system and to a system of k-automata, $A=\{A_j \mid j\in J, J=\{1, 2, ---, k\}\}$.

Definition 5.7. An *internal stimulus for a given jth automaton* operating over a given moment base M_j is a stimulus produced as a transformation response by that same automaton at some prior moment within M_j; if the stimulus of every jth automaton of a system of k-automata is an internal stimulus at some system moment $m+i=(m_j+i \mid j\in J)$, $J=\{1, 2, ---, k\}$, the k-tuple of stimuli at that moment is an *internal stimulus for the system of k-automata.*

Definition 5.8. An *external stimulus for a given jth automaton* operating over a given moment base M_j is a stimulus produced as a transformation response either by that automaton or some other automaton external to the system at some moment prior to the given moment base, or by another automaton of the system prior to or within the moment base; if the stimulus of every jth automaton of a system of k-automata at some system moment $m+i=(m_j+i\,|\,j\in J), J=\{1,2,---,k\}$ was produced as a transformation response by an automaton at some moment prior to the given moment base, the k-tuple of stimuli at that moment $m+i$ is an *external stimulus for the system of k-automata.*

Recall, a stimulus location encountered for the first time produces an external stimulus.

Theorem 5.4. *For a jth automaton, an initial stimulus is an external stimulus; for a system of k-automata,* $A=\{A_j \mid j\in J, J=\{1, 2, ---, k\}\}$*, an initial system stimulus is a k-tuple of external stimuli.*

Internal stimuli produced by successive encounters of a single location at successive moments correspond to a response containing a null spatial change response. Recall the definition of a null response.

Theorem 5.5. *For a jth automaton operating over a moment base* M_j*, any null response occurring without a change of state requires the given automaton to operate internally in a loop until an interaction occurs with another automaton to supply a change of stimulus. For a system of k-automata, any null system response occurring without a change of system state results in the system operating internally in an infinite loop.*

Definition 5.9. When all stimuli of a jth automaton defined over a moment base M_j are internal stimuli, that jth automaton is said to operate internally over M_j and is an observed automaton by those automata of the system sharing its environment.

Theorem 5.6. *A system of interactive k-automata,* $A=\{A_j\,|\,j\in J, J=\{1,2,---,k\}\}$*, defined for a moment base M, operates internally over that moment base.*

Suppose the system of interactive k-automata also received external stimuli excluding those obtained from environments accessed for the first time. Automata of the system would be interacting with automata not included in the system set $\{A_j \mid j \in J, J=\{1, 2, ---, k\}\}$, which contradicts the definition of a system of Interactive automata.

Definition 5.10. When all stimuli of a jth automaton are external stimuli and the automaton has accessed at least one of its environments more than once during the moment base M_j, that jth automaton is said to operate externally over M_j and is an observing automaton of other automata of the system sharing its environment.

Definition 5.11. When stimuli of a jth automaton are both internal and external within a given moment base excluding those obtained from environments accessed for the first time, that jth automaton is said to operate both internally and externally over M_j and is performing both as an observed and observing automaton in the system.

5.1.2 Microsystem Model

The two sets of functions $\{5.1F_j, 5.2F_j, ---, 5.14F_j\}$ and $\{5.1F_S, 5.2F_S, ---, 5.14F_S\}$ suggest two classes of models for a system of interactive k-automata. The graphs that are defined from the jth automaton functions will be established first and are useful in viewing a system on the component level. Such a set of graphs is a system model described as a *microsystem model.* The primitives for the graphs of the model are similar to those for the graphs of Section 2.1 for the general automaton. There are the three types of graphs, the processor, environment, and time graphs, and for each of two of the types, the environment and time graphs, there are two kinds, the principal and alternate graphs. The graph functions for defining the graphs must be derived from the functions which describe the operation of a jth automaton as a component of the system. Emphasis will be placed on the processor graph, the principal time graph, and the alternate environment graph.

Processor Graph. Consider Functions $5.1F_j$ through $5.5F_j$ and define a function Ω_j mapping a subset of the product space of internal states and stimuli of a jth automaton into the product space of responses and states. That is, Ω_j: $V_j \subseteq Q_j \times S_j \rightarrow R_j \times Q_j$, or more specifically

$$\Omega_j : V_j \subseteq Q_j \times S_j \rightarrow \sigma_j(V_j) \times \delta_j(V_j) \times \tau_j(V_j) \times \omega_j(V_j) \subseteq F_j \times D_j \times C_j \times Q_j \qquad 5.1\text{GFj}$$

where $\Omega_j(q_{ji}, s_{ji}) = (\sigma_j(q_{ji}, s_{ji}), \delta_j(q_{ji}, s_{ji}), \tau_j(q_{ji}, s_{ji}), \omega_j(q_{ji}, s_{ji}))$. Hence, the image under Ω_j, denoted by $\Omega_j(V_j)$, is a tetradic relation on $F_j \times D_j \times C_j \times Q_j$; i.e., $\Omega_j(V_j) \subseteq F_j \times D_j \times C_j \times Q_j$. This definition parallels the definition of Ω given in Section 2.1 for a general automaton. The function Ω_j describes the basic action of a jth automaton as a component of the system.

Given the graph function Ω_j of $5.1GF_j$ and the notion of a general graph, the following four primitives are established for defining the *processor graph* for the microsystem model:

Primitive Psi1. A family of sets $\{Q'_j\}$ and a set $\cup_j Q'_j$ of elements called points $\forall j \in J$ where J is an index set.

Primitive Psi2. A family of sets $\{S'_j, F'_j, D'_j, C'_j\}$, and a set $\cup_j U_j$ of elements called lines, where $U_j \subseteq S'_j \times (F'_j \times D'_j \times C'_j)$.

Primitive Psi3. A family of sets of functions $\{\{\sigma_j\}, \{\delta_j\}, \{\tau_j\}, \{\omega_j\}\}$ and a set of functions $\{\Omega_j\}$ whose domains are $V_j \subseteq Q_j \times S_j$, $Q_j \subseteq Q'_j$, $S_j \subseteq S'_j$, and whose range sets are $\Omega_j(V_j) = (\sigma_j(V_j) \times \delta_j(V_j) \times \tau_j(V_j) \times \omega_j(V_j) \subseteq F_j \times D_j \times C_j \times Q_j$, where $\sigma_j : V_j \subseteq Q_j \times S_j \to \sigma_j(V_j) = F_j \subseteq F'_j$, $\delta_j : V_j \to \delta_j(V_j) = D_j \subseteq D'_j$, $\tau_j : V_j \to \tau_j(V_j) = C_j \subseteq C'_j$, and $\omega_j : V_j \to \omega_j(V_j) \subseteq Q_j$. Furthermore, $Q_j = \{q_{ji} \mid (q_{ji}, s_{ji}) \in V_j\} \cup \{q_{j(i+1)} \mid q_{j(i+1)} \in \omega_j(V_j), (q_{j(i+1)}, s_{ji}) \notin V_j\}$, and Ω_j is defined by $\Omega_j(q_{ji}, s_{ji}) = (\sigma_j(q_{ji}, s_{ji}), \delta_j(q_{ji}, s_{ji}), \tau_j(q_{ji}, s_{ji}), \omega_j(q_{ji}, s_{ji})) = (f_{j(i+1)}, d_{j(i+1)}, c_{j(i+1)}, q_{j(i+1)})$. The set of lines $\cup_j U_j$ is the set of all the $(s_{ji}, (f_{j(i+1)}, d_{j(i+1)}, c_{j(i+1)}))$ in the sextuples $(q_{ji}, s_{ji}, f_{j(i+1)}, d_{j(i+1)}, c_{j(i+1)}, q_{j(i+1)})$, such that $\Omega_j(q_{ji}, s_{ji}) = (f_{j(i+1)}, d_{j(i+1)}, c_{j(i+1)}, q_{j(i+1)})$, and if $\cup_j Q'_j - \cup_j Q_j$ is not empty, the set of $\cup_j Q_{j0} = (\cup_j Q'_j - \cup_j Q_j)$ is the set of points with no lines.

Primitive Psi4. The subset $\cup_j U_{j0} \subseteq \cup_j U_j$ is a set of directed lines if, for $u_{j0} \in U_{j0}$, $u_{j0} = (s_{ji}, (f_{j(i+1)}, d_{j(i+1)}, c_{j(i+1)}))$, the point q_{ji} associated with the element (q_{ji}, s_{ji}) from the domain is not equal to the point $q_{j(i+1)}$ in the image quadruple $\Omega_j(q_{ji}, s_{ji}) = (f_{j(i+1)}, d_{j(i+1)}, c_{j(i+1)}, q_{j(i+1)})$. The line u_{j0} is said to be directed from point q_{ji} to point $q_{j(i+1)}$.

The details of the processor graph are presented more specifically in terms of the automaton system. The graph reflects some of the integral parts of the system as it operates over any moment base within its life. The set of points $\cup_j Q'_j$, $\forall j \in J$, $J = \{1, 2, ---, k\}$, for the graph corresponds to a set of internal states for the system processors. There is a directed line $(s_{ji}, (f_{j(i+1)}, d_{j(i+1)}, c_{j(i+1)}))$ in the graph from a point representing state q_{ji} to another point representing state $q_{j(i+1)}$ if, whenever a processor of a jth automaton is in state q_{ji} and receives a stimulus s_{ji}, it responds with $r_{j(i+1)} = (f_{j(i+1)}, d_{j(i+1)}, c_{j(i+1)})$ and changes to state $q_{j(i+1)}$. An action of a jth automaton not resulting in a change of state is reflected in the graph by a nondirected line called a loop; i.e., when $q_{ji} = q_{j(i+1)}$. Also, if the processor of a jth automaton is in state q_{ji} at different moments and therein receives different stimuli before changing to state $q_{j(i+1)}$, this is represented in the graph by parallel lines (or parallel loops).

Let M be an arbitrary system moment base, $M \subseteq L$, and denote the corresponding processor graph by $G_M(Psi)$. Then $G_M(Psi)$ is a subgraph of $G_L(Psi)$ for the life of the system since the points and lines in $G_M(Psi)$ are also points and lines in $G_L(Psi)$. Now let $G_{M''}(Psi)$ be the subgraph for moment base M', $M \subseteq M' \subseteq L$. Comparing $G_{M'}(Psi)$ with $G_M(Psi)$, any additional points in $G_{M'}(Psi)$ reflect latent states with respect to the moment base M. The subgraph $G_M(Psi)$ may be viewed as having the same number of points as $G_{M'}(Psi)$, where the latent states are represented by isolated points in $G_M(Psi)$. In a similar manner, additional lines in $G_{M'}(Psi)$ may reflect latent stimuli and responses for moment base M.

Theorem 5.7 (Psi). *Given a processor graph $G_L(Psi)$ of a microsystem model describing the operation of a system of interactive automata over the life of the system, every subgraph describing the operation of a component jth automaton is connected, and the graph of the system is a set of disconnected subgraphs.*

The proof of connectiveness of a subgraph for a *j*th automaton parallels that of Theorem 2.6(C) of Section 2.1.3 for a general automaton. If the subgraphs were connected, there would exist a state for which more than one component automaton could enter, but this cannot be the case since every component automaton is distinct with its own set of states. Thus, the subgraphs are disconnected.

Corollary 5.7 (Psi). *Let M_j be a moment base for a jth automaton of a system of interactive automata and $M_j \subseteq L_j$. Suppose no information is given for any other moment base M'_j properly containing M_j. Then every processor subgraph describing the operation of a component jth automaton is connected, and the graph of the system $G_M(Psi)$ is a set of disconnected subgraphs.*

This follows from Theorem 5.7 (*Psi*). Since the subgraph for any *j*th automaton is unknown for any moment base M'_j properly containing M_j, then M_j can be treated as L_j in the sense of the theorem.

Principal Environment Graph. Consider the above graph function Ω_j and Functions 5.8F$_j$ and 5.13F$_j$ and define Γ_j mapping a subset of the product space of environments, internal states, and stimuli of a *j*th automaton into the product space of recorded responses, responses, states, and environments. That is, $\Gamma_j: W_j \subseteq E_j \times V_j \rightarrow B_j \times \Omega_j(V_j) \times E_j$; more specifically

$$\Gamma_j: W_j \subseteq E_j \times V_j \rightarrow (\eta_j(W_j) \times \sigma_j(V_j) \times \delta_j(V_j) \times \tau_j(V_j) \times \omega_j(V_j) \times \gamma_j(W_j)) \subseteq B_j \times F_j \times D_j \times C_j \times Q_j \times E_j \qquad \text{5.2GF}_j$$

where $\Gamma_j(e_{ji}, (q_{ji}, s_{ji})) = (\eta_j(e_{ji}, q_{ji}, s_{ji}), (\sigma_j(q_{ji}, s_{ji}), \delta_j(q_{ji}, s_{ji}), \tau_j(q_{ji}, s_{ji}), \omega_j(q_{ji}, s_{ji})), \gamma_j(e_{ji}, q_{ji}, s_{ji}))$, $V_j \subseteq Q_j \times S_j$, $\Omega_j(V_j) \subseteq F_j \times D_j \times C_j \times Q_j$. The image under Γ_j, $\Gamma_j(W_j)$ is a hexadic relation on $B_j \times F_j \times D_j \times C_j \times Q_j \times E_j$; i.e., $\Gamma_j(W_j) \subseteq B_j \times \Omega_j(V_j) \times E_j$. This function Γ_j gives information on the environment of a *j*th automaton in addition to its basic action.

The graph using Γ_j parallels the principal environment graph for a general automaton detailed in Section 2.1.2 and is called the principal environment graph for the microsystem model. Briefly, the graph is defined as follows: An element $(e_{ji}, (q_{ji}, s_{ji}))$ from the domain and the image element $\Gamma_j(e_{ji}(q_{ji}, s_{ji})) = (b_{j(i+1)}, f_{j(i+1)}, d_{j(i+1)}, c_{j(i+1)}, q_{j(i+1)}, e_{j(i+1)})$ form a graph nine-tuple $(e_{ji}, q_{ji}, s_{ji}, b_{j(i+1)}, f_{j(i+1)}, d_{j(i+1)}, c_{j(i+1)}, q_{j(i+1)}, e_{j(i+1)})$, which is represented in the graph by two points e_{ji} and $e_{j(i+1)}$, $(e_{ji} \neq e_{j(i+1)})$, and a connecting line directed from e_{ji} to $e_{j(i+1)}$ identified by (q_{ji}, s_{ji}) at its origin and $(b_{j(i+1)}, f_{j(i+1)}, d_{j(i+1)}, c_{j(i+1)}, q_{j(i+1)})$ at its terminal. If $e_{ji}=e_{j(i+1)}$, e_{ji} and $e_{j(i+1)}$ are the same point, and the line is a nondirected loop with the line identification established on the basis of an assumed line direction.

Principal Time Graph. Consider the above graph function Γ_j and Function 5.10F$_j$, and define a function Λ_j as $\Lambda_j: Y_j \subseteq T_j \times W_j \rightarrow \Gamma_j(W_j) \times T_j$, more specifically

$$\begin{aligned} &\Lambda_j : Y_j \subseteq T_j \times W_j \subseteq T_j \times E_j \times V_j \rightarrow \\ &(\eta_j(W_j) \times (\sigma_j(V_j) \times \delta_j(V_j) \times \tau_j(V_j) \times \omega_j(V_j)) \times \pi_j(Y_j)) \\ &\subseteq B_j \times F_j \times D_j \times C_j \times Q_j \times E_j \times T_j \end{aligned} \qquad 5.3\text{GF}_j$$

where $\Lambda(t_{ji}, (e_{ji}, (q_{ji}, s_{ji}))) = (\eta_j(e_{ji}, q_{ji}, s_{ji}), (\sigma_j(q_{ji}, s_{ji}), \delta_j(q_{ji}, s_{ji}), \tau_j(q_{ji}, s_{ji}), \omega_j(q_{ji}, s_{ji})), \pi_j(t_{ji}, e_{ji}, q_{ji}, s_{ji}))$. This function Λ_j provides information on the operation times of the jth automaton in addition to the information given by the environment graph function Γ_j, and the graph to be defined using Λ_j is the *principal time graph* for the microsystem model.

Given the graph function Λ_j of 5.3GF$_j$ and the notion of a general graph, the following four primitives are established for defining the principal time graph:

Primitive Ts$_p$ i1. A family of sets $\{T'_j\}$ and a set $\cup_j T'_j$ of elements called points $\forall j \in J$, where J is an index set.

Primitive Ts$_p$ i2. A family of sets $\{E'_j, Q'_j, S'_j, B'_j, F'_j, D'_j, C'_j\}$ and a set $\cup_j Z_j$ of elements called lines, where $Z_j \subseteq (E'_j \times Q'_j \times S'_j) \times (B'_j \times F'_j \times D'_j \times C'_j \times Q'_j \times E'_j)$.

Primitive Ts$_p$ i3. A family of sets of functions $\{\{\eta_j\}, \{\sigma_j\}, \{\tau_j\}, \{\omega_j\}, \{\gamma_j\}, \{\lambda_j\}, \{\pi_j\}\}$ and a set of functions $\{\Lambda_j\}$ whose domains are $Y_j \subseteq T_j \times W_j \subseteq T_j \times E_j \times V_j \subseteq T_j \times E_j \times Q_j \times S_j$, $T_j \subseteq T'_j$, $E_j \subseteq E'_j$, $Q_j \subseteq Q'_j$, $S_j \subseteq S'_j$, and whose range sets are $\Lambda_j(Y_j) \subseteq (\eta_j(W_j) \times (\sigma_j(V_j) \times \delta_j(V_j) \times \tau_j(V_j) \times \omega_j(V_j)) \times \gamma_j(W_j) \times \lambda_j(N_j)) \subseteq B_j \times F_j \times D_j \times C_j \times Q_j \times E_j \times T_j$, where $\pi_j: Y_j \subseteq T_j \times W_j \rightarrow \pi_j(Y_j) \subseteq E_j \times T_j \subseteq E'_j \times T'_j$. The definitions for the remaining functions and for the sets Q_j, E_j, T_j are given in Primitive Es_a $i3$, and the definition for Λ_j is similar to that of function Λ'_j. The set of lines $\cup_j Z_j$ is the set of all the $((e_{ji}, q_{ji}, s_{ji}), (b_{j(i+1)}, f_{j(i+1)}, d_{j(i+1)}, c_{j(i+1)}, q_{j(i+1)}, e_{j(i+1)}))$ in $(t_{ji}, e_{ji}, q_{ji}, s_{ji}, b_{j(i+1)}, f_{j(i+1)}, d_{j(i+1)}, c_{j(i+1)}, q_{j(i+1)}, t_{j(i+1)}, e_{j(i+1)})$, such that $\Lambda_j(t_{ji}, (e_{ji}, (q_{ji}, s_{ji}))) = (b_{j(i+1)}, f_{j(i+1)}, d_{j(i+1)}, c_{j(i+1)}, q_{j(i+1)}, e_{j(i+1)}, t_{j(i+1)}$, and if $\cup_j T'_j - \cup_j T_j$ is not empty, the set $\cup_j T_{j0} = (\cup_j T'_j - \cup_j T_j)$ is the set of points with no lines.

Primitive Ts$_p$ i4. The subset $\cup_j Z_{j\,0} \subseteq \cup_j Z_j$ is a set of directed lines if, for $z_{j0} \in Z_{j0}$, $z_{j0} = ((e_{ji}, q_{ji}, s_{ji}), (b_{j(i+1)}, f_{j(i+1)}, d_{j(i+1)}, c_{j(i+1)}, q_{j(i+1)}, e_{j(i+1)}))$, the point t_{ji} associated with the element $(t_{ji}, (e_{ji}, (q_{ji}, s_{ji})))$ from the domain is not equal to the point $t_{j(i+1)}$ in the image $\Lambda_j(t_{ji}, (e_{ji}, (q_{ji}, s_{ji}))) = (b_{j(i+1)}, f_{j(i+1)}, d_{j(i+1)}, c_{j(i+1)}, q_{j(i+1)}, e_{j(i+1)}, t_{j(i+1)})$. The line z_{j0} is said to be directed from point t_{ji} to $t_{j(i+1)}$.

To describe these primitives in terms of a jth automaton, consider the 11-tuple $(t_{ji}, e_{ji}, q_{ji}, s_{ji}, b_{j(i+1)}, f_{j(i+1)}, d_{j(i+1)}, c_{j(i+1)}, q_{j(i+1)}, e_{j(i+1)}, t_{j(i+1)})$ associated with two points and a connecting line. This 11-tuple is represented in the graph by two points t_{ji} and $t_{j(i+1)}$, $(t_{j\,i} \neq t_{j(i+1)})$, and the connecting line carries the identification of the nine-tuple of the Γ_j function of 5.2GF$_j$. From the definition of the system moment base in Section 5.1, t_{ji} can never equal $t_{j(i+1)}$ and loops cannot be present.

Let $G_L(Ts_p\ i)$ denote the principal time graph of a microsystem model for the life of the system, and let $G_M(Ts_p\ i)$ be the graph for an arbitrary moment base $M \subseteq L$. Then $G_M(Ts_p\ i)$ is a subgraph of $G_L(Ts_p\ i)$ since the points and lines in $G_M(Ts_p\ i)$ are also points and lines in $G_L(Ts_p\ i)$.

Theorem 5.8 (Ts$_p$ i). *Given a time graph $G_L(Ts_p\ i)$ of a microsystem model describing the operation of a system of interactive automata over the life of the system, every subgraph describing the operation of a component jth automaton is connected, and the graph of the system is a set of disconnected subgraphs.*

The proof that a subgraph describing the operation of a *j*th automaton over its life is connected parallels that of Section 2.1 for a general automaton. Suppose the subgraphs are connected; there would exist a single absolute time for more than one component automaton. This contradicts the definition of the moment base, and the graph of a system is a set of disconnected subgraphs.

Corollary 5.8 (Ts$_p$ i). *Let M_j be a moment base for a jth automaton of a system of interactive automata, and $M_j \subseteq L_j$. Suppose no information is given for any other moment base M'_j properly containing M_j. Then, every time subgraph describing the operation of a component jth automaton is connected, and the graph of the system $G_M(Ts_p\ i)$ is a set of disconnected subgraphs.*

M_j can be treated as L_j in the sense of Theorem 5.8 $(Ts_p\ i)$.

Alternate Time Graph. Consider the Graph Function 5.1GF$_j$, Ω_j, and the Function 5.9F$_j$ and define Γ'_j, $\Gamma'_j : N_j \subseteq T_j \times V_j \rightarrow \Omega_j(V_j) \times T_j$. More specifically

$$\Gamma'_j : N_j \subseteq T_j \times V_j \subseteq T \times Q \times S_j \rightarrow (\sigma_j(V_j) \times \delta_j(V_j) \times \tau_j(V_j) \times \omega_j(V_j) \times \lambda_j(N_j)) \subseteq F_j \times D_j \times C_j \times Q_j \times T_j \qquad \text{5.4GF}_j$$

where $\Gamma'_j(t_{ji}, (q_{ji}, s_{ji})) = ((\sigma_j(q_{ji}, s_{ji}), \delta_j(q_{ji}, s_{ji}), \tau_j(q_{ji}, s_{ji}), \omega_j(q_{ji}, s_{ji})), \lambda_j(t_{ji}, q_{ji}, s_{ji}))$. This definition of Γ'_j parallels the definition of Γ' given in Section 2.1.2 for a general automaton. Γ'_j provides information on operating times of the *j*th automaton in addition to the information provided by the Graph Function Ω_j. This is the second function that can be used in defining a time graph. For the purpose of distinction, the previous function Λ_j is said to provide for a principal time graph and this function Γ'_j is said to provide for an alternate time graph. The principal time graph gives complete information.

Briefly, the alternate time graph is defined as follows: An element $(t_{ji}, (q_{ji}, s_{ji}))$ from the domain and the image element $\Gamma'_j(t_{ji}, (q_{ji}, s_{ji})) = (f_{j(i+1)}, d_{j(i+1)}, c_{j(i+1)}, q_{j(i+1)}, t_{j(i+1)})$ form a graph eight-tuple $(t_{ji}, q_{ji}, s_{ji}, f_{j(i+1)}, d_{j(i+1)}, c_{j(i+1)}, q_{j(i+1)}, t_{j(i+1)})$, which is represented in the graph by two points t_{ji} and $t_{j(i+1)}$, and a connecting line directed from t_{ji} to $t_{j(i+1)}$ identified by (q_{ji}, s_{ji}) at its origin and $(f_{j(i+1)}, d_{j(i+1)}, c_{j(i+1)}, q_{j(i+1)})$ at its terminal. If $t_{ji}=t_{j(i+1)}$, t_{ji} and $t_{j(i+1)}$ are the same point, and the line is a nondirected loop with the line identification established on the basis of an assumed line direction.

Alternate Environment Graph. Consider the above graph function Γ'_j and Functions 5.11F$_j$, 5.13F$_j$. A function Λ'_j is defined as $\Lambda'_j: Y'_j \subseteq E_j \times N_j \to B_j \times \Gamma'_j(N_j) \times E_j$, more specifically

$$\Lambda'_j: Y'_j \subseteq E_j \times N_j \subseteq E_j \times T_j \times V_j \to (\eta_j(W_j) \times (\sigma_j(V_j) \times \delta_j(V_j) \times \tau_j(V_j) \times \omega_j(V_j)) \times \mu_j(Y'_j)) \subseteq B_j \times F_j \times D_j \times C_j \times Q_j \times T_j \times E_j \qquad 5.5\text{GF}_j$$

where $\Lambda'_j(e_{ji}, (t_{ji}, (q_{ji}, s_{ji}))) = (\eta_j(e_{ji}, q_{ji}, s_{ji}), (\sigma_j(q_{ji}, s_{ji}), \delta_j(q_{ji}, s_{ji}), \tau_j(q_{ji}, s_{ji}), \omega_j(q_{ji}, s_{ji})), \mu_j(e_{ji}, t_{ji}, q_{ji}, s_{ji}))$. This function Λ'_j provides information on the environments and the recorded responses in addition to the information provided by the graph function Γ'_j. This is the second graph function that can be used in defining an environment graph. The previous function Γ_j is said to provide for a principal environment graph, and this function Λ'_j is said to provide for an alternate environment graph for the microsystem model.

Because the alternate environment graph provides maximum information, the primitives defining the graph will be written in detail.

Primitive Es$_a$ i1. A family of sets $\{E'_j\}$ and a set $\cup_j E'_j$ of elements called points $\forall j \in J$, where J is an index set.

Primitive Es$_a$ i2. A family of sets $\{T'_j, Q'_j, S'_j, B'_j, F'_j, D'_j, C'_j\}$ and a set $\cup_j X_j$ of elements called lines, where $X_j \subseteq (T'_j \times Q'_j \times S'_j) \times (B'_j \times F'_j \times D'_j \times C'_j \times Q'_j \times T'_j)$.

Primitive Es$_a$ i3. A family of sets of functions $\{\{\eta_j\}, \{\sigma_j\}, \{\delta_j\}, \{\tau_j\}, \{\omega_j\}, \{\lambda_j\}, \{\gamma_j\}, \{\mu_j\}\}$ and a set of functions $\{\Lambda'_j\}$ whose domains are $Y'_j \subseteq E_j \times N_j \subseteq E_j \times T_j \times V_j \subseteq E_j \times T_j \times Q_j \times S_j$, $E_j \subseteq E'_j$, $T_j \subseteq T'_j$, $Q_j \subseteq Q'_j$, $S_j \subseteq S'_j$, and whose range sets are $\Lambda'_j(Y'_j) = (\eta_j(W_j) \times (\sigma_j(V_j) \times \delta_j(V_j) \times \tau_j(V_j) \times \omega_j(V_j)) \times \mu_j(Y'_j)) \subseteq (\eta_j(W_j) \times (\sigma_j(V_j) \times \delta_j(V_j) \times \tau_j(V_j) \times \omega_j(V_j)) \times \lambda_j(N_j) \times \gamma_j(W_j)) \subseteq B_j \times F_j \times D_j \times C_j \times Q_j \times T_j \times E_j$, where $\eta_j: W_j \to \eta_j(W_j) = B_j \subseteq B'_j$, $\sigma_j: V_j \to \sigma_j(V_j) = F_j \subseteq F'_j$, $\delta_j: V_j \to \delta_j(V_j) = D_j \subseteq D'_j$, $\tau_j: V_j \to \tau_j(V_j) = C_j \subseteq C'_j$, $\omega_j: V_j \to \omega_j(V_j) \subseteq Q_j \subseteq Q'_j$, $\lambda_j: N_j \subseteq T_j \times V_j \to \lambda_j(N_j) \subseteq T_j \subseteq T'_j$, $\gamma_j: W_j \subseteq E_j \times V_j \to \gamma_j(W_j) \subseteq E_j \subseteq E'_j$, $\mu_j: Y'_j \subseteq E_j \times N_j \to \mu_j(Y'_j) \subseteq \lambda_j(N_j) \times \gamma_j(W_j) \subseteq T_j \times E_j \subseteq T'_j \times E'_j$. Furthermore, $Q_j = \{q_{ji} \mid (q_{ji}, s_{ji}) \in V_j\} \cup \{q_{j(i+1)} \mid q_{j(i+1)} \in \omega_j(V_j), (q_{j(i+1)}, s_{ji}) \notin V_j\}$, $T_j = \{t_{ji} \mid (t_{ji}, (q_{ji}, s_{ji})) \in N_j\} \cup \{t_{j(i+1)} \in \lambda_j(N_j), (t_{j(i+1)}, (q_{ji}, s_{ji})) \notin N_j\}$, $E_j = \{e_{ji} \mid (e_{ji}, (t_{ji}, (q_{ji}, s_{ji}))) \in Y'_j\} \cup \{e_{j(i+1)} \mid e_{j(i+1)} \in \gamma_j(W_j), (e_{j(i+1)}, (t_{ji}, (q_{ji}, s_{ji}))) \notin Y'_j\}$, and Λ'_j is defined by $\Lambda'_j(e_{ji}, (t_{ji}, (q_{ji}, s_{ji}))) = (\eta_j(e_{ji}, (q_{ji}, s_{ji})), \sigma_j(q_{ji}, s_{ji}), \delta_j(q_{ji}, s_{ji}), \tau_j(q_{ji}, s_{ji}), \omega_j(q_{ji}, s_{ji}), \lambda_j(t_{ji}, (q_{ji}, s_{ji})), \gamma_j(e_{ji}, (q_{ji}, s_{ji}))) = (b_{j(i+1)}, f_{j(i+1)}, d_{j(i+1)}, c_{j(i+1)}, q_{j(i+1)}, t_{j(i+1)}, e_{j(i+1)})$. The set of lines $\cup_j X_j$ is the set of all the $((t_{ji}, q_{ji}, s_{ji}), (b_{j(i+1)}, f_{j(i+1)}, d_{j(i+1)}, c_{j(i+1)}, q_{j(i+1)}, t_{j(i+1)}))$ in $(e_{ji}, t_{ji}, q_{ji}, s_{ji}, b_{j(i+1)}, f_{j(i+1)}, d_{j(i+1)}, c_{j(i+1)}, q_{j(i+1)}, t_{j(i+1)}, e_{j(i+1)})$, such that $\Lambda'_j(e_{ji}, (t_{ji}, (q_{ji}, s_{ji}))) = (b_{j(i+1)}, f_{j(i+1)}, d_{j(i+1)}, c_{j(i+1)}, q_{j(i+1)}, t_{j(i+1)},$

$e_{j(i+1)})$, and if $\cup_j E'_j - \cup_j E_j$ is not empty, the set $\cup_j E_{j0} = (\cup_j E'_j - \cup_j E_j)$ is the set of points with no lines.

Primitive Es$_a$ i4. The subset $\cup_j X_{j0} \subseteq \cup_j X_j$ is a set of directed lines if for $x_{j0} \in X_{j0}$, $x_{j0} = ((t_{ji}, q_{ji}, s_{ji}), (b_{j(i+1)}, f_{j(i+1)}, d_{j(i+1)}, c_{j(i+1)}, q_{j(i+1)}, t_{j(i+1)}))$, the point e_{ji} associated with the element $(e_{ji}, (t_{ji}, (q_{ji}, s_{ji})))$ from the domain is not equal to the point $e_{j(i+1)}$ in the image $\Lambda'_j(e_{ji}, (t_{ji}, (q_{ji}, s_{ji}))) = (b_{j(i+1)}, f_{j(i+1)}, d_{j(i+1)}, c_{j(i+1)}, q_{j(i+1)}, t_{j(i+1)}, e_{j(i+1)})$. The line x_{j0} is said to be directed from point e_{ji} to point $e_{j(i+1)}$.

The primitives and construction details for the graph may be stated more precisely in terms of the processor of a jth automaton and its environment. The 11-tuple associated with two points and a connecting line means that at some moment $m_j + k$ the environment e_{jk} supplies a stimulus s_{jk} to the processor in state q_{jk}, and at the next moment $m_j + (k+1)$, the processor is in state $q_{j(k+1)}$ and gives a transformation response $f_{j(k+1)}$, a spatial change response $d_{j(k+1)}$, and a time change response $c_{j(k+1)} \cdot d_{j(k+1)}$ provides the processor with a new environment $e_{j(k+1)} = e_{jk} + d_{j(k+1)}$ as the location of a next stimulus. For the case of a zero vector d_{j0}, the response is a partial null response, and the environment e_{jk} at moment $m_j + k$ and $e_{j(k+1)}$ at the next moment are the same. This condition corresponds to a nondirected loop on the graph.

Let M be any arbitrary system moment base for a system of interactive automata, $M \subseteq L$, and denote the corresponding alternate environment graph by $G_M(Es_a\ i)$. Then $G_M(Es_a\ i)$ is a subgraph of $G_L(Es_a\ i)$ for the life of the system since the points and lines in $G_M(Es_a\ i)$ are also points and lines in $G_L(Es_a\ i)$. Now let $G_M(Es_a\ i)$ be the subgraph for the system moment base M', $M \subseteq M' \subseteq L$. Comparing $G_{M'}(Es_a\ i)$ with $G_M(Es_a\ i)$, any additional points in $G_{M'}(Es_a\ i)$ reflect latent system environments (stimulus locations) with respect to the moment base M, which are represented by isolate points in $G_M(Es_a\ i)$. Additional lines in $G_{M'}(Es_a\ i)$ may reflect latent states, stimuli, and responses for a moment base M.

Theorem 5.9 (Es$_a$ i). *Given an alternate environment graph $G_L(Es_a\ i)$ of a microsystem model describing the operation of a system of interactive automata over the life of the system, every subgraph describing the operation of a component jth automaton is connected, and the graph of the system is a set of connected subgraphs.*

The proof that every subgraph describing the operation of a component jth automaton is connected follows that of a general automaton, Section 2.1. For proof that the subgraphs are connected, consider environments e_{ju}, and e_{np} of a jth automaton and an nth automaton of the system; if $u \neq p$, the points representing these environments are distinct. If $u = p$, e_{ju}, and e_{np} are the same point representing a shared environment, and by definition of a system of interactive automata, the graph of the system is a set of connected subgraphs.

Corollary 5.9 (Es_a i). *Let M_j be a moment base for a jth automaton of a system of interactive automata, and $M_j \subseteq L_j$. Suppose no information is given for any other moment base M'_j properly containing M_j. Then every alternate environment subgraph describing the operation of a component jth automaton is connected, and the graph of the system $G_M(Es_a\ i)$ is a set of connected subgraphs.*

M_j can be treated as L_j in the sense of Theorem 5.9 ($Es_a\ i$).

5.1.3 Macrosystem Model

The set of functions {5.1F_S, 5.2F_S, ---, 5.14F_S} provides for a *Macrosystem Model* for a system of interactive k-automata. The graphs that can be defined from these functions are useful when a system is viewed as a single automaton.

For a k-automaton system, let the processor graph defined by the graph function Ω be denoted by $G_M(Psa)$; let the principal and alternate environment graphs defined by the graph functions Γ and Λ' be denoted by $G_M(Es_p\ a)$ and $G_M(Es_a\ a)$; and let the principal and alternate time graphs for Λ and Γ' be denoted by $G_M(Ts_p\ a)$ and $G_M(Ts_a\ a)$.

Processor Graph. Consider Functions 5.1F_S through 5.5F_S and define processor graph function Ω of the k-automata.

$$\Omega: V \subseteq Q \times S \rightarrow (\sigma(V) \times \delta(V) \times \tau(V) \times \omega(V)) \subseteq F \times D \times C \times Q \qquad 5.1\mathrm{GF_S}$$

with $\Omega(q, s) = (\sigma(q, s), \delta(q, s), \tau(q, s), \omega(q, s))$. Let Z denote any of the functions σ, δ, τ, ω. $Z = (Z_j \mid j \in J)$, $J = \{1, 2, ---, k\}$, is defined by $Z(q, s) = (Z_j(p_j(q), p_j(s)) \mid j \in J)$, where p_j is the jth projection function and $Z_j: V_j \subseteq Q_j \times S_j \rightarrow Z_j(V_j)$.

Given the graph function Ω of 5.1$\mathrm{GF_S}$ and the notion of a general graph, the following four primitives are established for defining the *processor graph* for the macrosystem model:

Primitive Psa1. A subset $Q' \subseteq \Pi Q'_j$ called points $\forall j \in J$, where J is an index set.

Primitive Psa2. A family of sets $\{S', F', D', C'\}$ and a subset $U \subseteq S' \times (F' \times D' \times C')$ called lines, where for any member of $\{S', F', D', C'\}$ denoted by X', $X' \subseteq \Pi_j X'_j$.

Primitive Psa3. A set of functions $\{\sigma, \delta, \tau, \omega\}$ and a function Ω whose domain is $V \subseteq Q \times S$, $Q \subseteq Q'$, $S \subseteq S'$ and whose range is $\Omega(V) \subseteq (\sigma(V) \times \delta(V) \times \tau(V) \times \omega(V)) \subseteq F \times D \times C \times Q$, where $\sigma: V \subseteq Q \times S \rightarrow \sigma(V) = F \subseteq F'$, $\delta: V \rightarrow \delta(V) = D \subseteq D'$, $\tau: V \rightarrow \tau(V) = C \subseteq C'$, and $\omega: V \rightarrow \omega(V) \subseteq Q$. Let Z

denote any of the functions belonging to $\{\sigma, \delta, \tau, \omega\}$, and $Z=(Z_j \mid j\in J)$ is defined by $Z(q, s)=(Z_j(p_j(q), p_j(s)) \mid j\in J)$ where p_j is the jth projection function and $Z_j: V_j \subseteq Q_j \times S_j \rightarrow Z_j(V_j)$. Furthermore, $Q=\{q_i \mid (q_i, s_i)\in V\} \cup \{q_{i+1} \mid q_{i+1}\in \omega(V), (q_{i+1}, s_i)\notin V\}$ and Ω is defined by $\Omega(q_i, s_i)=(\sigma(q_i, s_i), \delta(q_i, s_i), \tau(q_i, s_i), \omega(q_i, s_i))=(f_{i+1}, d_{i+1}, c_{i+1}, q_{i+1})$. The set of lines U is all the $(s_i, (f_{i+1}, d_{i+1}, c_{i+1}))$ in $(q_i, s_i, f_{i+1}, d_{i+1}, c_{i+1}, q_{i+1})$, such that $\Omega(q_i, s_i)=(f_{i+1}, d_{i+1}, c_{i+1}, q_{i+1})$, and if $Q'-Q$ is not empty, the $q_0\in(Q'-Q)$ are points with no lines.

Primitive Psa4. The subset $U_0 \subseteq U$ is a set of directed lines if for $u\in U_0$, $u=(s_i(f_{i+1}, d_{i+1}, c_{i+1}))$, the point q_i associated with the element (q_i, s_i) from the domain is not equal to the point q_{i+1} in the image $\Omega(q_i, s_i)=(f_{i+1}, d_{i+1}, c_{i+1}, q_{i+1})$. The line u is said to be directed from q_i to point q_{i+1}.

The processor graph views the system of k-processors as a single processor; i.e., as a system processor. A point of the graph corresponds to a system state q_i in the form of a k-tuple, $q_i=(q_{ji} \mid j\in J)$, $J=\{1, 2, ---, k\}$, representing the internal states of all k-processors for some system moment $m+i=(m_j+i \mid j\in J)$. Any element of $(s_i, (f_{i+1}, d_{i+1}, c_{i+1}))$ identifying a line is also in the form of a k-tuple. A line is directed in the graph from a point representing a system state q_i to another point if, whenever the system processor is in system state q_i and receives a k-tuple of stimuli s_i, the processor responds with a k-tuple of responses $r_{i+1}=(f_{i+1}, d_{i+1}, c_{i+1})$ and changes to a system state q_{i+1}. An action of the system of automata that does not result in a change in any of the elements in the k-tuple of states is reflected by a nondirected loop.

Let $G_M(Psa)$ for a moment base M be a subgraph of $G_{M'}(Psa)$ for a moment base M'; $M\subseteq M'\subseteq L$. Comparing $G_{M'}(Psa)$ with $G_M(Psa)$, any additional points in $G_{M'}(Psa)$ reflect latent system states in the form of k-tuples with respect to the moment base M. The subgraph $G_M(Psa)$ may be viewed as having the same number of points as $G_{M'}(Psa)$ where the latent system states are represented by isolate points in $G_M(Psa)$. Additional lines in $G_{M'}(Psa)$ may reflect latent stimuli and responses for moment base M.

Theorem 5.10 (Psa). *The processor graph $G_L(Psa)$ of a macrosystem model describing the operation of a system of interactive automata over the life of the system is a connected graph.*

All sets associated with the life of a system contain only active elements. If $G_L(Psa)$ were not connected, there would not exist an active system state $q_i = \bar{Q}$ for which the system processor could not enter and leave. This is contrary to the definition of $\bar{Q}$. A parallel proof is given in Section 2.1 for a general automaton.

Corollary 5.10 (Psa). *Let M be a moment base for a system of interactive automata, and $M\subseteq L$. Suppose no information is given for any other moment base M' properly containing M. Then $G_M(Psa)$ is a connected graph.*

This follows immediately from Theorem 5.10 (*Psa*) in the sense that M can be treated as L.

Principal Environment Graph. Consider the above graph function Ω and Functions 5.8F$_S$ and 5.13F$_S$, and define the principal graph function Γ for k-automata

$$\Gamma: W \subseteq E \times V \to (\eta(W) \times (\sigma(V) \times \delta(V) \times \tau(V) \times \omega(V) \times \gamma(W))) \subseteq B \times F \times D \times C \times Q \times E \qquad 5.2\mathrm{GF_S}$$

with $\Gamma(e, (q, s)) = (\eta(e, q, s), (\sigma(q, s), \delta(q, s), \tau(q, s), \omega(q, s), \gamma(e, q, s)))$. Let G denote either of the functions η, γ. $G = (G_L \mid j \in J)$ is defined by $G(e, (q, s)) = (G_j(p_j(e), p_j(q), p_j(s)) \mid j \in J)$, where $G_j: W_j \subseteq E_j \times V_j \to G_j(W_j)$.

Briefly, the principal environment graph is defined as follows: A nine-tuple $(e_i, q_i, s_i, b_{i+1}, f_{i+1}, d_{i+1}, c_{i+1}, q_{i+1}, e_{i+1})$ is represented in the graph by two points, e_i and e_{i+1}, and a connecting line directed from e_i to e_{i+1} identified by (q_i, s_i) at its origin and $(b_{i+1}, f_{i+1}, d_{i+1}, c_{i+1}, q_{i+1})$ at its terminal. If $e_i = e_{i+1}$, e_i and e_{i+1} are the same point, and the line is a nondirected loop. Any element of the nine-tuple denoted by x_i is in the form of a k-tuple, $x_i = (x_{ji} \mid j \in J)$, $J = \{1, 2, ---, k\}$.

Principal Time Graph. Consider the above graph function Γ and Function 5.10F$_S$, and define the principal time graph function Λ of k-automata

$$\Lambda: Y \subseteq T \times W \subseteq T \times E \times V \to (\eta(W) \times (\sigma(V) \times \delta(V) \times \tau(V) \times \omega(V)) \times \pi(V)) \subseteq B \times F \times D \times C \times Q \times E \times T \qquad 5.3\mathrm{GF_S}$$

with $\Lambda(t, (e, (q, s))) = (\eta(e, q, s), (\sigma(q, s), \delta(q, s), \tau(q, s), \omega(q, s)), \pi(t, e, q, s))$. $\pi(\pi_j \mid j \in J)$ is defined by $\pi(t, (e, (q, s))) = (\pi_j(p_j(t), p_j(e), p_j(q), p_j(s)) \mid j \in J)$ and $\pi_j: Y_j \subseteq T_j \times W_j \to \pi_j(Y_j) \subseteq E_j \times T_j$.

Given the graph function Λ of 5.3GF$_S$ and the notion of a general graph, the following four primitives are established for defining the time graph of the macrosystem model:

Primitive Ts$_p$ a1. A subset $T' \subseteq \Pi_j T'_j$ called points, $\forall j \in J$, where J is an index set.

Primitive Ts$_p$ a2. A family of sets $\{E', Q', S', B', D', C'\}$ and a subset $Z \subseteq (E' \times Q' \times S') \times (B' \times F' \times D' \times C' \times Q' \times E')$ called lines, where $k = |J|$ and, for any member of the family of sets denoted by X', $X' \subseteq \Pi_j X'_j$.

Primitive Ts$_p$ a3. A set of functions $\{\eta, \sigma, \delta, \tau, \omega, \gamma, \lambda, \pi\}$ and a function Λ whose domain is $Y \subseteq T \times W \subseteq T \times E \times V \subseteq T \times E \times Q \times S$, $T \subseteq T'$, $E \subseteq E'$, $Q \subseteq Q'$, $S \subseteq S'$, and whose range is $\Lambda(Y) = (\eta(W) \times (\sigma(V) \times \delta(V) \times \tau(V) \times \omega(V)) \times \pi(Y)) \subseteq (\eta(W) \times (\sigma(V) \times \delta(V) \times \tau(V) \times \omega(V)) \times \gamma(W) \times \lambda(N)) \subseteq B \times F \times D \times C \times Q \times E \times T$, where $\eta: W \to \eta(W) = B \subseteq B'$, $\sigma: V \to \sigma(V) = F \subseteq F'$, $\delta: V \to \delta(V) = D \subseteq D'$, $\tau: V \to \tau(V) = C \subseteq C'$, $\omega: V \to \omega(V) \subseteq Q \subseteq Q'$, $\gamma: W \subseteq E \times V \to \gamma(W) \subseteq E \subseteq E'$, $\lambda: N$

$\subseteq T\times V\rightarrow\lambda(N)\subseteq T\subseteq T'$, $\pi: Y\subseteq T\times W'\rightarrow\pi(Y)\subseteq\gamma(W)\times\lambda(N)\subseteq E\times T\subseteq E'\times T'$. The definitions for the functions belonging to $\{\eta, \sigma, \delta, \tau, \omega, \gamma, \lambda, \pi\}$ are similar to the definitions for the functions given in Primitive $Es_a\, a3$ for the alternate environment graph of the macromodel. The definitions for the sets Q, E, T are also given in Primitive $Es_a\, a3$. The set of lines Z is all the $((e_i, q_i, s_i), (b_{i+1}, f_{i+1}, d_{i+1}, c_{i+1}, q_{i+1}, e_{i+1}))$ in $(t_i, e_i, q_i, s_i, b_{i+1}, f_{i+1}, d_{i+1}, c_{i+1}, q_{i+1}, e_{i+1}, t_{i+1})$, such that $\Lambda(t_i, (e_i, (q_i, s_i)))=(f_{i+1}, d_{i+1}, c_{i+1}, q_{i+1}, e_{i+1}, t_{i+1})$. If $T'-T$ is not empty, the $t_0\in(T'-T)$ are points with no lines.

Primitive $\mathbf{Ts_p}$ a4. The subset $Z_0\subseteq Z$ is a set of directed lines if for $z\in Z_0$, $z=((e_i, q_i, s_i), (b_{i+1}, f_{i+1}, d_{i+1}, c_{i+1}, q_{i+1}, e_{i+1}))$, the point t_i associated with the element $(t_i, (e_i, (q_i, s_i)))$ from the domain is not equal to the point t_{i+1} in the image $\Lambda(t_i, (e_i, (q_i, s_i)))=(b_{i+1}, f_{i+1}, d_{i+1}, c_{i+1}, q_{i+1}, e_{i+1}, t_{i+1})$. The line z is said to be directed from t_i to point t_{i+1}.

By definition of the system moment, t_i cannot equal t_{i+1} to provide loops in the graph, and the system time change response cannot include the zero vector.

Let $G_L(Ts_p\ a)$ denote the time graph of a system defined by its respective sets for the life of the system, and let $G_M(Ts_p\ a)$ be the graph for an arbitrary moment base $M\subseteq L$. Then $G_M(Ts_p\ a)$ is a subgraph of $G_L(Ts_p\ a)$ since the points and lines in $G_M(Ts_p\ a)$ are also the points and lines in $G_L(Ts_p\ a)$.

Theorem 5.11 ($\mathbf{Ts_p}$ a). *The time graph $G_L(Ts_p\ a)$ of a macrosystem model describing the operation of a system of interactive automata over the life of the system is a connected graph.*

Paralleling the proof of Section 2.1 for a general automaton, $G_L(Ts_p\ a)$ is connected since the points correspond to system absolute times from the beginning of the system's operation throughout its life, and lines must necessarily join these points to reflect the passing from one time to the next.

Corollary 5.11 ($\mathbf{Ts_p}$ a). *Let M be a moment base for a system of interactive automata, and $M\subseteq L$. Suppose no information is given for any other moment base M' properly containing M. Then $G_M(Ts_p\ a)$ is a connected graph.*

Since the graphs are unknown for any moment base M' properly containing M, then M can be treated as L in the sense of Theorem 5.11 ($Ts_p\ a$).

Alternate Time Graph. Consider the Graph Function 5.1GF$_S$, Ω and the Function 5.9F$_S$, and define the alternate time graph function Γ' for k-automata

$$\Gamma': N\subseteq T\times V\subseteq T\times Q\times S\rightarrow((\sigma(V)\times\delta(V)\times\tau(V)\times\omega(V))\times\lambda(N))\subseteq F\times D\times C\times Q\times T \qquad 5.4\text{GF}_S$$

with $\Gamma'(t, (q, s))=((\sigma(q, s), \delta(q, s), \tau(q, s), \omega(q, s)), \lambda(t, q, s))$ where $\lambda=(\lambda_j \mid j\in J)$ is defined by $\lambda(t, (q, s))=(\lambda_j(p_j(t), p_j(q), p_j(s)) \mid j\in J)$ and $\lambda_j:N_j\subseteq T_j\times V_j\rightarrow\lambda_j(N_j)\subseteq T_j$.

Briefly, the alternate time graph is defined as follows: An eight-tuple $(t_i, q_i, s_i, f_{i+1}, d_{i+1}, c_{i+1}, q_{i+1}, t_{i+1})$ is represented in the graph by two points t_i and t_{i+1} and a connecting line directed from t_i to t_{i+1} identified by (q_i, s_i) at its origin and $(f_{i+1}, d_{i+1}, c_{i+1}, q_{i+1})$ at its terminal. If $t_i=t_{i+1}$, t_i and t_{i+1} are the same point, and the line is a nondirected loop. Any element of the eight-tuple denoted by x_i is in the form of a k-tuple, $x_i=(x_{ji} \mid j\in J)$, $J=\{1, 2, ---, k\}$.

Alternate Environment Graph. Consider the above graph function Γ' and Functions 5.11F$_S$, 5.13F$_S$. A graph function Λ' for the alternate environment graph for k-automata is defined by

$$\Lambda':Y'\subseteq E\times N\subseteq E\times T\times V\subseteq E\times T\times Q\times S\rightarrow(\eta(W)\times(\sigma(V)\times\delta(V)\times\tau(V)\times\omega(V))\times\mu(Y'))\subseteq(\eta(W)\times\delta(V)\times\tau(V)\times\omega(V)\times\lambda(N)\times\gamma(W))\subseteq B\times F\times D\times C\times Q\times T\times E \qquad 5.5\text{GF}_S$$

with $\Lambda'(e, (t, (q, s)))=(\eta(e, q, s), (\sigma(q, s), \delta(q, s), \tau(q, s), \omega(q, s)), \mu(e, t, q, s))$. $\mu=(\mu_j \mid j\in J)$ is defined by $\mu(e, (t, (q, s)))=(\mu_j(p_j(e), p_j(t), p_j(q), p_j(s)) \mid j\in J)$ and $\mu_j: Y_j'\subseteq E_j\times N_j\rightarrow\mu_j(Y_j')\subseteq(\lambda_j(N_j)\times\gamma_j(W_j))\subseteq T_j\times E_j$.

The graph function Λ' of 5.5GF$_S$ is employed in describing the alternate environment graph $G_M(Es_a\ a)$ of the macrosystem model by the following primitives:

Primitive Es$_a$ a1. A subset $E'\subseteq\Pi_j E_j'$ called points, $\forall j\in J$, where J is an index set.

Primitive Es$_a$ a2. A family of sets $\{T', Q', S', B', F', D', C'\}$ and a subset $X\subseteq(T'\times Q'\times S'\times(B'\times F'\times D'\times C'\times Q'\times T'))$ called lines, where $k=|J|$, and, for any member of the family of sets denoted by U', $U'\subseteq\Pi_j U_j'$.

Primitive Es$_a$ a3. A set of functions $\{\eta, \sigma, \delta, \tau, \omega, \lambda, \gamma, \mu\}$ and a function Λ' whose domain is $Y'\subseteq E\times N\subseteq E\times T\times V\subseteq E\times T\times Q\times S$, $E\subseteq E'$, $T\subseteq T'$, $Q\subseteq Q'$, $S\subseteq S'$, whose range is $\Lambda'(Y')=(\eta(W)\times(\sigma(V)\times\delta(V)\times\tau(V)\times\omega(V))\times\mu(Y'))\subseteq(\eta(W)\times(\sigma(V)\times\delta(V)\times\tau(V)\times\omega(V))\times\lambda(N)\times\gamma(W))\subseteq B\times F\times D\times V\times Q\times T\times E$, where $\eta:W\rightarrow\eta(W)=B\subseteq B'$, $\sigma:V\rightarrow\sigma(V)=F\subseteq F'$, $\delta:V\rightarrow\delta(V)=D\subseteq D'$, $\tau:V\rightarrow\tau(V)=C\subseteq C'$, $\omega:V\rightarrow\omega(V)\subseteq Q\subseteq Q'$, $\lambda:N\subseteq T\times V\rightarrow\lambda(N)\subseteq T\subseteq T'$, $\gamma:W\subseteq E\times V\rightarrow\gamma(W)\subseteq E\subseteq E'$, $\mu:Y'\subseteq E\times N\rightarrow\mu(Y')\subseteq\lambda(N)\times\gamma(W)\subseteq T\times E\subseteq T'\times E'$. Let Z denote any of the functions belonging to $\{\sigma, \delta, \tau, \omega\}$ and G any of the functions belonging to $\{\eta, \gamma\}$, $Z=(Z_j \mid j\in J)$, $G=(G_j \mid j\in J)$, $\lambda=(\lambda_j \mid j\in J)$ are defined by $Z(q, s)=(Z_j(p_j(q), p_j(s)) \mid j\in J)$, $G(e, (q, s))=(G_j(p_j(e), p_j(q), p_j(s)) \mid j\in J)$ and $\lambda(t, (q, s))=(\lambda_j(p_j(t), p_j(q), p_j(s)) \mid j\in J)$, where p_j is the jth projection function and $Z_j: V_j\subseteq Q_j\times S_j\rightarrow Z_j(V_j)$, $G_j: W_j\subseteq E_j\times V_j\rightarrow G_j(W_j)$, $\lambda_j: N_j\subseteq T_j\times V_j\rightarrow\lambda_j(N_j)\subseteq T_j$. Furthermore, $Q=\{q_i \mid (q_i, s_i)\in V\}\cup\{q_{i+1} \mid q_{i+1}\in\omega(V), (q_{i+1}, s_i)\notin V\}$, $T=\{t_i \mid (t_i, (q_i, s_i))\in N\}\cup\{t_{i+1} \mid t_{i+1}\in\lambda(N), (t_{i+1}, (q_i, s_i))\notin N\}$, $E=\{e_i \mid (e_i, (t_i, (q_i, s_i)))\in Y'\}\cup\{e_{i+1} \mid e_{i+1}\in\gamma(W), (e_{i+1}, (t_i, (q_i, s_i)))\notin Y'\}$, and Λ' is defined by $\Lambda'(e_i, (t_i, (q_i, s_i))) = (\eta(e_i, (q_i, s_i)), \sigma(q_i, s_i), \delta(q_i, s_i), \tau(q_i, s_i), \omega(q_i, s_i), \lambda(t_i, (q_i, s_i)), \gamma(e_i, (q_i, s_i))) = (b_{i+1}, f_{i+1}, d_{i+1}, c_{i+1}, q_{i+1}, t_{i+1}, e_{i+1})$. The set of lines X is all the $((t_i, q_i, s_i), b_{i+1}, f_{i+1}, d_{i+1}, c_{i+1}, q_{i+1}, t_{i+1})$ in $(e_i, t_i, q_i, s_i, b_{i+1}, f_{i+1}, d_{i+1}, c_{i+1}, q_{i+1}, t_{i+1}, e_{i+1})$, such that $\Lambda'(e_i, (t_i, (q_i, s_i))) = (b_{i+1}, f_{i+1}, d_{i+1}, c_{i+1}, q_{i+1}, t_{i+1}, e_{i+1})$, and if $E' - E$ is not empty, the $e_0 \in (E' - E)$ are points with no lines.

Primitive Es$_a$ a4. The subset $X_0 \subseteq X$ is a set of directed lines for $x \in X_0$, $x = ((t_i, q_i, s_i), (b_{i+1}, f_{i+1}, d_{i+1}, c_{i+1}, q_{i+1}, t_{i+1}))$, the point e_i associated with the element $(e_i, (t_i, (q_i, s_i)))$ from the domain is not equal to the point e_{i+1} in the image $\Lambda'(e_i, (t_i, (q_i, s_i))) = (b_{i+1}, f_{i+1}, d_{i+1}, c_{i+1}, q_{i+1}, t_{i+1}, e_{i+1})$. The line x is said to be directed from e_i to point e_{i+1}.

For a given 11-tuple, $(e_i, t_i, q_i, s_i, b_{i+1}, f_{i+1}, d_{i+1}, c_{i+1}, q_{i+1}, t_{i+1}, e_{i+1})$, at some system moment $m+i = (m_j + i \mid j \in J)$ corresponding to a k-tuple of absolute times t_i, the system environment e_i in the form of a k-tuple supplies a k-tuple of stimuli s_i to the system processor in a system state q_i and at the next moment $m+(i+1)$ the system processor is in system state q_{i+1} and gives a k-tuple of transformation responses f_{i+1}, a k-tuple of spatial change responses d_{i+1}, and a k-tuple of time change responses c_{i+1}. The system transformation response f_{i+1} is recorded as a k-tuple of recorded responses b_{i+1}; d_{i+1} provides the system processor with a new system environment $e_{i+1} = e_i + d_{i+1}$; and c_{i+1} provides for the k-tuple of absolute times at the system moment $m+(i+1)$, $t_{i+1} = t_i + c_{i+1}$.

For $M \subseteq L$, $G_M(Es_a\ a)$ is a subgraph of $G_L(Es_a\ a)$ since the points and lines in $G_M(Es_a\ a)$ are also points and lines in $G_L(Es_a\ a)$. For $M \subseteq M' \subseteq L$, $G_{M'}(Es_a\ a)$ is also a subgraph. Comparing $G_{M'}(Es_a\ a)$ with $G_M(Es_a\ a)$, any additional points in $G_{M'}(Es_a\ a)$ reflect latent system environments with respect to the system moment base M, which are represented by isolated points in $G_M(Es_a\ a)$. Additional lines in $G_{M'}(Es_a\ a)$ may reflect latent system states, stimuli, and responses for M.

Theorem 5.12 (Es$_a$ a). *The alternate environment graph $G_L(Es_a\ a)$ of a macrosystem model describing the operation of a system of interactive automata over the life of the system is a connected graph.*

Suppose $G_L(Es_a\ a)$ is not connected. Then there are system environments in $\bar{E}$ which the system cannot encounter during L. But this contradicts the definition of $\bar{E}$ which is taken to be all the environments which the system accesses during its life. This proof parallels the proof in Section 2.1 for a general automaton.

Corollary 5.12 (Es$_a$ a). *Let M be a moment base for a system of interactive automata, and $M \subseteq L$. Suppose no information is given for any other moment base M' properly containing M. Then $G_M(Es_a\ a)$ is a connected graph.*

This follows immediately from Theorem 5.12 ($Es_a\ a$) in the sense that M can be treated as L.

Appendix B covers the development and application of recursive methods and graph-theoretic principles for determining the effective operation of the general system of interactive automata through select properties of the graphs in the model (Frederick, Koenig 1971a, 1971b).

5.2 EXAMPLE APPLICATIONS

Primitives were established in the most detail for the processor graph, principal time graph, and alternate environment graph for each of the two models, the microsystem model and the macrosystem model. These three graphs for each model contain maximum information. Of these three, the alternate environment graph $G_M(Es_a\ i)$ for the micromodel and $G_M(Es_a\ a)$ for the macromodel are the most important to the study of interactive systems. For this graph, the points represent environments, and, since interactive automata share environments, the graph best demonstrates interaction. On this basis, the alternate environment graph of both models has been chosen to demonstrate the use of the functions and the primitives in establishing a graph and the use of the graph in the study of interaction (Koenig 1986b).

5.2.1 A Two-Component System

The system of interactive automata chosen is a two-component system $A=\{A_j \mid j \in J, J=\{1, 2\}\}$. The automaton component A_1 is identified with a man, and A_2 is identified with a computer. A_1 and A_2 share a display and a keyboard as shared environments as shown in Figure 5.1. The nonshared environment of A_1 is man's mind, and the nonshared environment of A_2 is memory. The only system functions defined by formulas for the example are 5.6F$_j$, 5.7F$_j$ and 5.6F$_S$, 5.7F$_S$, hence, most of the required information for the graphs must be observed, and very little can be calculated. The observing automaton that obtains the observed information has been excluded as a part of the system. The observations were made over the system moment base $M=\{m+i \mid i \in I, (m+i)=(m_j+i) \mid j \in J\}, I=\{0, 1, 2, ---, 8\}, J=\{1, 2\}$; i.e., $M=\{(m_1, m_2), (m_1+1, m_2+1), ---, (m_1+8, m_2+8)\}$. The physical interpretations of the symbols for states, stimuli, and environments are given in Figure 5.1.

The man A_1 will be considered first as an observed automaton as he interacts with the second automaton A_2. At m_1, the following information was obtained:

1. The *initial absolute time* was observed to be $\theta_{T1}\ (m_1)=t_{10} \in T_1$ (from 5.8M$_j$), where the second subscript of t establishes the value given by an integer vector with this entry.
2. The *initial environment* was observed to be $\theta_{E1}\ (m_1)=e_{10} \in E_1$ (from 5.7M$_j$), where the second subscript of e identifies the stimulus location given by an integer vector with this entry.

3. The *initial state* was observed to be $\theta_{Q1}(m_1)=q_{10}\in Q_1$ (from 5.1M$_j$).
4. The *initial stimulus* was observed to be $\theta_{S1}(m_1)=\alpha_{1x}(b_{x0})=s_{10}\in S_1$ (from 5.2M$_j$, 5.14F$_j$). The subscript x of b denotes an unknown automaton that supplied the recorded transformation response prior to the given moment base and applies to the recorded responses of environments accessed for the first time within the moment base and also to those that are not changed by null recorded responses after the first accesses. There is a corresponding x for the subscript of the interpreting function α.

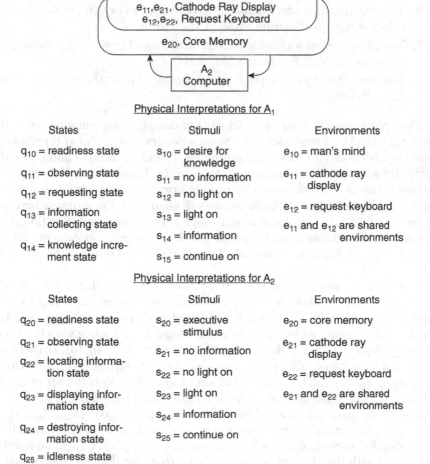

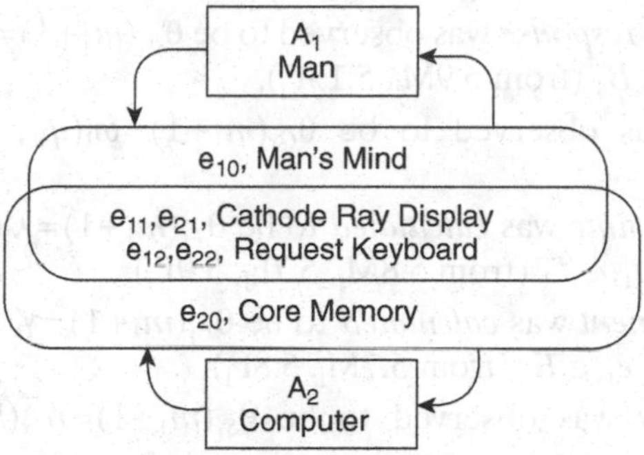

Physical Interpretations for A_1

States	Stimuli	Environments
q_{10} = readiness state	s_{10} = desire for knowledge	e_{10} = man's mind
q_{11} = observing state	s_{11} = no information	e_{11} = cathode ray display
q_{12} = requesting state	s_{12} = no light on	e_{12} = request keyboard
q_{13} = information collecting state	s_{13} = light on	e_{11} and e_{12} are shared environments
q_{14} = knowledge increment state	s_{14} = information	
	s_{15} = continue on	

Physical Interpretations for A_2

States	Stimuli	Environments
q_{20} = readiness state	s_{20} = executive stimulus	e_{20} = core memory
q_{21} = observing state	s_{21} = no information	e_{21} = cathode ray display
q_{22} = locating information state	s_{22} = no light on	e_{22} = request keyboard
q_{23} = displaying information state	s_{23} = light on	e_{21} and e_{22} are shared environments
q_{24} = destroying information state	s_{24} = information	
q_{25} = idleness state	s_{25} = continue on	

Figure 5.1 An interactive human-computer system.

At moment m_1+1, the following information was obtained:

1. The *transformation response* was observed to be $\theta_{F_1}(m_1+1)=\sigma_1(q_{10}, s_{10})=f_{15}\in F_1$ (from 5.4M$_j$, 5.3F$_j$).
2. The *spatial change response* was observed to be $\theta_{D_1}(m_1+1)=\delta_1(q_{10}, s_{10})=d_{11}\in D_1$ (from 5.5M$_j$, 5.4F$_j$), where the second subscript of d establishes the value given by an integer vector.
3. The *time change response* was observed to be $\theta_{C_1}(m_1+1)=\tau_1(q_{10}, s_{10})=c_{13}\in C_1$ (from 5.6M$_j$, 5.5F$_j$), where the second subscript of c establishes the value given by an integer vector.
4. The *recorded response* was observed to be $\theta_{B_1}(m_1+1)=\eta_1'(e_{10}, f_{15})=\eta_1'(e_{10}, q_{10}, s_{10})=b_{15}\in B_1$ (from 5.9M$_j$, 5.13F$_j$).
5. The *state* was observed to be $\theta_{Q_1}(m_1+1)=\omega_1(q_{10}, s_{10})=q_{11}\in Q_1$(from 5.1M$_j$, 5.1F$_j$).
6. The *absolute time* was *calculated* to be $\theta_{T_1}(m_1+1)=\lambda_1'(t_{10}, c_{13})=\lambda_1(t_{10}, q_{10}, s_{10})=t_{10}+c_{13}=t_{13}\in T_1$ (from 5.8M$_j$, 5.7F$_j$, 5.9F$_j$).
7. The *environment* was *calculated* to be $\theta_{E_1}(m_1+1)=\gamma_1'(e_{10}, d_{11})=\gamma_1(e_{10}, q_{10}, s_{10})=e_{10}+d_{11}=e_{11}\in E_1$ (from 5.7M$_j$, 5.8F$_j$).
8. The *stimulus* was observed to be $\theta_{S_1}(m_1+1)=\alpha_{1x}(b_{x1})=s_{11}\in S_1$ (from 5.2M$_j$, 5.14F$_j$).

This procedure is continued to obtain complete information over the moment base M_1 of automaton A_1. Since a system moment $m+i$ represents a time slice necessary to view both automata A_1 and A_2 at m_1+i and m_2+i, the computer as the automaton component A_2 is also being observed during each system moment. The complete observed information on A_1 and A_2 is given in Tables 5.1(a), 5.1(b), and the active sets defined over the moment base are given in Table 5.2. The domain and range sets of the graph function Λ_j' of 5.5GF$_j$ for the alternate environment graph of the microsystem model are given in Table 5.3 and were obtained from the observed information of Tables 5.1(a), 5.1(b).

From the information on the sets of Table 5.2, Λ_j' of Table 5.3, and the primitives $Es_a\,i1$ through $Es_a\,i4$, the alternate environment graph $G_M(Es_a\,i)$ of Figure 5.2 was obtained. The set $E_1\cup E_2$ of elements is the set of points. The set $X_1\cup X_2$ of elements is the set of lines, where $X_j\subseteq(T_j\times Q_j\times S_j)\times(B_j\times F_j\times D_j\times C_j\times Q_j\times T_j)$, $J=\{1, 2\}$. For the function Λ_j' defined by $\Lambda_j'(e_{ji}, (t_{ji}, (q_{ji}, s_{ji})))=(\eta_j(e_{ji}, (q_{ji}, s_{ji})), \sigma_j(q_{ji}, s_{ji}), \delta_j(q_{ji}, s_{ji}), \tau_j(q_{ji}, s_{ji}), \omega_j(q_{ji}, s_{ji}), \lambda_j(t_{ji}, (q_{ji}, s_{ji})), \gamma_j(e_{ji}, (q_{ji}, s_{ji}))=(b_{j(i+1)}, f_{j(i+1)}, d_{j(i+1)}, c_{j(i+1)}, q_{j(i+1)}, t_{j(i+1)}, e_{j(i+1)})$, the set of lines $X_1\cup X_2$ is the set of all the $((t_{ji}, q_{ji}, s_{ji}), (b_{j(i+1)}, f_{j(i+1)}, d_{j(i+1)}, c_{j(i+1)}, q_{j(i+1)}, t_{j(i+1)}))$ in the 11-tuple $(e_{ji}, t_{ji}, q_{ji}, s_{ji}, b_{j(i+1)}, f_{j(i+1)}, d_{j(i+1)}, c_{j(i+1)}, q_{j(i+1)}, t_{j(i+1)}, e_{j(i+1)})$, such that $\Lambda_j'(e_{ji}, (t_{ji}, (q_{ji}, s_{ji})))=(b_{j(i+1)}, f_{j(i+1)}, d_{j(i+1)}, c_{j(i+1)}, q_{j(i+1)}, t_{j(i+1)}, e_{j(i+1)})$. The subset $X_{10}\cup X_{20}\subseteq X_1\cup X_2$ is a set of directed lines if for $x_{j0}\in X_{j0}$, $x_{j0}=((t_{ji}, q_{ji}, s_{ji}), (b_{j(i+1)}, f_{j(i+1)}, d_{j(i+1)}, c_{j(i+1)}, q_{j(i+1)}, t_{j(i+1)}))$, the point e_{ji} associated with the element $(e_{ji}, (t_{ji}, (q_{ji}, s_{ji})))$ from the domain is not equal to the point $e_{j(i+1)}$ in the image $\Lambda_j'(e_{ji}, (t_{ji}, (q_{ji}, s_{ji})))=(b_{j(i+1)}, f_{j(i+1)}, d_{j(i+1)}, c_{j(i+1)}, q_{j(i+1)}, t_{j(i+1)}, e_{j(i+1)})$. The line is directed from e_{ji} to point $e_{j(i+1)}$.

TABLE 5.1.

(a) Observed Information on the Human as an Automaton Component A_1

*$(\theta_{x_1}(m_1+i) \mid i \varepsilon I)$	F_1	D_1	C_1	B_1
$\theta_{x_1}(m_1)$	—	—	—	—
$\theta_{x_1}(m_1+1)$	$\sigma_1(q_{10}, s_{10})=f_{15}$	$\delta_1(q_{10}, s_{10})=d_{11}$	$\tau_1(q_{10}, s_{10})=c_{13}$	$\eta_1(e_{10}, q_{10}, s_{10})=b_{15}$
$\theta_{x_1}(m_1+2)$	$\sigma_1(q_{11}, s_{11})=f_{10}$	$\delta_1(q_{11}, s_{11})=d_{11}$	$\tau_1(q_{11}, s_{11})=c_{12}$	$\eta_1(e_{11}, q_{11}, s_{11})=b_{10}$
$\theta_{x_1}(m_1+3)$	$\sigma_1(q_{12}, s_{12})=f_{13}$	$\delta_1(q_{12}, s_{12})=d_{10}$	$\tau_1(q_{12}, s_{12})=c_{11}$	$\eta_1(e_{12}, q_{12}, s_{12})=b_{13}$
$\theta_{x_1}(m_1+4)$	$\sigma_1(q_{12}, s_{13})=f_{10}$	$\delta_1(q_{12}, s_{13})=d_{1\text{-}1}$	$\tau_1(q_{12}, s_{13})=c_{13}$	$\eta_1(e_{12}, q_{12}, s_{13})=b_{10}$
$\theta_{x_1}(m_1+5)$	$\sigma_1(q_{13}, s_{11})=f_{10}$	$\delta_1(q_{13}, s_{11})=d_{10}$	$\tau_1(q_{13}, s_{11})=c_{12}$	$\eta_1(e_{11}, q_{13}, s_{11})=b_{10}$
$\theta_{x_1}(m_1+6)$	$\sigma_1(q_{13}, s_{11})=f_{10}$	$\delta_1(q_{13}, s_{11})=d_{10}$	$\tau_1(q_{13}, s_{11})=c_{12}$	$\eta_1(e_{11}, q_{13}, s_{11})=b_{10}$
$\theta_{x_1}(m_1+7)$	$\sigma_1(q_{13}, s_{14})=f_{10}$	$\delta_1(q_{13}, s_{14})=d_{1\text{-}1}$	$\tau_1(q_{13}, s_{14})=c_{12}$	$\eta_1(e_{11}, q_{13}, s_{14})=b_{10}$
$\theta_{x_1}(m_1+8)$	$\sigma_1(q_{14}, s_{15})=f_{10}$	$\delta_1(q_{14}, s_{15})=d_{11}$	$\tau_1(q_{14}, s_{15})=c_{12}$	$\eta_1(e_{10}, q_{14}, s_{15})=b_{10}$
$(\theta_{x_1}(m_1+i) \mid i \varepsilon I)$	Q_1	T_1	E_1	S_1
$\theta_{x_1}(m_1)$	q_{10}	t_{10}	e_{10}	$\alpha_{1x}(b_{x0})=s_{10}$
$\theta_{x_1}(m_1+1)$	$\omega_1(q_{10}, s_{10})=q_{11}$	$\lambda_1(t_{10}, q_{10}, s_{10}) \quad =t_{13}$	$\gamma_1(e_{10}, q_{10}, s_{10})=e_{11}$	$\alpha_{1x}(b_{x1})=s_{11}$
$\theta_{x_1}(m_1+2)$	$\omega_1(q_{11}, s_{11})=q_{12}$	$\lambda_1(t_{13}, q_{11}, s_{11}) \quad =t_{15}$	$\gamma_1(e_{11}, q_{11}, s_{11})=e_{12}$	$\alpha_{1x}(b_{x2})=s_{12}$
$\theta_{x_1}(m_1+3)$	$\omega_1(q_{12}, s_{12})=q_{12}$	$\lambda_1(t_{15}, q_{12}, s_{12}) \quad =t_{16}$	$\gamma_1(e_{12}, q_{12}, s_{12})=e_{12}$	$\alpha_{11}(b_{13})=s_{13}$
$\theta_{x_1}(m_1+4)$	$\omega_1(q_{12}, s_{13})=q_{13}$	$\lambda_1(t_{16}, q_{12}, s_{13}) \quad =t_{19}$	$\gamma_1(e_{12}, q_{12}, s_{13})=e_{11}$	$\alpha_{1x}(b_{x1})=s_{11}$
$\theta_{x_1}(m_1+5)$	$\omega_1(q_{13}, s_{11})=q_{13}$	$\lambda_1(t_{19}, q_{13}, s_{11}) \quad =t_{1,11}$	$\gamma_1(e_{11}, q_{13}, s_{11})=e_{11}$	$\alpha_{1x}(b_{x1})=s_{11}$
$\theta_{x_1}(m_1+6)$	$\omega_1(q_{13}, s_{11})=q_{13}$	$\lambda_1(t_{1,11}, q_{13}, s_{11}) \quad =t_{1,13}$	$\gamma_1(e_{11}, q_{13}, s_{11})=e_{11}$	$\alpha_{12}(b_{24})=s_{14}$
$\theta_{x_1}(m_1+7)$	$\omega_1(q_{13}, s_{14})=q_{14}$	$\lambda_1(t_{1,13}, q_{13}, s_{14}) \quad =t_{1,15}$	$\gamma_1(e_{11}, q_{13}, s_{14})=e_{10}$	$\alpha_{11}(b_{15})=s_{15}$
$\theta_{x_1}(m_1+8)$	$\omega_1(q_{14}, s_{15})=q_{11}$	$\lambda_1(t_{1,15}, q_{14}, s_{15}) \quad =t_{1,17}$	$\gamma_1(e_{10}, q_{14}, s_{15})=e_{11}$	

* x_1 denotes any of the sets F_1, D_1, C_1, B_1, Q_1, T_1, E_1, S_1, $I=\{0, 1, 2, \ldots, 8\}$

TABLE 5.1. *Continued*

(b) Observed Information of the Computer as an Automaton Component A_2

$*(\theta_{x_2}(m_2+i)\vert i\varepsilon I)$	F_2	D_2	C_2	B_2
$\theta_{x_2}(m_2)$	—	—	—	—
$\theta_{x_2}(m_2+1)$	$\sigma_2(q_{20}, s_{20})=f_{25}$	$\delta_2(q_{20}, s_{20})=d_{22}$	$\tau_2(q_{20}, s_{20})=c_{21}$	$\eta_2(e_{20}, q_{20}, s_{20})=b_{25}$
$\theta_{x_2}(m_2+2)$	$\sigma_2(q_{21}, s_{22})=f_{20}$	$\delta_2(q_{21}, s_{22})=d_{20}$	$\tau_2(q_{21}, s_{22})=c_{22}$	$\eta_2(e_{22}, q_{21}, s_{22})=b_{20}$
$\theta_{x_2}(m_2+3)$	$\sigma_2(q_{25}, s_{22})=f_{20}$	$\delta_2(q_{25}, s_{22})=d_{20}$	$\tau_2(q_{25}, s_{22})=c_{23}$	$\eta_2(e_{22}, q_{25}, s_{22})=b_{20}$
$\theta_{x_2}(m_2+4)$	$\sigma_2(q_{21}, s_{23})=f_{22}$	$\delta_2(q_{21}, s_{23})=d_{2\text{-}2}$	$\tau_2(q_{21}, s_{23})=c_{21}$	$\eta_2(e_{22}, q_{21}, s_{23})=b_{22}$
$\theta_{x_2}(m_2+5)$	$\sigma_2(q_{22}, s_{25})=f_{20}$	$\delta_2(q_{22}, s_{25})=d_{21}$	$\tau_2(q_{22}, s_{25})=c_{22}$	$\eta_2(e_{20}, q_{22}, s_{25})=b_{20}$
$\theta_{x_2}(m_2+6)$	$\sigma_2(q_{23}, s_{21})=f_{24}$	$\delta_2(q_{23}, s_{21})=d_{20}$	$\tau_2(q_{23}, s_{21})=c_{22}$	$\eta_2(e_{21}, q_{23}, s_{21})=b_{24}$
$\theta_{x_2}(m_2+7)$	$\sigma_2(q_{24}, s_{24})=f_{21}$	$\delta_2(q_{24}, s_{24})=d_{21}$	$\tau_2(q_{24}, s_{24})=c_{22}$	$\eta_2(e_{21}, q_{24}, s_{24})=b_{21}$
$\theta_{x_2}(m_2+8)$	$\sigma_2(q_{21}, s_{22})=f_{20}$	$\delta_2(q_{21}, s_{22})=d_{20}$	$\tau_2(q_{21}, s_{22})=c_{22}$	$\eta_2(e_{22}, q_{21}, s_{22})=b_{20}$
$(\theta_{x_2}(m_2+i)\vert i\varepsilon I)$	Q_2	T_2	E_2	S_2
$\theta_{x_2}(m_2)$	q_{20}	t_{21}	e_{20}	$\alpha_{2x}(b_{x0})=s_{20}$
$\theta_{x_2}(m_2+1)$	$\omega_2(q_{20}, s_{20})=q_{21}$	$\lambda_2(t_{21}, q_{20}, s_{20}) \ =t_{22}$	$\gamma_2(e_{20}, q_{20}, s_{20})=e_{22}$	$\alpha_{2x}(b_{x2})=s_{22}$
$\theta_{x_2}(m_2+2)$	$\omega_2(q_{21}, s_{22})=q_{25}$	$\lambda_2(t_{22}, q_{21}, s_{22}) \ =t_{24}$	$\gamma_2(e_{22}, q_{21}, s_{22})=e_{22}$	$\alpha_{2x}(b_{x2})=s_{22}$
$\theta_{x_2}(m_2+3)$	$\omega_2(q_{25}, s_{22})=q_{21}$	$\lambda_2(t_{24}, q_{25}, s_{22}) \ =t_{27}$	$\gamma_2(e_{22}, q_{25}, s_{22})=e_{22}$	$\alpha_{21}(b_{13})=s_{23}$
$\theta_{x_2}(m_2+4)$	$\omega_2(q_{21}, s_{23})=q_{22}$	$\lambda_2(t_{27}, q_{21}, s_{23}) \ =t_{28}$	$\gamma_2(e_{22}, q_{21}, s_{23})=e_{20}$	$\alpha_{22}(b_{25})=s_{25}$
$\theta_{x_2}(m_2+5)$	$\omega_2(q_{22}, s_{25})=q_{23}$	$\lambda_2(t_{28}, q_{22}, s_{25}) \ =t_{2,10}$	$\gamma_2(e_{20}, q_{22}, s_{25})=e_{21}$	$\alpha_{2x}(b_{x1})=s_{21}$
$\theta_{x_2}(m_2+6)$	$\omega_2(q_{23}, s_{21})=q_{24}$	$\lambda_2(t_{2,10}, q_{23}, s_{21}) =t_{2,12}$	$\gamma_2(e_{21}, q_{23}, s_{21})=e_{21}$	$\alpha_{22}(b_{24})=s_{24}$
$\theta_{x_2}(m_2+7)$	$\omega_2(q_{24}, s_{24})=q_{21}$	$\lambda_2(t_{2,12}, q_{24}, s_{24}) =t_{2,14}$	$\gamma_2(e_{21}, q_{24}, s_{24})=e_{22}$	$\alpha_{22}(b_{22})=s_{22}$
$\theta_{x_2}(m_2+8)$	$\omega_2(q_{21}, s_{22})=q_{25}$	$\lambda_2(t_{2,14}, q_{21}, s_{22}) =t_{2,16}$	$\gamma_2(e_{22}, q_{21}, s_{22})=e_{22}$	—

* x_2 denotes any of the sets F_2, D_2, C_2, B_2, Q_2, T_2, E_2, S_2, $I=\{0, 1, 2, \ldots, 8\}$

TABLE 5.2. Active sets determined by observation over the system moment base M

Active Sets for A_1 Determined by Observation Over the Moment Base $M_1=\{m_1=i \mid i\varepsilon I,\ I=\{0, 1, 2, \ldots, 8\}\}$		
$Q_1=\{q_{10}, q_{11}, q_{12}, q_{13}, q_{14}\}$	$F_1=\{f_{15}, f_{10}, f_{13}\}$	$T_1=\{t_{10}, t_{13}, t_{15}, t_{16}, t_{19}, t_{1,11}, t_{1,13}, t_{1,15}, t_{1,17}\}$
$S_1=\{s_{10}, s_{11}, s_{12}, s_{13}, s_{14}, s_{15}\}$	$D_1=\{d_{11}, d_{10}, d_{1\text{-}1}\}$	
$B_1=\{b_{15}, b_{10}, b_{13}\}$	$C_1=\{c_{13}, c_{12}, c_{11}\}$	$E_1=\{e_{10}, e_{11}, e_{12}\}$
Active Sets for A_2 Determined by Observation Over the Moment Base $M_1=\{m_1+i \mid i\varepsilon I,\ I=\{0, 1, 2, \ldots, 8\}\}$		
$Q_2=\{q_{20}, q_{21}, q_{25}, q_{22}, q_{23}, q_{24}\}$	$F_2=\{f_{25}, f_{20}, f_{22}, f_{24}, f_{21}\}$	$T_2=\{t_{21}, t_{22}, t_{24}, t_{27}, t_{28}, t_{2,10}, t_{2,12}, t_{2,14}, t_{2,16}\}$
$S_2=\{s_{20}, s_{22}, s_{23}, s_{25}, s_{21}, s_{24}\}$	$D_2=\{d_{22}, d_{20}, d_{2\text{-}2}, d_{21}\}$	
$B_2=\{b_{25}, b_{20}, b_{24}, b_{21}, b_{22}\}$	$C_2=\{c_{21}, c_{22}, c_{23}\}$	$E_2=\{e_{20}, e_{22}, e_{21}\}$

Active System Sets Over the System Moment Base M

$Q=\{(q_{10}, q_{20}), (q_{11}, q_{21}), (q_{12}, q_{25}), (q_{12}, q_{21}), (q_{13}, q_{22}), (q_{13}, q_{23}), (q_{13}, q_{24}), (q_{14}, q_{21}), (q_{11}, q_{25})\}$

$S=\{(s_{10}, s_{20}), (s_{11}, s_{22}), (s_{12}, s_{22}), (s_{13}, s_{23}), (s_{11}, s_{25}), (s_{11}, s_{21}), (s_{14}, s_{24}), (s_{15}, s_{22})\}$

$F=\{(f_{15}, f_{25}), (f_{10}, s_{20}), (f_{13}, f_{20}), (f_{10}, f_{22}), (f_{10}, f_{24}), (f_{10}, f_{21})\}$

$D=\{(d_{11}, d_{22}), (d_{11}, d_{20}), (d_{10}, d_{20}), (d_{1\text{-}1}, d_{2\text{-}2}), (d_{10}, d_{21}), (d_{1\text{-}1}, d_{21}), (d_{1\text{-}1}, d_{20})\}$

$C=\{(c_{13}, c_{21}), (c_{12}, c_{22}), (c_{11}, c_{23})\}$

$B=\{(b_{15}, b_{25}), (b_{10}, b_{20}), (b_{13}, b_{20}), (b_{10}, b_{22}), (b_{10}, b_{24}), (b_{10}, b_{21})\}$

$T=\{(t_{10}, t_{21}), (t_{13}, t_{22}), (t_{15}, t_{24}), (t_{16}, t_{27}), (t_{19}, t_{28}), (t_{1,11}, t_{2,10}), (t_{1,13}, t_{2,12}), (t_{1,15}, t_{2,14}), (t_{1,17}, t_{2,16})\}$

$E=\{(e_{10}, e_{20}), (e_{11}, e_{22}), (e_{12}, e_{22}), (e_{11}, e_{20}), (e_{11}, e_{21}), (e_{10}, e_{22})\}$

TABLE 5.3. Domain and corresponding range sets of the graph function Λ'_j for the alternate environment graph of the microsystem model

Operation	Y'_1	$\Lambda'_1(Y'_1)$
1_1	$e_{10}, t_{10}, q_{10}, s_{10}$	$b_{15}, f_{15}, d_{11}, c_{13}, q_{11}, t_{13}, e_{11}$
2_1	$e_{11}, t_{13}, q_{11}, s_{11}$	$b_{10}, f_{10}, d_{11}, c_{12}, q_{12}, t_{15}, e_{12}$
3_1	$e_{12}, t_{15}, q_{12}, s_{12}$	$b_{13}, f_{13}, d_{10}, c_{11}, q_{12}, t_{16}, e_{12}$
4_1	$e_{12}, t_{16}, q_{12}, s_{13}$	$b_{10}, f_{10}, d_{1\text{-}1}, c_{13}, q_{13}, t_{19}, e_{11}$
5_1	$e_{11}, t_{19}, q_{13}, s_{11}$	$b_{10}, f_{10}, d_{10}, c_{12}, q_{13}, t_{1,11}, e_{11}$
6_1	$e_{11}, t_{1,11}, q_{13}, s_{11}$	$b_{10}, f_{10}, d_{10}, c_{12}, q_{13}, t_{1,13}, e_{11}$
7_1	$e_{11}, t_{1,13}, q_{13}, s_{14}$	$b_{10}, f_{10}, d_{1\text{-}1}, c_{12}, q_{14}, t_{1,15}, e_{10}$
8_1	$e_{10}, t_{1,15}, q_{14}, s_{15}$	$b_{10}, f_{10}, d_{11}, c_{12}, q_{11}, t_{1,17}, e_{11}$
Operation	Y'_2	$\Lambda'_2(Y'_2)$
1_2	$e_{20}, t_{21}, q_{20}, s_{20}$	$b_{25}, f_{25}, d_{22}, c_{21}, q_{21}, t_{22}, e_{22}$
2_2	$e_{22}, t_{22}, q_{21}, s_{22}$	$b_{20}, f_{20}, d_{20}, c_{22}, q_{25}, t_{24}, e_{22}$
3_2	$e_{22}, t_{24}, q_{25}, s_{22}$	$b_{20}, f_{20}, d_{20}, c_{23}, q_{21}, t_{27}, e_{22}$
4_2	$e_{22}, t_{27}, q_{21}, s_{23}$	$b_{22}, f_{22}, d_{2\text{-}2}, c_{21}, q_{22}, t_{28}, e_{20}$
5_2	$e_{20}, t_{28}, q_{22}, s_{25}$	$b_{20}, f_{20}, d_{21}, c_{22}, q_{23}, t_{2,10}, e_{21}$
6_2	$e_{21}, t_{2,10}, q_{23}, s_{21}$	$b_{24}, f_{24}, d_{20}, c_{22}, q_{24}, t_{2,12}, e_{21}$
7_2	$e_{21}, t_{2,12}, q_{24}, s_{24}$	$b_{21}, f_{21}, d_{21}, c_{22}, q_{21}, t_{2,14}, e_{22}$
8_2	$e_{22}, t_{2,14}, q_{21}, s_{22}$	$b_{20}, f_{20}, d_{20}, c_{22}, q_{25}, t_{2,16}, e_{22}$

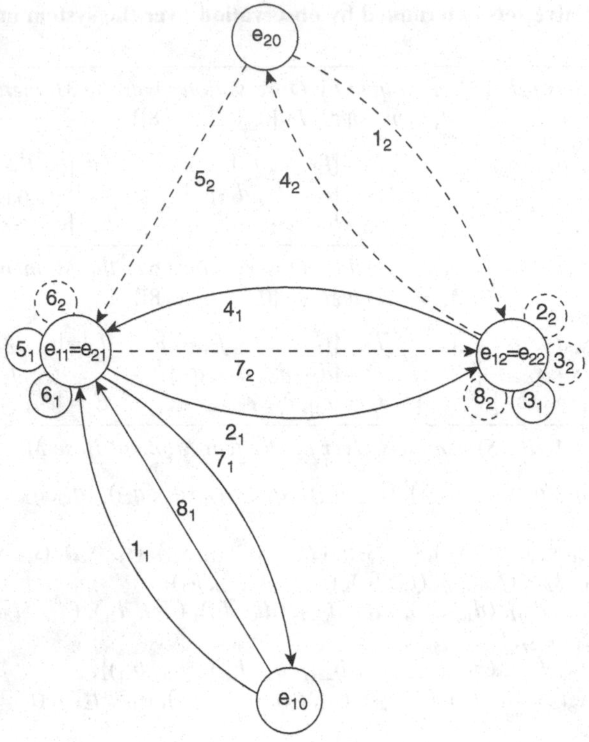

Figure 5.2 The alternate Environment Graph $G_M(Es_a\, i)$ of the microsystem model for a human-computer system obtained from Table 5.3.

An effective operation of a system, in this case the observed operation, can be represented by paths through the graph. The paths are given by alternating sequences of points and lines. The sequence of numbers 1_1, 2_1, ---, 8_1 on the graph represents the operation of A_1, and the path given by the sequence 1_2, 2_2, ---, 8_2 represents the operation of A_2. By following these sequences on the graph and by giving a physical interpretation of the graph information, the effective operation can be clearly described.

For 1_1. A_1 (man) interprets a recorded response b_{x0} as a stimulus s_{10} (desire for knowledge) at environment e_{10} (his mind) while in state q_{10} (readiness state) at time t_{10} (value of 0) and goes into state q_{11} (observing state), leaves a recorded response b_{15} at e_{10}, and completes a move to environment e_{ji} (display) at time t_{13} (value of 3).

For 2_1. A_1 interprets a recorded response b_{x1} as a stimulus s_{11} (no information) at e_{j1} while in state q_{11} at time t_{13} (value of 3) and goes into state q_{12} (requesting state), leaves a null recorded response b_{10} at e_{j1} (does nothing to change the existing recorded response), and completes a move to environment e_{j2} (keyboard) at time t_{15} (value of 5).

For 3_1. A_1 interprets a recorded response b_{x2} as a stimulus s_{12} (no light on) at e_{j2} while in state q_{12} at time t_{15}, and remains in the same state q_{12}, leaves a recorded response b_{13} at e_{j2}, and remains at the environment e_{j2}. The time is t_{16} (value of 6).

In the meantime, A_2 has been operating.

For 1_2. A_2 (computer) interprets a recorded response b_{x0} as a stimulus s_{20} (executive stimulus) at environment e_{20} (memory) while in state q_{20} (readiness state) at time t_{21} (value of 1) and goes into state q_{21} (observing state), leaves a recorded response b_{25} at e_{20}, and completes a move to e_{j2} (keyboard) at time t_{22}.

For 2_2. A_2 interprets a recorded response b_{x2} as a stimulus s_{22} (no light on) at e_{j2} while in state q_{21} at time t_{22} (value of 2) and goes into state q_{25} (idleness state), leaves a null recorded response b_{20} at e_{j2} (does nothing to change the existing recorded response) and remains at environment e_{j2}. The time is t_{24} (value of 4).

For 3_2. A_2 interprets a recorded response b_{x2} as a stimulus s_{22} (no light on) at e_{j2} while in state q_{25} at time t_{24} and goes into state q_{21} (observing state), leaves a recorded response b_{20} at e_{j2} (does nothing to change the existing recorded response) and remains at environment e_{j2}. The time is t_{27} (value of 7).

For 4_2. A_2 interprets a recorded response b_{13} (a recorded response of A_1 at t_{16}) as a stimulus s_{23} (light on) at e_{j2} while in state q_{21} at time t_{27} and goes into state q_{22} (locating information state), leaves a recorded response b_{22} at e_{j2} and completes a move to e_{20} (memory) at time t_{28}.

This procedure may be continued to provide a word description of the complete effective operation over the moment base M.

Recall that a macrosystem model is useful when a system of interactive automata is viewed as a single automaton. The alternative environment graph $G_M(Es_a\ a)$ for a macrosystem model will now be established for the example system. From the observed information of Tables 5.1(a), 5.1(b) and the functions given by 5.1F$_S$ through 5.14F$_S$, the domain and range sets of Table 5.4 are established. Table 5.4, in turn, supplies information for the domain and range sets of the graph function Λ' given in Table 5.5.

From the information on the sets of Table 5.2, Λ' of Table 5.5, and the Primitives $Es_a\ a1$ through $Es_a\ a4$, the alternate environment graph $G_M(Es_a\ a)$ of Figure 5.3 was obtained. The subset $E \subseteq E_1 \times E_2$ of elements are called points, and a subset $X \subseteq (T \times Q \times S) \times (B \times F \times D \times C \times Q \times T)$ are called lines. For the function Λ' defined by $\Lambda'(e_i, (t_i, (q_i, s_i))) = (\eta(e_i, (q_i, s_i)), \sigma(q_i, s_i), \delta(q_i, s_i), \tau(q_i, s_i), \omega(q_i, s_i), \lambda(t_i, (q_i, s_i)), \gamma(e_i, (q_i, s_i))) = (b_{i+1}, f_{i+1}, d_{i+1}, c_{i+1}, q_{i+1}, t_{i+1}, e_{i+1})$, the set of lines X is all the $((t_i, q_i, s_i), (b_{i+1}, f_{i+1}, d_{i+1}, c_{i+1}, q_{i+1}, t_{i+1}))$ in the 11-tuple $(e_i, t_i, q_i, s_i, b_{i+1}, f_{i+1}, d_{i+1}, c_{i+1}, q_{i+1}, t_{i+1}, e_{i+1})$. The subset $X_0 \subseteq X$ is a set of directed lines for $x \in X_0$, $x = ((t_i, q_i, s_i), (b_{i+1}, f_{i+1}, d_{i+1}, c_{i+1}, q_{i+1}, t_{i+1}))$, the point e_i associated with the

TABLE 5.4. Domain and range sets for the system functions

V	$\sigma(V)$	$d(V)$	$\tau(V)$	$\omega(V)$
$((q_{10}, q_{20}), (s_{10}, s_{20}))$	(f_{15}, f_{25})	(d_{11}, d_{22})	(c_{13}, c_{21})	(q_{11}, q_{21})
$((q_{11}, q_{21}), (s_{11}, s_{22}))$	(f_{10}, f_{20})	(d_{11}, d_{20})	(c_{12}, c_{22})	(q_{12}, q_{25})
$((q_{12}, q_{25}), (s_{12}, s_{22}))$	(f_{13}, f_{20})	(d_{10}, d_{20})	(c_{11}, c_{23})	(q_{12}, q_{21})
$((q_{12}, q_{21}), (s_{13}, s_{23}))$	(f_{10}, f_{22})	$(d_{1\text{-}1}, d_{2\text{-}2})$	(c_{13}, c_{21})	(q_{13}, q_{22})
$((q_{13}, q_{22}), (s_{11}, s_{25}))$	(f_{10}, f_{20})	(d_{10}, d_{21})	(c_{12}, c_{22})	(q_{13}, q_{23})
$((q_{13}, q_{23}), (s_{11}, s_{21}))$	(f_{10}, f_{24})	(d_{10}, d_{20})	(c_{12}, c_{22})	(q_{13}, q_{24})
$((q_{13}, q_{24}), (s_{14}, s_{24}))$	(f_{10}, f_{21})	$(d_{1\text{-}1}, d_{21})$	(c_{12}, c_{22})	(q_{14}, q_{21})
$((q_{14}, q_{21}), (s_{15}, s_{22}))$	(f_{11}, f_{20})	(d_{11}, d_{20})	(c_{12}, c_{22})	(q_{11}, q_{25})

W	$\eta(W)$	$\lambda(W)$
$((e_{10}, e_{20}), (q_{10}, q_{20}), (s_{10}, s_{20}))$	(b_{15}, b_{25})	(e_{11}, e_{22})
$((e_{11}, e_{22}), (q_{11}, q_{21}), (s_{11}, s_{22}))$	(b_{10}, b_{20})	(e_{12}, e_{22})
$((e_{12}, e_{22}), (q_{12}, q_{25}), (s_{12}, s_{22}))$	(b_{13}, b_{20})	(e_{12}, e_{22})
$((e_{12}, e_{22}), (q_{12}, q_{21}), (s_{13}, s_{23}))$	(b_{10}, b_{22})	(e_{11}, e_{20})
$((e_{11}, e_{20}), (q_{13}, q_{22}), (s_{11}, s_{25}))$	(b_{10}, b_{20})	(e_{11}, e_{21})
$((e_{11}, e_{21}), (q_{13}, q_{23}), (s_{11}, s_{21}))$	(b_{10}, b_{24})	(e_{11}, e_{21})
$((e_{11}, e_{21}), (q_{13}, q_{24}), (s_{14}, s_{24}))$	(b_{10}, b_{21})	(e_{10}, e_{22})
$((e_{10}, e_{22}), (q_{14}, q_{21}), (s_{15}, s_{22}))$	(b_{10}, b_{20})	(e_{11}, e_{22})

N	$\lambda(N)$
$((t_{10}, t_{21}), (q_{10}, q_{20}), (s_{10}, s_{20}))$	(t_{13}, t_{22})
$((t_{13}, t_{22}), (q_{11}, q_{21}), (s_{11}, s_{22}))$	(t_{15}, t_{24})
$((t_{15}, t_{24}), (q_{12}, q_{25}), (s_{12}, s_{22}))$	(t_{16}, t_{27})
$((t_{16}, t_{27}), (q_{12}, q_{21}), (s_{13}, s_{23}))$	(t_{19}, t_{28})
$((t_{19}, t_{28}), (q_{13}, q_{22}), (s_{11}, s_{25}))$	$(t_{1,11}, t_{2,10})$
$((t_{1,11}, t_{2,10}), (q_{13}, q_{23}), (s_{11}, s_{21}))$	$(t_{1,13}, t_{2,12})$
$((t_{1,13}, t_{2,12}), (q_{13}, q_{24}), (s_{14}, s_{24}))$	$(t_{1,15}, t_{2,14})$
$((t_{1,15}, t_{2,14}), (q_{14}, q_{21}), (s_{15}, s_{22}))$	$(t_{1,17}, t_{2,16})$

domain element $(e_i, (t_i, (q_i, s_i)))$ is not equal to the point e_{i+1} in the image Λ' $(e_i, (t_i, (q_i, s_i))) = (b_{i+1}, f_{i+1}, d_{i+1}, c_{i+1}, q_{i+1}, t_{i+1}, e_{i+1})$. The line x is directed from e_i to point e_{i+1}.

The graph (Figure 5.3) views the system as a single automaton and, in this case, the effective operation path through the graph is obvious. For a brief physical interpretation of the graph, consider only the environments of each system moment. A_1 and A_2 are initially at environments e_{10}, e_{20} (man's mind and memory). Each then access different shared environments e_{11}, e_{12} (display and keyboard). They then access at two successive system moments a single shared environment e_{12}, e_{22} (keyboard) at different absolute times. e_{11} and e_{20} (display and memory) are next accessed followed by two successive accesses of a single shared environment e_{11}, e_{21} (display). Accesses of e_{10}, e_{22} and then a second access of e_{11}, e_{22} completes the effective operation path over the system moment base M. The graphs of the macrosystem model, in general, give

Table 5.5. Domain and range sets of the graph function Λ' for the alternate environment graph of the macrosystem model

Operation	Y', Domain Set
(a)	$(e_{10}, e_{20}), (t_{10}, t_{21}), (q_{10}, q_{20}), (s_{10}, s_{20})$
(b)	$(e_{11}, e_{22}), (t_{13}, t_{22}), (q_{11}, q_{21}), (s_{11}, s_{22})$
(c)	$(e_{12}, e_{22}), (t_{15}, t_{24}), (q_{12}, q_{25}), (s_{12}, s_{22})$
(d)	$(e_{12}, e_{22}), (t_{16}, t_{27}), (q_{12}, q_{21}), (s_{13}, s_{23})$
(e)	$(e_{11}, e_{20}), (t_{19}, t_{28}), (q_{13}, q_{22}), (s_{11}, s_{25})$
(f)	$(e_{11}, e_{21}), (t_{1,11}, t_{2,10}), (q_{13}, q_{23}), (s_{11}, s_{21})$
(g)	$(e_{11}, e_{21}), (t_{1,13}, t_{2,12}), (q_{13}, q_{24}), (s_{14}, s_{24})$
(h)	$(e_{10}, e_{22}), (t_{1,15}, t_{2,14}), (q_{14}, q_{21}), (s_{15}, s_{22})$
Operation	**$\Lambda'(Y')$, Corresponding Range Set**
(a)	$(b_{15}, b_{25}), (f_{15}, f_{25}), (d_{11}, d_{22}), (c_{13}, c_{21}), (q_{11}, q_{21}), (t_{13}, t_{22}), (e_{11}, e_{22})$
(b)	$(b_{10}, b_{20}), (f_{10}, f_{20}), (d_{11}, d_{20}), (c_{12}, c_{22}), (q_{12}, q_{25}), (t_{15}, t_{24}), (e_{12}, e_{22})$
(c)	$(b_{13}, b_{20}), (f_{13}, f_{20}), (d_{10}, d_{20}), (c_{11}, c_{23}), (q_{12}, q_{21}), (t_{16}, t_{27}), (e_{12}, e_{22})$
(d)	$(b_{10}, b_{22}), (f_{10}, f_{22}), (d_{1-1}, d_{2-2}), (c_{13}, c_{21}), (q_{13}, q_{22}), (t_{19}, t_{28}), (e_{11}, e_{20})$
(e)	$(b_{10}, b_{20}), (f_{10}, f_{20}), (d_{10}, d_{21}), (c_{12}, c_{22}), (q_{13}, q_{23}), (t_{1,11}, t_{2,10}), (e_{11}, e_{21})$
(f)	$(b_{10}, b_{24}), (f_{10}, f_{24}), (d_{10}, d_{20}), (c_{12}, c_{22}), (q_{13}, q_{24}), (t_{1,13}, t_{2,12}), (e_{11}, e_{21})$
(g)	$(b_{10}, b_{21}), (f_{10}, f_{21}), (d_{1-1}, d_{21}), (c_{12}, c_{22}), (q_{14}, q_{21}), (t_{1,15}, t_{2,14}), (e_{10}, e_{22})$
(h)	$(b_{10}, b_{20}), (f_{10}, f_{20}), (d_{11}, d_{20}), (c_{12}, c_{22}), (q_{11}, q_{25}), (t_{1,17}, t_{2,16}), (e_{11}, e_{22})$

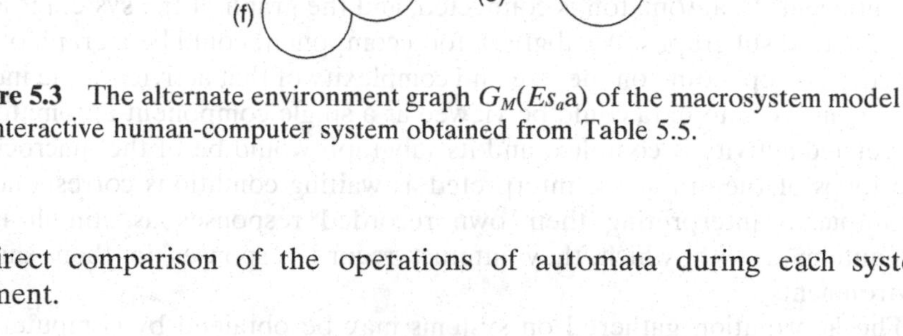

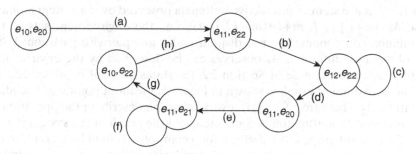

Figure 5.3 The alternate environment graph $G_M(Es_a a)$ of the macrosystem model for an interactive human-computer system obtained from Table 5.5.

a direct comparison of the operations of automata during each system moment.

5.2.2 A System of Many Components

Consider now a system of many components (Chrystal et al. 1962a, 1962b), (Koenig 1973–74). There are given a process industry, a supplier industry, and a customer industry, shown in Figure 5.4, and an observer. Over a system

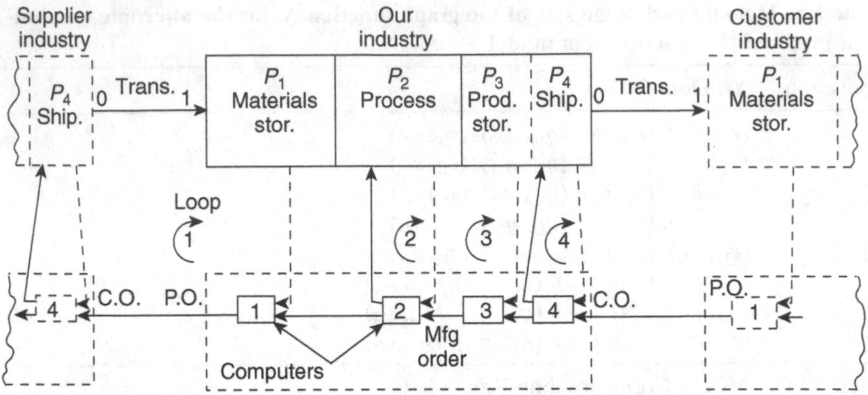

Figure 5.4 A process industry and its supplier and customer industry.

moment base M, the observer observes the process industry receiving material from the supplier industry, processing the material, and shipping the product to a customer industry. Let the observer be described as an automaton A_0. Let $A_1, A_2, ---, A_{10}$ denote the automata of the system, input inventory, process, product storage, shipping, transportation, postal service, purchasing, input inventory control, process control, product inventory, and customer order automaton. Let the system of 10 automata $A=\{A_j \mid j\in J=\{1, 2, ---, 10\}\}$ be described as a system of interactive automata observed over a system moment base $M=\{m+i \mid i\in I,\ m+i=(m_j+i \mid j\in J)\}$; i.e., the environment sets of the automaton components are not disjoint under any pairwise partition. What the observing automaton A_0 observes can be described by the environment graph. As in the example of Section 2.2, the states may not be observed.

The environment graph is shown in Figure 5.5 without point and line identification. By Theorem 5.9 ($Es_a\ i$), every subgraph describing the operation of a component jth automaton is connected, and the graph of the system is a set of connected subgraphs. A subgraph for a component could be a graph of the macroclass, depending on the size and complexity of that activity of the industry; i.e., many automata could be viewed as a single component automaton if the related activity is complex, and its subgraph would be of the macroclass. The loops of the graph are interpreted as waiting conditions corresponding to automata interpreting their own recorded responses as stimuli until the automata with which they interact record responses in their shared environments.

The information gathered on systems may be obtained by computers as observing automata, in which case, large complex systems of interactive automata may be observed. The amount of information required by observation for a given system is dependent on the number of functions defined by formulas, i.e., on the amount that can be calculated. When large amounts of information are stored and organized in computers representing complex

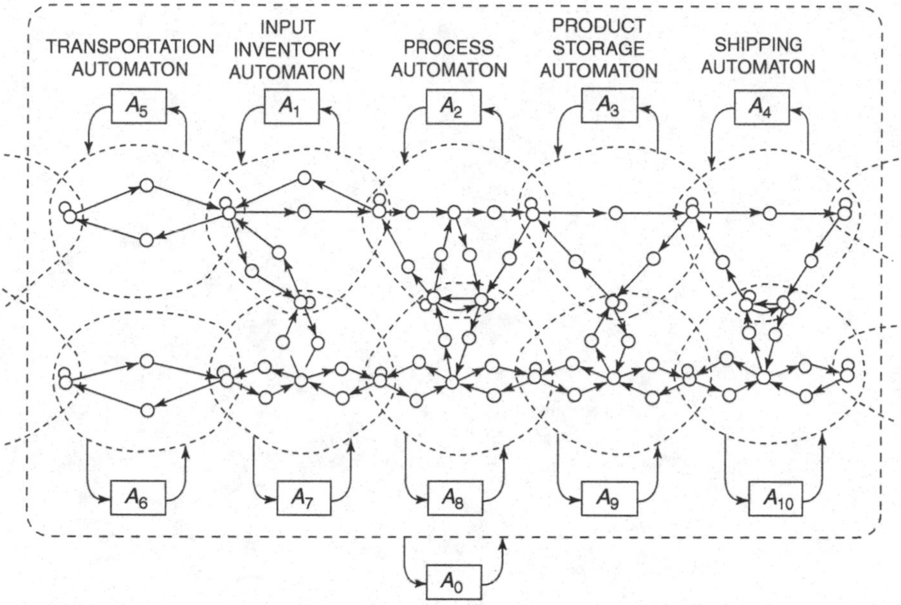

Figure 5.5 Environment graph for an interactive system of 10 automata describing the activity of an industry of Figure 5.4.

graph structures, it is very important to be able to determine effective operations by computer methods. Appendix B presents the development of important recursive methods and graph-theoretic principles for determining the effective operation of a general system of interactive automata.

EXERCISES

5.1 Establish, in detail, the four primitives for the a) principal environment graph of the microsystem model, $G_M(Es_p\, i)$, and b) alternate time graph of the microsystem model, $G_M(Ts_a\, i)$.

5.2 Establish, in detail, the four primitives for the a) principal environment graph of the macrosystem model, $G_M(Es_p\, a)$, and b) alternate time graph of the macrosystem model, $G_M(Ts_a\, a)$.

CHAPTER 6

Processing Knowledge About Systems of Automata

6.1 A GENERAL SYSTEM OF INTERACTIVE AUTOMATA: DETAILED ANALYSIS

In Section 5.1, a formal analysis was made of a general system of interactive automata, and graph primitives were established for the processor, environment, and time graphs for both the microsystem and macrosystem models. A detailed analysis of a general automaton was made in Chapter 3 to make the receptors and effectors distinguishable and to establish transformation response components. A summary of the results of the analysis of a general system of interactive automata is given here when the components of the system are the general automata described in the detail of Chapter 3 (Koenig 1986b, 1987, 1997a).

System Principle. Given a discrete system moment $m+i=(m_j+i \mid j\in J)$, $m+i\in M$, each component automaton A_j is accessing one and only one environment from a set of nonhomogeneous environments $E_j \subseteq R^n$ of the environment space, receiving one and only one stimulus from a set S_j of potential stimuli through one and only one auxiliary receptor and one and only one principal receptor from sets X_j'' and X_j' of receptors, is in one and only one state from a set Q_j^* of possible states, and has one and only one absolute time associated with m_j+1 from a set T of times. Multiple accesses of a same environment during M can be made by different components of the system. Furthermore, during the next system moment $m+(i+1)=(m_j+(i+1) \mid j\in J)$, each component automaton A_j produces one and only one triadic response from a set R_j of possible responses through one and only one principal effector and one and only one auxiliary effector from sets Z_j' and Z_j'' of effectors based on the aforementioned conditions. The triadic response consists of an ordered triple from a set F_j of transformation responses, a set D_j of

Knowledge Structures for Communications in Human-Computer Systems: General Automata-Based, by Eldo C. Koenig

spatial change responses, and a set C_j of time change responses. The pentadic transformation response at $(m_j+(i+1) \mid j \in J)$ consists of the ordered quintuple from sets FS_j, FM_j, FH_j, FT_j, FA_j of responses interpreted as stimuli for sight, smell, hearing, touch, and taste and is recorded as a response from the set B_j of recorded responses at the environment accessed at moment $(m_j+i \mid j \in J)$.

6.1.1 The Microsystem Model

Recall, the microsystem model is useful in viewing a system on the component level while the macrosystem model is useful when a system is viewed as a single automaton. As was the case of the general automaton, the alternate environment graph, principal time graph, and processor graph are chosen for the graph model of a general system of interactive automata for both the microsystem and macrosystem models.

First, define the alternate environment graph of the microsystem model. In summary, the set of four primitives of the jth automaton of a system gives a graph (15+1)-tuple associated with two points (not necessarily distinct) and a connecting line

$$TsE_a i(m_j+i, m_j+i+1)=(e_{ji}, t_{ji}, q^*_{ji}, x''_{ji}, x'_{ji}, z'_{ji}, z''_{ji}, s_{ji}, a_j, b_{j(i+1)}, (fs, fm, fh, ft, fa)_{j(i+1)}, d_{j(i+1)}, c_{j(i+1)}, q^*_{j(i+1)}, t_{j(i+1)}, e_{j(i+1)})$$

such that $\Lambda^{*\prime}_j(e_{ji}, t_{ji}, q^*_{ji}, x''_{ji}, x'_{ji}, z'_{ji}, z''_{ji}, s_{ji})=(b_{j(i+1)}, (fs, fm, fh, ft, fa)_{j(i+1)}, d_{j(i+1)}, c_{j(i+1)}, q^*_{j(i+1)}, t_{j(i+1)}, e_{j(i+1)})$, and the line is directed from e_{ji} of the domain element to $e_{j(i+1)}$ of the image.

The set of primitives defining the principal time graph of the microsystem model also gives a graph (15+1)-tuple associated with two points and a connecting line. It is the same as that of the alternate environment graph except the elements e_{ji} and t_{ji} and the elements $e_{j(i+1)}$ and $t_{j(i+1)}$ are interchanged, and the function $\Lambda^{*\prime}_j$ is replaced with the graph function Λ^*_j.

The set of primitives defining the processor graph of the microsystem model gives a graph 10-tuple associated with two points and a connecting line

$$TsQi(m_j+i, m_j+i+1)=(q^*_{ji}, x''_{ji}, x'_{ji}, z'_{ji}, z''_{ji}, s_{ji}, a_j, (fs, fm, fh, ft, fa)_{j(i+1)}, d_{j(i+1)}, c_{j(i+1)}, q^*_{j(i+1)})$$

such that $\Omega^*_j(q^*_{ji}, x''_{ji}, x'_{ji}, z'_{ji}, z''_{ji}, s_{ji})=((fs, fm, fh, ft, fa)_{j(i+1)}, d_{j(i+1)}, c_{j(i+1)}, q^*_{j(i+1)})$, and the line is directed from q^*_{ji} of the domain element to $q^*_{j(i+1)}$ of the image.

6.1.2 The Macrosystem Model

The macrosystem model is useful when a system is viewed as a single automaton. For any element u of the graph $(n+1)$-tuple for the macrosystem model, $u=(u_j \mid j \in J)$, $J=\{1, 2, ---, k\}$; i.e., the system element u is in the form of a k-tuple.

First, define the alternate environment graph of the macrosystem model. In summary, the set of four primitives of the system of k-automata gives a graph (15+1)-tuple associated with two points (not necessarily distinct) and a connecting line

$$TsE_a a(m+i, m+i+1) = (e_i, t_i, q_i^*, x_i'', x_i', z_i', z_i'', s_i, a, b_{i+1}, (fs, fm, fh, ft, fa)_{i+1}, d_{i+1}, c_{i+1}, q_{i+1}^*, t_{i+1}, e_{i+1})$$

such that $\Lambda^{*\prime}(e_i, t_i, q_i^*, x_i'', x_i', z_i', z_i'', s_i) = (b_{i+1}, (fs, fm, fh, ft, fa)_{i+1}, d_{i+1}, c_{i+1}, q_{i+1}^*, t_{i+1}, e_{i+1})$, and the line is directed from e_i of the domain element to e_{i+1} of the image.

The principal time graph of the macrosystem model is the same as that of the alternate environment graph except the elements e_i and t_i and the elements e_{i+1} and t_{i+1} are interchanged, and the function $\Lambda^{*\prime}$ is replaced with the graph function Λ^*.

The set of primitives defining the processor graph of the macrosystem model gives a graph 10-tuple associated with two points and a connecting line

$$TsQa(m+i, m+i+1) = (q_i^*, x_i'', x_i', z_i', z_i'', s_i, a, (fs, fm, fh, ft, fa)_{i+1}, d_{i+1}, c_{i+1}, q_{i+1}^*)$$

such that $\Omega^*(q_i^*, x_i'', x_i', z_i', z_i'', s_i) = ((fs, fm, fh, ft, fa)_{i+1}, d_{i+1}, c_{i+1}, q_{i+1}^*)$, and the line is directed from q^*_i of the domain element to q_{i+1}^* of the image.

6.2 KNOWLEDGE STRUCTURES FOR SENTENCES DESCRIBING SYSTEMS OF INTERACTIVE AUTOMATA

Interactions between automata over some system moment base can be very complex. Knowledge structures for human automata who have had experience in the interactions must reflect these complexities, and a sequence of sentences that describes the meaning of the knowledge structures may be difficult to construct. Thus, the one who says

"John and Jim played a game of chess."

is likely to expect the receiver of this knowledge to have played the game and to associate past knowledge obtained by experience. *Democracy*, *socialism*, *academic freedom*, *manufacturing company*, all suggest types of interactive behavior among the automata components that may require volumes of printed material to describe.

Consider some of the basic interactive behaviors that make up the above more complex behaviors. An example of a simple interaction is described by the sentence

"John gave a book to Jim."

The equivalent sentence is

"Jim received a book from John."

Two knowledge structures sharing a single point are required to store the meaning of either of the two sentences. A first, consisting of two points and an adjacent line, is required for the giver, and a second, also consisting of two points and an adjacent line, is required by the receiver. An environment is shared by the two-component automaton of the system, which is required if one (the receiver) is to interpret the recorded response of the other (the giver), i.e., if the definition of a system of interactive automata is to be satisfied. Referring to the primitives of Section 5.1.2, the two tuples for the alternate environment graph become

$$TsE_ai(m_j+i+1, m_j+i+2)=(e_{j(i+1)}=(\text{-, -, -}, E'_{j(i+1)}),$$
$$\text{-, -, -, -, -, -, -}, a_j=\text{John}, b_{j(i+2)}=+\text{book}, f_{j(i+2)}=(\text{gave, -, -, -, -}),$$
$$\text{-, -, -}, t_{j(i+2)}, e_{j(i+2)}=(\text{-, -, -, -}))$$

$$TsE_ai(m_h+k+1, m_h+k+2)=(e_{h(k+1)}=(\text{book, -, -}, E'_{h(k+1)}=E'_{j(i+1)}),$$
$$\text{-, -, -, -, -, -, -}, a_h=\text{Jim}, b_{h(k+2)}=-\text{book}, f_{h(k+2)}=(\text{received, -, -, -, -})$$
$$\text{, -, -, -}, t_{h(k+2)}, e_{h(k+2)}=(\text{-, -, -, -}))$$

Figure 6.1 shows the knowledge structure for the meaning of either of the two example sentences to demonstrate the basic interaction of either *give* or *receive*. For the proper interaction $t_{j(i+2)}<t_{h(k+1)}$. Following the effective operation path through the structure for a_j (following the solid lines of the graph) for the pair of moments (m_j+i+1, m_j+i+2), the knowledge is expressed by the sentence

"John gave a book to Jim."

John must take a book from some environment e_{ji} before he can give the book to Jim, and this knowledge is stored in a third tuple shown in the Figure as

$$TsE_ai(m_j+i, m_j+i+1)=(e_{ji}=(\text{book, -, -, -}), \text{-, -, -, -, -, -, -},$$
$$a_j=\text{John}, b_{j(i+1)}=-\text{book}, f_{j(i+1)}=(\text{took, -, -, -, -}), \text{-, -, -},$$
$$t_{j(i+1)}, e_{j(i+1)}=(\text{-, -, -}, E'_{j(i+1)}))$$

Following the effective operation path through the structure for a_h (following the broken lines of the graph) for the pair of moments (m_h+k+1, m_h+k+2), the knowledge is expressed by the sentence

"Jim received a book from John."

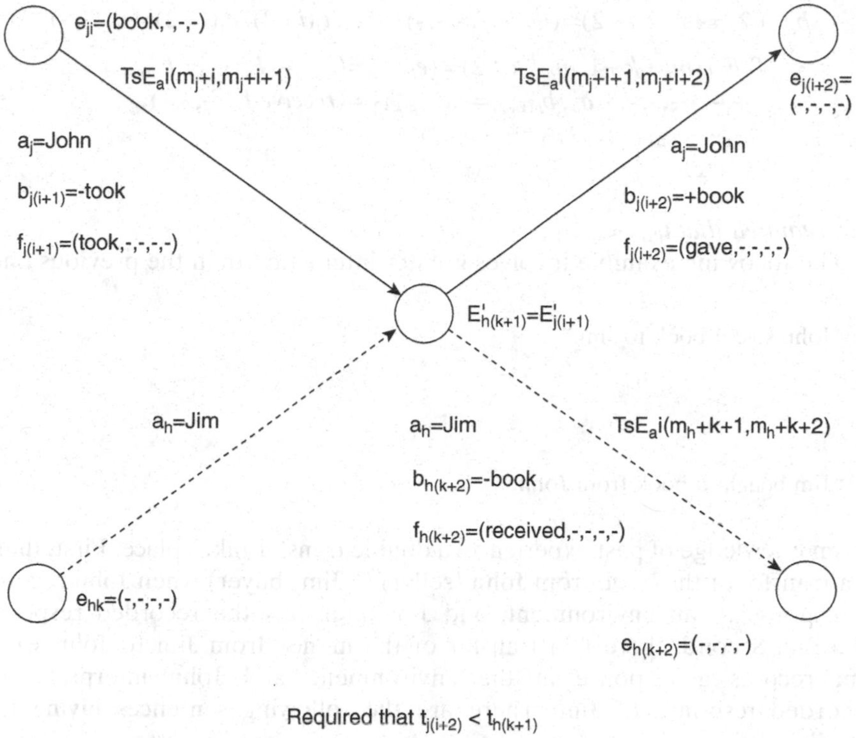

Figure 6.1 Knowledge structure for the meaning of either of the example sentences

"John gave a book to Jim."
"Jim received a book from John."

to demonstrate the basic interaction of either "give" or "receive" ($t_{j(i+2)} < t_{h(k+1)}$).

The above discussion establishes the following theorem (Koenig 1979b):

Theorem 6.1. *Given either of the following sentences:*

"a_j gave x to a_h."
"a_h received x from a_j.

then the tuples for the knowledge structure for storing the meaning are

$TsE_a i(m_j+i, m_j+i+1)=(e_{ji}=(x, \text{-}, \text{-}, \text{-}), \text{-}, \text{-}, \text{-}, \text{-}, \text{-}, \text{-}, \text{-}, a_j, b_{j(i+1)}=-x, f_{j(i+1)}=$ $(took, \text{-}, \text{-}, \text{-}, \text{-}), \text{-}, \text{-}, \text{-}, t_{j(i+1)}, e_{j(i+1)}=(\text{-}, \text{-}, \text{-}, E'_{j(i+1)}))$

$TsE_a i(m_j+i+1, m_j+i+2)=(e_{j(i+1)}=(\text{-}, \text{-}, \text{-},) E'_{j(i+1)}), \text{-}, \text{-}, \text{-}, \text{-}, \text{-}, \text{-}, \text{-}, a_j,$

$$b_j(i+2)=+x,\ f_j(i+2)=(gave, \text{-}, \text{-}, \text{-}, \text{-}), \text{-}, \text{-}, \text{-}, t_j(i+2), e_j(i+2)=(\text{-}, \text{-}, \text{-}, \text{-})$$

$$T_S E_a i(m_h+k+1, m_h+k+2)=(e_{h(k+1)}=(x, \text{-}, \text{-}, E'_{h(k+1)}=E'_{j(i+1)}), \text{-}, \text{-}, \text{-}, \text{-}, \text{-}, \text{-}, \text{-}, a_h, b_{h(k+2)}=-x, f_{h(k+2)}=(received, \text{-}, \text{-}, \text{-}, \text{-}), \text{-}, \text{-}, \text{-}, t_{h(k+2)}, e_{h(k+2)}=(\text{-}, \text{-}, \text{-}, \text{-}))$$

It is required that $t_{j(i+2)}<t_{h(k+1)}$.

The following example involves greater interaction than the previous one:

"John sold a book to Jim."

or

"Jim bought a book from John."

From knowledge of past experience, a double transfer takes place. First, there is a transfer of the book from John (seller) to Jim (buyer) when John records a response at an environment, and Jim interprets the recorded response of John. Second, there is a transfer of the money from Jim to John when Jim records a response at the environment, and John interprets the recorded response of Jim. There are the following sentences giving the detail:

"John took the book from an environment,
and Jim took the money from a different environment."
"John put the book in an environment,
and Jim put the money in the same environment."
"John waited while Jim took the book from the environment."
"Then John took the money from the environment."

The sentences giving detail are examined, and the (15+1)-tuples are established for storing their meanings.

"John took the book from an environment,
and Jim took the money from a different environment."

$$T'sE_a i(m_j+i, m_j+i+1)=(e_{ji}=(\text{book}, \text{-}, \text{-}, \text{-}), \text{-}, \text{-}, \text{-}, \text{-}, \text{-}, \text{-}, \text{-}, a_j=\text{John}, b_{j(i+1)}=-\text{book}, f_{j(i+1)}=(\text{took}, \text{-}, \text{-}, \text{-}, \text{-}), \text{-}, \text{-}, \text{-}, t_{j(i+1)}, e_{j(i+1)}=(\text{-}, \text{-}, \text{-}, E'_{j(i+1)}))$$

$$T'sE_a i(m_h+i, m_h+i+1)=(e_{hi}=(\text{money}, \text{-}, \text{-}, \text{-}), \text{-}, \text{-}, \text{-}, \text{-}, \text{-}, \text{-}, \text{-}, a_h=\text{Jim}, b_{h(i+1)}=-\text{money}, f_{h(i+1)}=(\text{took}, \text{-}, \text{-}, \text{-}, \text{-}), \text{-}, \text{-}, \text{-}, t_{h(i+1)}, e_{h(i+1)}=(\text{-}, \text{-}, \text{-}, E'_{h(i+1)}))$$

"John put the book in an environment,
and Jim put the money in the same environment."

$$T'sE_ai(m_j+i+1, m_j+i+2)=(e_{j(i+1)}=(\text{-, -, -}, E'_{j(i+1)}), \text{-, -, -, -, -, -, -}, a_j=\text{John}, b_{j(i+2)}=+\text{book}, f_{j(i+2)}=(\text{put, -, -, -, -}), \text{-, -, -}, t_{j(i+2)}, e_{j(i+2)}=(\text{-, -, -}, E'_{j(i+2)}))$$

$$T'sE_ai(m_h+i+1, m_h+i+2)=(e_{h(i+1)}=(\text{-, -, -}, E'_{h(i+1)}=E'_{j(i+1)}), \text{-, -, -, -, -, -, -}, a_h=\text{Jim}, b_{h(i+2)}=+\text{money}, f_{h(i+2)}=(\text{put, -, -, -, -}), \text{-, -, -}, t_{h(i+2)}, e_{h(i+2)}=(\text{-, -, -}, E'_{h(i+2)}))$$

The time $t_{j(i+2)}$ must be less than $t_{h(i+2)}$ in order to have the interaction described by the next sentence

"John waited while Jim took the book from the environment."

$$T'sE_ai(m_j+i+2, m_j+i+3)=(e_{j(i+2)}=(\text{-, -, -}, E'_{j(i+2)}), \text{-, -, -, -, -, -, -}, a_j=\text{John}, b_{j(i+3)}=0, f_{j(i+3)}=(\text{waited, -, -, -, -}), d_{j(i+3)}=0, \text{-, -}, t_{j(i+3)}, e_{j(i+3)}=(\text{-, -, -}, E'_{j(i+3)}=E'_{j(i+2)}))$$

$$T'sE_ai(m_h+i+2, m_h+i+3)=(e_{h(i+2)}=((\text{book, -, -}), (\text{money, -, -}), E'_{h(i+2)}=E'_{j(i+1)}), \text{-, -, -, -, -, -, -}, a_h=\text{Jim}, b_{h(i+3)}=-\text{book}, f_{h(i+3)}=(\text{took, -, -, -, -}), \text{-, -, -}, t_{h(i+3)}, e_{h(i+3)}=(\text{-, -, -, -}))$$

Suppose John did not wait, then, for a_h to interpret the recorded response of a_j, $t_{j(i+2)}$ must be less than $t_{h(i+2)}$, and, for a_j to interpret the recorded response of a_h, $t_{h(i+2)}$ must be less than $t_{j(i+2)}$. Thus, there is a conflict. Now consider a_j to be waiting over the period (m_j+i+2, m_j+i+3). For a_h to interpret the recorded response of a_j, $t_{j(i+2)}$ must be less than $t_{h(i+2)}$ as before. Now, with a_j having waited, a_j at $t_{j(i+3)}$ can interpret the response of a_h recorded at $t_{h(i+2)}$, for $t_{h(i+2)}<t_{j(i+3)}$.

Finally, there is the sentence

"John took the money from the environment."

$$T'sE_ai(m_j+i+3, m_j+i+4)=(e_{j(i+3)}=(\text{money, -, -}, E'_{j(i+3)}=E'_{h(i+2)}), \text{-, -, -, -, -, -, -}, a_j=\text{John}, b_{h(i+4)}=-\text{money}, f_{j(i+4)}=(\text{took, -, -, -, -}), \text{-, -, -}, t_{j(i+4)}, e_{j(i+4)}=(\text{-, -, -, -}))$$

Figure 6.2 shows the knowledge structure for the meaning of either of the two example sentences to demonstrate the basic interaction of either **sold** or **bought**. For the proper interaction, $t_{j(i+2)}<t_{h(i+2)}<t_{j(i+3)}$. The loop described by $T'sE_ai(m_j+i+2, m_j+i+3)$ helps to accomplish this.

It should be apparent that the results are the same as the above when **John** and **Jim**, and **money** and **book** are interchanged in the above set of sentences giving detail and in the corresponding tuples. The sentences become

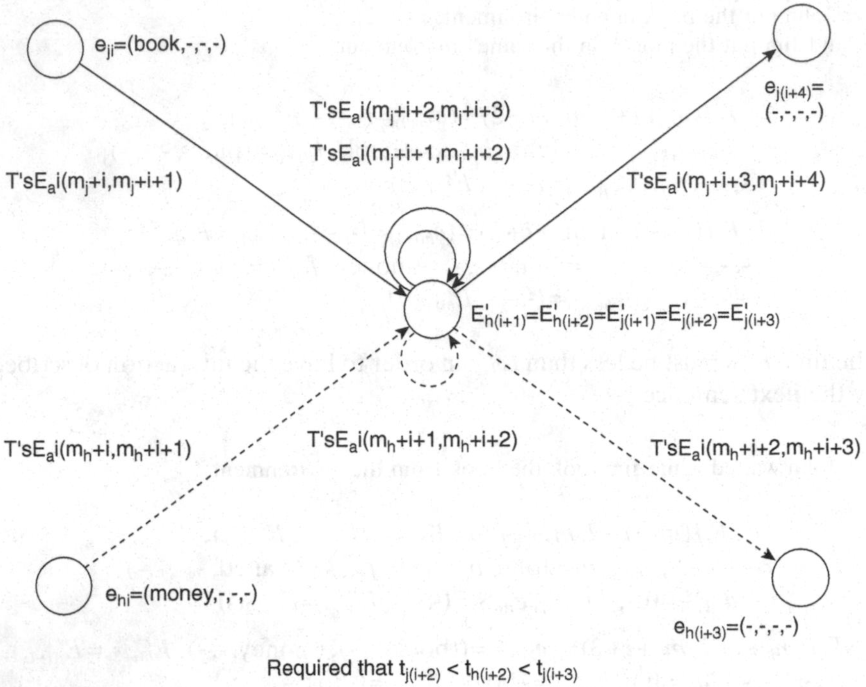

Figure 6.2 Knowledge structure for the meaning of either of the two example sentences

"John sold a book to Jim."
"Jim bought a book from John."

to demonstrate the basic interaction of either "sold" or "bought."

"John sold a book to Jim"
(or "Jim bought a book from John.")

"Jim took the money from an environment,
and John took the book from a different environment."

"Jim put the money in an environment,
and John put the book in the same environment."

"Jim waited while John took the money from the environment.
Then Jim took the book from the environment."

The above discussion establishes the following theorem (Koenig 1979c):

Theorem 6.2. *Given either of the following sentences*:

"*a_j sold x to a_h.*"
"*a_h bought x from a_j.*"

defined in greater detail by the following set of sentences:

"*a_j took x from an environment,*
and a_h took money from a different environment."

"*a_j put x in an environment,*
and a_h put the money in the same environment."

"*a_j waited while a_h took x from the environment.*
Then a_j took the money from the environment."

then the following (15+1)-*tuples for a Knowledge Structure stores the meaning*

"*a_j took x from an environment,*
and a_h took money from a different environment."

$$T'sE_ai(m_j+i, m_j+i+1)=(e_{ji}=(x, -, -, -), -, -, -, -, -, -, -, a_j, b_{j(i+1)}=-x, f_{j(i+1)}=(took, -, -, -, -), -, -, -, t_{j(i+1)}, e_{j(i+1)}=(-, -, -, E'_{j(i+1)}))$$

$$T'sE_ai(m_h+i, m_h+i+1)=(e_{hi}=(money, -, -, -), -, -, -, -, -, -, -, a_h, b_{h(i+1)}=-money, f_{h(i+1)}=(took, -, -, -, -), -, -, -, t_{h(i+1)}, e_{h(i+1)}=(-, -, -, E'_{h(i+1)}))$$

"*a_j put x in an environment,*
and a_h put the money in the same environment."

$$T'sE_ai(m_j+i+1, m_j+i+2)=(e_{j(i+1)}=(-, -, -, E'_{j(i+1)}), -, -, -, -, -, -, -, a_j, b_{j(i+2)}=+x, f_{j(i+2)}=(put, -, -, -, -), -, -, -, t_{j(i+2)}, e_{j(i+2)}=(-, -, -, E'_{j(i+2)}))$$

$$T'sE_ai(m_h+i+1, m_h+i+2)=(e_{h(i+1)}=(-, -, -, E'_{h(i+1)}=E'_{j(i+1)}), -, -, -, -, -, -, -, a_h, b_{h(i+2)}=+money, f_{h(i+2)}=(put, -, -, -, -), -, -, -, t_{h(i+2)}, e_{h(i+2)}=(-, -, -, E'_{h(i+2)}))$$

"*a_j waited while a_h took x from the environment.*"

$$T'sE_ai(m_j+i+2, m_j+i+3)=(e_{j(i+2)}=(-, -, -, E'_{j(i+2)}), -, -, -, -, -, -, -, a_j, b_{j(i+3)}=0, f_{j(i+3)}=(waited, -, -, -, -), d_{j(i+3)}=0, -, -, t_{j(i+3)}, e_{j(i+3)}=(-, -, -, E'_{j(i+3)}=E'_{j(i+2)}))$$

$$T'sE_ai(m_h+i+2, m_h+i+3)=(e_{h(i+2)}=((x, -, -), (money, -, -), E'_{h(i+2)}=E'_{j(i+1)}), -, -, -, -, -, -, -, a_h, b_{h(i+3)}=-x, f_{h(i+3)}=(took, -, -, -, -), -, -, -, t_{h(i+3)}, e_{h(i+3)}=(-, -, -, -))$$

"*Then a_j took the money from the environment.*"

$$T'sE_ai(m_j+i+3, m_j+i+4)=(e_{j(i+3)}=(money, -, -, E'_{j(i+3)}=E'_{h(i+2)}), -, -, -, -, -, -, -, a_j, b_{j(i+4)}=-money, f_{j(i+4)}=(took, -, -, -, -), -, -, -, t_{j(i+4)}, e_{j(i+4)}=(-, -, -, -))$$

It is required that $t_{j(i+2)}<t_{h(i+2)}<t_{j(i+3)}$. For a_j and a_h, and x and money interchanged in the set of sentences giving detail, the same interchanges made

in the (15+1)-*tuples also describe a knowledge structure for storing the meaning.*

Storing the meanings of sentences involving quantifiers, *all*, *some*, and group words like **they**, **the club**, can be handled by the graph macromodels. For example, the sentence

"They struck their tables with hammers."

or

"All struck their tables with hammers."

means the knowledge structure is equivalent to that for a single automaton. If by **they** is meant two automata a_j and a_h, then

$$TsE_a a(m+i, m+i+1)=(((\text{table}, -, -, -)_j, (\text{table}, -, -, -)_h), (-, -), (-, -), (-, -), (-, -), (-, -), (\text{hammer}_j, \text{hammer}_h), (-, -), (a_j, a_h), (-, -), (((\text{struck}, -, -, -, -)_j, (\text{struck}, -, -, -, -)_h), (-, -), (-, -), (-, -), (-, -), ((-, -, -, -), (-, -, -, -)))$$

where m is the system moment.

Now, suppose the above sentence was modified to read

"They struck the table with a hammer."

The analysis is an exercise at the end of this chapter.

EXERCISES

6.1 Discuss the conditions for interaction when $t_{h(i+1)} < t_{j(i+1)}$ for the example, "John sold a book to Jim," and draw the corresponding graph of Figure 6.2.

6.2 Review Figures 6.1 and 6.2, and draw the alternate environment graph that describes the meaning of the sentence, "Jim bought a book from John to give to Jane."

6.3 Discuss the conditions when the example sentence of Section 6.2, "They struck their tables with hammers," is changed to, "They struck the table with a hammer," and write the tuple for the alternate environment graph of the macromodel.

CHAPTER 7

Changing Expressions of Knowledge for Communication from One Form and Style to Another

7.1 INTRODUCTION

Important needs exist for electronic translations. Clearly, there is an important need for electronic translations of messages on the World-Wide Web when interactions of the users take place in different languages. For interactions in the conversational mode, the speed of translation requires special consideration (Bulkeley 1998). Published results of electronic translations suggest that much remains to be done in achieving satisfactory translations. Translations of some of the literary works appear to receive the most criticism (Bulkeley 1998, Immergut 1996).

The effort here addresses problems of electronic translation and establishes computer procedures that are general purpose and flexible in application. For example, literary forms and styles of languages can be treated as separate from the common forms and styles, and optional amounts of expert human participation can be engaged in the preparation of data files that are addressed by the procedures.

An expression of knowledge for communication in the most fundamental sense is considered to be a sequence of stimuli of the senses of the knowledge receiver, such as a sequence of vocal sounds, gestures, signs, signals, or the written symbols for them, and the sequences of stimuli are sequences of knowledge elements. Thus, an expression of knowledge for communication is considered a sequence of knowledge elements.

The changing of expressions of knowledge for communication from one form and style (fs) to another requires that, after the changes, the expressions in the one fs communicate the same (or nearly the same) knowledge as the

Knowledge Structures for Communications in Human-Computer Systems: General Automata-Based, by Eldo C. Koenig

expressions in the other fs. For a given single expression of knowledge in one fs, the knowledge communicated by that expression is said to be communicated by a set of one or more expressions in another fs. For example, a knowledge expression (k-exp), (He, buys, umbrella), in the language of English is changed to a k-exp, (Er, kaupt, Regenrehirm), in the language of German (Rosenberg 1962), and the two expressions are accepted as conveying the same knowledge.

The effort here utilizes open expressions of knowledge for communication. The variables of the open expressions are the variables of the knowledge base under consideration. That is, the variables of the knowledge base categorize the stimuli of the senses of the knowledge receiver. For example, words of a human language as stimuli of the senses of the knowledge receiver are categorized by the variables related to the knowledge base. For this example, each expression of knowledge represents the knowledge communicated by one type of sentence for that language. A particular knowledge base may not require as large a number of variables as another, and the set of values of each variable may not be as large.

For the above example k-exp (He, buys, umbrella), the open expression (o-exp) of three variables is established to be say, (a, f, e). For a particular knowledge base, only people may be categorized by the variable a and used as values for a. The variable f may be limited in values to either buy or sell, and variable e may be limited in values to consumer products. Only certain combinations of three values will yield expressions that convey knowledge.

Expressions of knowledge for communication are said to be changed from a fs, α, to any other fs, β. A single expression of knowledge in α is considered to be a sequence of knowledge elements, $(u1, u2, \text{---}, un)_\alpha$, n=1,2,---, and is to be changed to a set of one or more expressions in β. Each expression of the set in β is considered to be a sequence of knowledge elements in β, such that the knowledge contained in the set of expressions in β is taken to be the same knowledge as that of the single expression in α.

A knowledge element (k-ele) is classified as either a basic k-ele or a supporting k-ele. The basic k-eles are considered the elements of the basic sets of a knowledge base, and the supporting k-eles are considered the elements of the supporting sets of the knowledge base. And these supporting elements are said to support the basic elements. For the above example k-exp, (He, buys, umbrella), a supporting k-ele that supports the basic k-ele, umbrella, could be the word **folding**.

Thirteen sets of basic k-eles and associated sets of supporting k-eles describe a knowledge base for the communication of knowledge about automata and histories of automata (Section 4.1). Included in the sets of supporting k-eles are descriptive and relationship sets (Koenig 1997a). The results of previous work on general automata will be followed in illustrating procedures for changing expressions of knowledge from one fs to another and will provide a means for including other facets of a broad based knowledge system.

7.2 SETS AND RELATIONS

Let U be a family of sets of k-eles with members $Uj, j=1,2,\text{--}, m$. And let $ui \in Ui$, $i=1,2,\text{--},n$, be the k-eles of an expression, $(u1, u2, \text{--}, un)$. The family of 13 sets of basic k-eles for communicating knowledge about automata and histories of automata is

$$U=\{A, E, T, Q, X2, X1, Z1, Z2, S, B, F, D, C\}.$$

The members are defined by describing the operation of a general automaton (Sections 2.1, 3.1). At a discrete moment $m+i \in M$, at time $t_i \in T$, a general automaton $a \in A$ is accessing one and only one environment $e_i \in E$, receiving one and only one stimulus $s_i \in S$, through one and only one auxiliary sensor $x2_i \in X2$, and one and only one principal sensor $x1_i \in X1$, and is in one and only one state $q_i \in Q$. Multiple accesses of a same environment during a period of consideration can be made only by the subject automaton. (For a system of interacting automata, multiple accesses of a same environment can be made by different component automata of the system.) Furthermore, during the next moment $m+i+1 \in M$, at the time $t_{i+1} \in T$, the automaton a produces one and only one response $r_{i+1} \in R$ through one and only one principal effector $z1_i \in Z1$ and one and only one auxiliary effector $z2_i \in Z2$ based on the aforementioned conditions. The response r_{i+1} consists of a transformation response $f_{i+1} \in F$, a spatial change response $d_{i+1} \in D$ and a time change response $c_{i+1} \in C$. The transformation response is recorded as a response $b_{i+1} \in B$ at the environment e_i accessed at the first moment $m+i$. At moment $m+i+1$, the automaton a is in state $q_{i+1} \in Q$.

Upon completion of the analysis and mathematical modeling, an environment graph is established. A line of the graph is directed from point e_i to point e_{i+1}. The corresponding graph tuple is

$$TE(m+i+1)=((e, t, q, x2, x1, z1, z2, s)_i, a(b, f, d, c, q, t, e)_{i+1}).$$

This tuple is considered a basic graph tuple and identifies all the basic knowledge elements.

The basic graph tuple is extended for including knowledge elements that support the environment and transformation response (Section 4.1)

$$(((e, ya, yn, E1), t, q, x2, x1, z1, z2, s)_i, a(b(f1, fad), d, c, q, t(e, ya, yn, E1))_{i+1}),$$

where

$ya \in Ya$ describes $e \in E$ by general appearance; e.g., (car, red, -, -);

$yn \in Yn$ describes $e \in E$ by function; e.g., (car, -, sports, -);

$E1 \in E$ is the environment of which $e \in E$ is a part; e.g., (car, -, -, garage);

$fad \in FAD$ describes $f1 \in F1$; e.g., (strikes, forcefully), and $F \subseteq F1 \times FAD$.

The family of sets of automaton k-eles will include the sets of supporting k-eles, *Ya*, *Yn*, *FAD*. Example k-eles of some of the automaton sets are given in Table 7.1 in two fss—English, and German designated by α and β, respectively (Koenig 1986a).

***Relation* Rn.** For $(u1, u2, \text{--}, un) \in Rn \subseteq U1 \times U2 \times\text{---}\times Un$, the expression $(u1, u2, \text{--}, un)$ is an element of the relation *Rn* whenever the expression is accepted as communicating knowledge. Then the expression, $(u1, u2, \text{--}, un) \in Rn$ is called a k-exp. Consider the example expression, (He, buys, umbrella), in relation to Table 7.1. He $\in A\alpha$, buys $\in F1\alpha$, umbrella $\in E\alpha$, and (He, buys, um brella) $\in R3 \subseteq A\alpha \times F1\alpha \times E\alpha$. Since the expression is accepted as conveying knowledge, it is an element of *R*3 and is called a k-exp.

***Relation* Rn.** With $ui\alpha \in Ui\alpha$ and $ui\beta \in Ui\beta$, for $(ui\alpha, ui\beta) \in Ru \subseteq Ui\alpha \times Ui\beta$, $(ui\alpha, ui\beta)$ is an element of *Ru* whenever $ui\beta$ is equivalent in meaning to $ui\alpha$. Consider the k-exps (He, buys, umbrella)$_\alpha$ and (Er, kauft, Regenscherm)$_\beta$. With, for example, He $\in A\alpha$ as $u1\alpha \in U1\alpha$ and Er $\in A\beta$ as $u1\beta \in U1\beta$, (He, Er) is an element of *Ru* since **Er** is equivalent in meaning to **He**.

Note also in the table that $F1Ya\alpha$ is a subset of $F1\alpha \times Ya\alpha$, and that $ui\alpha \in F1Ya\alpha$ has the same meaning as $ui\beta \in F1Ya\beta$. Similarly, $F1E\alpha$ is a subset of $F1\alpha \times E\alpha$, and $ui\alpha \in F1E\alpha$ has the same meaning as $ui\beta \in F1E\beta$. That is, the meaning of a single element of a set in β may have the same meaning as a combination of elements from more than one set in α. For example, **klingelt** as an element of $F1E\beta$ has the same meaning as the combination of elements rings $\in F1\alpha$ and bell $\in E\alpha$ which appears as a corresponding dual element, **rings bell**, of $F1E\alpha$.

***Relation* RUv.** Let *V* be a set of variables, $vj \in V$, $j=1,2,\text{--},m$. For $(Uj, vj) \in RUv \subseteq \mathsf{U} \times V$, (Uj, vj) is an element of the relation *RUv* whenever variable *vj* takes on an element of *Uj* as a value. For an illustration, let the basic automaton sets *A*, *E*, *T*, *Q*, *X*2, *X*1, *Z*1, *Z*2, *S*, *B*, *F*, *D*, *C* be the member sets

TABLE 7.1. Example K-eles of Automaton Sets in English and in German

$A\alpha$={we, He, teacher, Mr. Brown, ---}
$A\beta$={wer, Er, Lehrer, Herr Brown, ---}
$F1E\alpha$={rings bell, ---}
$F1E\beta$={klingelt, ---}
$E\alpha$={parcel, Berlin, umbrella, book, bell, bottle, table, ---}
$E\beta$={Packet, Berlin, Regenscherm, Buch, __, Flasche, Tische, ---}
$F1\alpha$={send, buys, gives, puts, rings, ---}
$F1\beta$={senden, kauft, gebt, __, __, ---}
$Ya\alpha$={upright position, horizontal position, ---}
$Ya\beta$={__, __, ---}
$F1Ya\alpha$={puts upright position, puts horizontal position, ---}
$F1Ya\beta$={stellt, legt, ---}

Uj, $j=1,2,\text{--},13$ of the family U, and let the set of variables $V=\{a,e,t,q,x2,x1,z1,z2,s,b,f,d,c\}$. For $j=2$, E is $U2$ and e is $v2$. (E, e) is an element of the relation RUv whenever the variable e takes on an element of the set E as a value.

***Relation* Ruv.** For $(ui, vi)\in Ruv\subseteq Ui\times V$, (ui, vi) is an element of Ruv whenever variable vi in an o-exp takes on a value ui as an element of the set Ui. Consider (He, buys, umbrella) as an example of a k-exp for $(u1, u2, u3)$, and consider $(a, f1, e)$ as an example of an o-exp for $(v1, v2, v3)$. For i=1, **He** is $u1$ in the k-exp, and a is $v1$ in the o-exp. (He, a) is an element of the relation Ruv since the variable $v1$ in the o-exp takes on a value $u1$ which is an element of the set $U1$.

***Relation* RU.** Knowledge communicated by a k-exp in α, $(u1, u2, \text{--}, un)_\alpha$ is to be communicated by a set of k-exps in β. Let Uα be a family of sets of k-eles in α with members $Uj\alpha, j=1,2,\text{--},m$. Let Uβ be a family of sets of k-eles in β with members $Uj\beta$. For $(Uj\alpha, Uj\beta)\in RU\subseteq U\alpha\times U\beta$, $(Uj\alpha, Uj\beta)$ is an element of RU whenever $Uj\beta$ is equivalent in meaning to $Uj\alpha$. Consider the automaton sets as member sets of Uα and Uβ. For example, for $A\alpha$ as $U1\alpha$ and $A\beta$ as $U1\beta$, $(A\alpha, A\beta)$ is an element of RU since $A\beta$ is made equivalent in meaning to $A\alpha$.

Relation Rv. With $vi\alpha\in V\alpha$ and $vi\beta\in V\beta$, for $(vi\alpha, vi\beta)\in Rv\subseteq V\alpha\times V\beta$, $(vi\alpha, vi\beta)$ is an element of Rv whenever $vi\beta$ is equivalent in meaning to $vi\alpha$. Consider, for example, $a\alpha\in V\alpha$ and $a\beta\in V\beta$. $(a\alpha, a\beta)$ is an element of Rv whenever $a\beta$ is equivalent in meaning to $a\alpha$.

7.3 ESTABLISHING OPEN EXPRESSIONS AND OPEN SENTENCES

Given, a family of sets of k-eles, U, whose member sets are Uj. Also given is a k-exp, $(u1, u2, \text{--}, un)\in Rn$, whose fs and knowledge base are the same as those of the sets Uj.

Required is an o-exp such that when its variables, vi, are provided with values, $ui\in Ui$, the resulting k-exp will communicate the same knowledge as the given k-exp.

For the given k-exp, $(u1, u2, \text{--}, un)\in Rn$, the k-eles ui are found to be elements of the member sets Ui, $ui\in Ui$, so that

$$(u1, u2, \text{--}, un)\in Rn\subseteq U1\times U2\times\text{---}\times Un.$$

The variables of an o-exp identify with member sets Ui, i.e., $(Ui, vi)\in RUv$. And values for vi are selected elements ui from Ui, i.e., $(ui, vi)\in Ruv$. Therefore, the o-exp that yields the k-exp, $(u1, u2, \text{--}, un)$, is $(v1, v2, \text{--}, vn)$. This established o-exp is added to a set of o-exps, OE, if it is not already an element.

Consider, for example, a k-exp, ($k1$, $k3$, $k2$, $k4$, $k5$, $k3$) $\in R6$, which is to provide an o-exp as an element of OE. The k-eles ki are found to be elements of the member sets Ui, $ki \in Ui$, so that

$$(k1, k3, k2, k4, k5, k3) \in R6 \subseteq U1 \times U3 \times U2 \times U4 \times U5 \times U3.$$

The o-exp that yields the k-exp, ($k1$, $k3$, $k2$, $k4$, $k5$, $k3$), is ($v1$, $v3$, $v2$, $v4$, $v5$, $v3$), which is added to the set OE if not already an element.

Consider the automaton sets as member sets of the family U in the fs, α, and establish an open expression for a knowledge sentence (k-sent). Given the sets of k-eles in α in Table 7.1 and the k-sent,

"He buys an umbrella."

From the above family of automaton sets, it is found that He $\in A\alpha$, buys $\in F1\alpha$, umbrella $\in E\alpha$. Then the triadic relation for this type of sentence is

$$R3\alpha \subseteq (A \times F1 \times E)_\alpha,$$

and (He, buys, umbrella) $\in R3\alpha$. The corresponding variable tuple that stores the knowledge of this type of sentence is

$$(((e,-,-,-),-,-,-,-,-,-,-)_i, a(-(f1,-),-,-,-,-,(-,-,-,))_{i+1}).$$

The symbol,-, indicates unknown values for the variable. The knowledge tuple (k-tuple) containing the knowledge of the given sentence is

$$(((\text{umbrella},-,-,-),-,-,-,-,-,-,-)_i, \text{He}, (-, (\text{buys},-),-,-,-,-,(-,-,-,-))_{i+1}).$$

The o-exp is

$$(a, f1, e),$$

and the sequence of variables with the filler words of the given k-exp inserted gives the o-sent,

$$a\,f1(\text{an})e.$$

The filler word (an) may be replaced by (a) when the variables take on other values to conform with the rules of grammar. A three-part result of the analysis of this example k-exp is summarized:

1. $(A \times F1 \times E)_\alpha$
2. $(a, f1, e)$
3. $a\,f1$ (an) e.

This three-part result becomes an entry in a three-part list if it is not already in the list.

The three parts of an entry in a list for any fs of k-exp have the following general descriptions with regard to their functions:

1. a product of k-ele sets for matching that of a k-exp of a given k-sent,
2. an o-exp whose variables take on k-eles as values,
3. an o-sent consisting of a sequence of variables with intermingled filler words for communicating knowledge when the variables take on values; (the filler words are subject to change depending on the set of values given the variables).

Table 7.2 gives a partial three-part list in English and Table 7.3 gives a partial three-part list in German.

In regard to number ***n*+2** of the list in English, if two different environment values appear in a given k-exp, a distinction between the two is maintained by labeling one of the two *E* sets of the product set, *E*2. The corresponding variable is labeled *e*2.

In regard to number ***n*+3**, if two different automaton values appear twice in a given k-exp (labeled *A* and *A*2 in the product set), interaction is indicated between the two automaton by two variable tuples, one for each of the two automaton variables *a* and *a*2. The distinction between *a* and *a*2 is carried by a single o-exp and o-sent.

TABLE 7.2. A partial three-part list in English

***n*+1** $A \times F1 \times E$
(*a*, *fl*, *e*)
a fl (an) *e*.
***n*+2** $A \times F1 \times E \times E2$
(*a*, *fl*, *e*, *e2*)
a fl (a) *e* (to) *e2*.
***n*+3** $A \times F1 \times A2 \times E$
(*a*, *fl*, *a2*, *e*)
a fl (the) *a2* (a) *e*.
***n*+4** $A \times F1 \times E \times Ya \times E2$ or $A \times F1 \times E \times E2 \times Ya$
(*a*, *fl*, *e*, *ya*, *e2*)
a fl (the) *e* (in an) *ya* (on the) *e2*
(*a*, *fl*, *e*, *e2*, *ya*)
a fl (the) *e* (on the) *e2* (in an) *ya*.
***n*+5** *A*
(*a*)
(There are) *a*
***n*+6** $FAD \times F1$
(*fad*, *fl*)
(There is a) *fad fl*

TABLE 7.3. A partial three-part list in German

m+1 $A \times F1E$
$(a, f1e)$
$a\ f1e.$
m+2 $A \times F1 \times E$
$(a, f1, e)$
a f1 (einen) *e*.
m+3 $A \times F1 \times E2 \times E$
$(a, f1, e2, e)$
a f1 e2 (ein) *e*
m+4 $A \times F1Ya \times E \times E2$
$(a, f1ya, e, e2)$
a f1ya (die) *e e2*
m+5 $A \times F1 \times A2 \times E$
$(a, f1, a2, e)$
a f1 (dem) *a2* (ein) *e*.

For number ***n*+4** of the list in English, same sets appear in different sequences in the two product sets. A given k-exp can be an element of either of the two product sets. The two o-exps and o-sents, when given the same set of values, will yield k-exps and sentences conveying the same knowledge.

In regard to ***m*+1** of the list in German, the value of $f1e \in F1E$ of a k-exp is a composite of the value $f1 \in F1$ and the value $e \in E$. The composite variable *f1e* appears in the o-exp and o-sent. Examples of composite values previously presented are the elements of automaton sets $F1E\alpha$ and $F1E\beta$:

$$(\text{rings bell, klingelt}) \in Ru \subseteq F1E\alpha \times F1E\beta.$$

That is, **rings bell** in α conveys the same knowledge as **klingelt** in β.

For ***m*+4** of the list in German, the explanation of set F1*Ya* is similar to that of *F1E* of ***m*+1**.

For ***m*+3**, ***m*+4**, ***m*+5** of the list in German, the discussions of ***n*+2**, ***n*+3**, of the list in English apply.

The three-part list illustrated by Tables 7.2 and 7.3 has utility in achieving other facets of knowledge based systems when the systems perform under the principles of general automata. These other facets include the following:

a. extracting and storing the knowledge of sentences (Sections 1.2, 4.2),
b. associating knowledge (Sections 1.2, 4.3),
c. communicating stored knowledge (generating sentences) (Koenig 1997a),
d. obtaining effective operation paths in knowledge structures (for finding and sequencing knowledge) (Appendixes A, B),
e. establishing intelligent-like features (reasoning, generating inferences) (Sections 4.4, 4.5).

7.4 SELECTING SUBSETS OF OPEN EXPRESSIONS

Given, an o-exp in α which was established from a single k-exp in α. Also given, an established set OEβ of o-exps in β that yield k-exp in β.

Required is a subset of o-exps of OEβ that is equivalent to the given o-exp in α in that equal knowledge may be communicated by the k-exps that they yield. For a selected o-exp from OEβ to qualify, four requirements must be met. The presentations and discussions of the requirements follow. In the discussions, $ui\beta$ and $ui\alpha$ as k-eles are considered to be equivalent in meaning, i.e., $(ui\alpha, ui\beta) \in Ru$. A k-ele $ui\beta$ is an element of a set $Ui\beta$ and is also a value for a variable $vi\beta$. Also, a k-ele $ui\alpha$ is an element of a set $Ui\alpha$ and is also a value for a variable $vi\alpha$. Then, $vi\beta$, $Ui\beta$, and $vi\alpha$, $Ui\alpha$, may also be considered equivalent in meaning, respectively, i.e., $(vi\alpha, vi\beta) \in Rv$ and $(Ui\alpha, Ui\beta) \in RU$. They are treated accordingly in the discussion.

Requirement 1. The set of variables of a selected o-exp in β must be a subset of the set of variables of the o-exp in α that yields a given k-exp in α.

A selected o-exp in β is to yield a k-exp that contains only knowledge contained in a given k-exp in α. This requires that the k-exp provided by a current o-exp in β contains only k-eles contained in the given k-exp in α, i.e., that the selected o-exp in β contains only the variables contained in the o-exp in α that yields the given k-exp in α. Then, the set of variables of a selected o-exp in β is a subset of the set of variables of the o-exp in α.

Requirement (Req) 1 is illustrated by an example: A selected o-exp in β, $(v1, v3, v5, v2)_\beta$ is to yield a k-exp that contains only knowledge contained in a given k-exp in α, $(k1, k3, k2, k4, k5, k3)_\alpha$. This requires that the k-exp, $(k1, k3, k5, k2)_\beta$ provided by the currently selected o-exp in β contain only k-eles contained in $(k1, k3, k2, k4, k5, k3)_\alpha$, i.e., that the selected o-exp, $(v1, v3, v5, v2)_\beta$ contain only the variables contained in the o-exp in α, $(v1, v3, v2, v4, v5, v3)_\alpha$. And Req 1 is satisfied:

$$\{v1, v2, v3, v5\}_\beta \subseteq \{v1, v2, v3, v4, v5\}_\alpha.$$

Requirement 2. The set of variables of a currently selected o-exp in β must not be a subset of the union of the sets of variables of all previously selected o-exps in β.

A currently selected o-exp in β is to provide for additional knowledge beyond what is provided by previously selected o-exps in β. This requires that the k-exp provided by the currently selected o-exp contain some k-eles not contained in the k-exps provided by previously selected o-exps in β. Then, the set of variables of the currently selected o-exp cannot be a subset of the union of the sets of variables of all previously selected o-exps in β.

The example introduced for illustrating Req 1 will be extended for illustrating Req 2. A currently selected o-exp, $(v1, v4, v5)_\beta$ is to provide for additional knowledge in β beyond what is provided by a previously selected o-exp, which is $(v1, v3, v5, v2)_\beta$. The k-exp, $(k1, k4, k5)_\beta$, provided by $(v1, v4, v5)_\beta$ contains a k-ele, $k4\beta$, which is not contained in the k-exp, $(k1, k3, k5, k2)_\beta$. Then, $\{v1, v4, v5\}_\beta \not\subseteq \{v1, v3, v5, v2\}_\beta$, and Req 2 is satisfied.

Requirement 3. The intersection of the set of variables of the currently selected o-exp in β and the union of the sets of variables of all previously selected o-exps in β must not be empty. This Req 3 does not apply to the first selection.

A currently selected o-exp in β is to provide knowledge linked to knowledge provided by previously selected o-exps in β. This requires that the k-exp, provided by the currently selected o-exp, contain k-eles that are also contained in the k-exps, provided by the previously selected o-exps in β. Then, the intersection of the set of variables of the currently selected o-exp in β, and the union of the sets of variables of all previously selected o-exps in β cannot be empty.

It will be determined whether the selected set of o-exps of the example illustrating Reqs 1 and 2 satisfies Req 3. The set of variables of the currently selected o-exp in β is $\{v1, v4, v5\}_\beta$. The union of the sets of variables of all previously selected o-exps in β is $\{v1, v3, v5, v2\}_\beta$. The intersection of the set of variables of the currently selected o-exp in β and the union of the sets of variables of all previously selected o-exps in β is not empty:

$$\{v1, v4, v5\}_\beta \cap \{v1, v3, v5, v2\}_\beta = \{v1, v5\}_\beta .$$

And the currently selected o-exp in β satisfies Req 3.

Requirement 4. The union of the sets of variables of a complete set of selected o-exps in β equals the set of variables of the o-exp in α that yields a given k-exp in α.

A complete set of selected o-exps in β is to yield a set of k-exps that communicates the same knowledge that is communicated by a given k-exp in α. This requires that the k-exps provided by the complete set of selected o-exps in β contain the same k-eles that are contained in the given k-exp in α, i.e., that the complete set of o-exps in β contain the same variables that are contained in the o-exp in α. Then, the union of the sets of variables of all the selected o-exps in β equals the set of variables of the o-exp in α, and the set of selected o-exps in β is complete.

The example that was introduced to illustrate the previous requirements serves also to illustrate Req 4. The union of the sets of variables of all the o-exps in β is:

$$\{v1, v4, v5\}_\beta \cup \{v1, v3, v5, v2\}_\beta = \{v1, v2, v3, v4, v5\}_\beta .$$

And the set equals the set of variables of the o-exp in α that yields the given k-exp in α:

$$\{v1, v2, v3, v4, v5\}_\beta = \{v1, v2, v3, v4, v5\}_\alpha .$$

Then, the set of two selected o-exps in β is a complete set, and no further selections of o-exps in β are required. That is to say, with $(ui\alpha, ui\beta\} \in Ru$ and $(Ui\alpha, Ui\beta) \in RU$, the resulting set of k-exps,

$$\{(k1, k4, k5), (k1, k3, k5, k2)\}_\beta .$$

communicates knowledge in the fs of β equivalent to the knowledge communicated by the given k-exp,

$$(k1, k3, k2, k4, k5, k3)_\alpha$$

in the fs of α.

7.5 APPLYING THE RESULTS OF THE ABOVE ANALYSIS

The results of the above analysis will be applied in achieving the objective of changing expressions of knowledge for communication from one fs to another.

Given:

a. a k-exp in α. $(u1, u2, \text{--}, un)_\alpha \in Rn\alpha$, $n=1,2,\text{--}$,
b. a family of sets of k-eles in α, Uα, with member sets $Uj\alpha$, $j=1,2,\text{--},m$,
c. a family of sets of k-eles in β, Uβ, with member sets $Uj\beta$, and with $(ui\alpha, ui\beta) \in Ru \subseteq Ui\alpha \times Ui\beta$,
d. a set $OE\beta$ of o-exps in β.

To achieve the objective:

1. Find the o-exp in α that yields the given k-exp in α: $(u1, u2, \text{--}, un)_\alpha \in Rn\alpha \subseteq (U1 \times U2 \times \text{—} Un)_\alpha$ (from (a) and (b)). Then, with $(Ui, vi)_\alpha \in RUv\alpha$, the o-exp that yields the given k-exp is $(v1, v2, \text{--}, vn)_\alpha$.
2. Select the o-exps of a set in β from $OE\beta$ of (d) based on the variables of the established o-exp in α of (1) and the four previously established requirements.
3. Give values, $ui\beta$, to the variables, $vi\beta$, of the o-exp of the established set of (2). Values are taken from member sets $Ui\beta$, in accord with $(ui\alpha, ui\beta) \in Ru \subseteq Ui\alpha \times Ui\beta$, (from (c)).

Then, the knowledge communicated by the established set of k-exps in the fs of β is considered to be the knowledge communicated by the given k-exp in the fs of α.

Applying the results of the analysis is demonstrated by expanding on the example previously introduced:

Given:

a. a k-exp, $(k1, k3, k2, k4, k5, k3)_\alpha \in R6$,
b. a family of sets of k-eles in α, Uα, with member sets $Uj\alpha$, $j=1,2,\text{--},m$,
c. a family of sets of k-eles in β, Uβ, with member sets $Uj\beta$ and with $(ki\alpha, ki\beta) \in Ru \subseteq Ui\alpha \times Ui\beta$, $i=1,2,\text{--},6$,
d. a set $OE\beta$ of o-exps in β, $OE\beta=\{\text{---},(v1, v4, v5)_\beta,\text{---},(v1, v3, v5, v2)_\beta,\text{---},(v3, v2)_\beta,\text{---}\}$ with $(Uj, vj)_\beta \in RUv\beta \subseteq U\beta \times V\beta$.

To achieve the objective:

1. Find the o-exp in α that yields the given k-exp in α: $(k1, k3, k2, k4, k5, k3)_\alpha \in R6\alpha \subseteq (U1 \times U3 \times U2 \times U4 \times U5 \times U3)_\alpha$, (from (a) and (b)). Then, with $(Ui, vi)_\alpha \in RUv\alpha$, the o-exp that yields the k-exp is $(v1, v3, v2, v4, v5, v3)_\alpha$.
2. Select the o-exps for a set in β from $OE\beta$ of (d).
   A first random selection of an o-exp is $(v1, v3, v5, v2)_\beta$.
   With $\{v1, v2, v3, v5\}_\beta \subseteq \{v1, v2, v3, v4, v5\}_\alpha$, Req 1 is satisfied.
   With $\{v1, v2, v3, v5\}_\beta \not\subseteq \varnothing_\beta$, Req 2 is satisfied.
   With $\{v1, v2, v3, v5\}_\beta$ a first selection, Req 3 does not apply.
   With $\{v1, v2, v3, v5\}_\beta \neq \{v1, v2, v3, v4, v5\}_\alpha$, Req 4 is not satisfied; that is, this first selection does not complete a set, and another selection is required.
   A second selection of an o-exp is $(v1, v4, v5)_\beta$.
   With $\{v1, v4, v5\}_\beta \subseteq \{v1, v2, v3, v4, v5\}_\alpha$, Req 1 is satisfied.
   With $\{v1, v4, v5\}_\beta \not\subseteq \{v1, v2, v3, v5\}_\beta$, Req 2 is satisfied.
   With $\{v1, v4, v5\}_\beta \cap \{v1, v2, v3, v5\}_\beta \neq \varnothing$, Req 3 is satisfied.
   With $\{v1, v4, v5\}_\beta \cup \{v1, v2, v3, v5\}_\beta = \{v1, v2, v3, v4, v5\}_\alpha$, Req 4 is satisfied, that is, the two o-exps make a complete set of o-exps in β.
3. Give values, $ui\beta$, to the variables, $vi\beta$, of the o-exps of the established set of (2). Values are taken from member sets $Ui\beta$ with $(ui\alpha, ui\beta) \in Ru$. And the given expression of knowledge for communication, $(k1, k3, k2, k4, k5, k3)_\alpha$ in the fs of α, is changed to the set $\{(k1, k3, k5, k2)_\beta, (k1, k4, k5)_\beta\}$ in the fs of β.

The changing of k-exps in one fs, α, to k-exps in another fs, β, is illustrated by three examples. For the first two examples, the k-eles of the k-exps in both α and β are in the languages of humans. For the third example, the k-eles of the k-exps in α are other than in the language of humans, and the k-eles of the k-exps in β are in the language of humans.

For Examples 7.1 and 7.2 that follow, the fs, α, is English, and the fs, β, is German. Sets of general automata of Table 7.1 categorize k-eles of English and German.

Example 7.1

Given:

a. a k-sent in α, "He buys an umbrella.",
b. a family of sets of k-eles in α (from Table 7.1), $U\alpha=\{A, E, F1, Ya, F1Ya, F1E\}_{\alpha}$,
c. a family of sets of k-eles in β (from Table 7.1), $U\beta=\{A, E, F1, Ya, F1Ya, F1E\}_{\beta}$,
d. a set OEβ of o-exps in β (from Table 7.3), $OE\beta=\{(a, f1e), (a, f1, e), (a, f1, e2, e), (a, f1ya, e, e2), (a, f1, a2, e)\}_{\beta}$.

To achieve the objective:

1. Find the o-exp in α that yields the given k-exp in α. (He, buys, umbrella) $\in R3\subseteq(A\times F1\times E)_{\alpha}$, (from (a) and (b)). Then with $(Ui, vi)_{\alpha}\in RUv\alpha$, the o-exp that yields the k-exp is $(a, f1, e)_{\alpha}$.
2. Select an o-exp as an element of a set in β from OEβ of (d). A first random selection of an o-exp is $(a, f1, e)_{\beta}$.
   With $\{a, f1, e\}_{\beta}\subseteq\{a, f1, e\}_{\alpha}$, Req 1 is satisfied.
   With $\{a, f1, e\}_{\beta}\not\subseteq\varnothing_{\beta}$, Req 2 is satisfied.
   With $\{a, f1, e\}_{\beta}$ a first selection, Req 3 does not apply.
   With $\{a, f1, e\}_{\beta}=\{a, f1, e\}_{\alpha}$, Req 4 is satisfied; that is, the single o-exp makes a complete set of o-exps in β.
3. Give values, $ui\beta$, to the variables, $vi\beta$, of the o-exps of the established set of (2). Values are taken from member sets $Ui\beta$ with $(ui\alpha, ui\beta)\in Ru$. And the given expression of knowledge for communication, $(\text{He, buys, umbrella})_{\alpha}$, in the fs, α, is changed to $(\text{Er, kauft, Regenscherm})_{\beta}$, in the fs, β.

Example 7.2

Given:

a. a k-sent in α, "He puts the bottle in an upright position on the table."
b.,c.,d. see Example 7.1.

To achieve the objective:

1. Find the o-exp in α that yields the given k-exp in α. $(\text{He, puts, bottle, upright position, table})_{\alpha}\in R5\alpha\subseteq(A\times F1\times E\times Ya\times E2)_{\alpha}$.
   Then with $(Ui, vi)_{\alpha}\in RUv\alpha$, $(a, f1, e, ya, e2)_{\alpha}$.

2. Select an o-exp as an element of a set in β from OEβ of (d). A first random selection of an o-exp is $(a, f1ya, e, e2)_\beta$; (The composite variables of o-exps are identified by their single variables in determining whether the Requirements are met.)
   With $\{a, f1, ya, e, e2\}_\beta \subseteq \{a, f1, e, ya, e2\}_\alpha$, Req 1 is satisfied.
   Similar to Example 7.1, the other Requirements are met, and the o-exp, $(a, f1ya, e, e2)_\beta$ is a single element in the set of o-exps in β.
3. Give values, *ui*β, to the variables *vi*β of the o-exps of the established set of (2). Values are taken from member sets *Ui*β of Table 2.1 with (*ui*α, *ui*β) ∈ *Ru*. The composite member sets *Uj*β determine the composite member sets *Uj*α. That is, for example, *F1Ya*β of the o-exp determines the composite member set *F1Ya*α, and the value, **stellt**, for the variable *f1ya*β is taken from *F1Ya*β. Then, with (puts upright position, stellt) ∈ *Ru*, the given expression of knowledge

   $$(\text{He, puts, bottle, upright position, table})_\alpha.$$

   is changed to

   $$(\text{Er, stellt, Flasche, Tisch})_\beta.$$

It is evident that Tables 7.1, 7.2, and 7.3 apply also to the changing of k-exps in German to k-exps in English. It is also evident that the tables can be extended to include other languages of humans and that any pair of languages might be designated as α and β.

For the Example 7.3 that follows, the k-eles of the k-exps in α are other than in the language of humans, and the k-eles of the k-exps in β are in the languages of humans. The k-exps in the language of birds (α) are changed to k-exps in the language of humans (β). Tables 7.4 and 7.5 were prepared to accomplish the changes and could be extended to accomplish additional changes in k-exps. The previous explanation of Table 7.1 applies to Table 7.4, and previous explanations associated with Tables 7.2 and 7.3 apply to Table 7.5. Table 7.2 in English will be used directly in Example 7.3 to follow. The

TABLE 7.4. Example K-eles of automaton sets in bird language and in English

*A*α={shrill sounds, melodious sounds, ---}
*A*β={bird predators, bird friends, ---}

TABLE 7.5. Partial 3-part list in bird language

p+1 *A*
(*a*)
a

k-eles representing bird language in Table 7.4 are described in the English language.

Example 7.3
Given:

a. oral sounds of a bird (recorded). "Shrill sounds of a bird."
b. a family of sets of k-eles in bird language, α, (from Table 7.4), $U\alpha=\{A,\text{--}\}_{\alpha}$,
c. a family of sets of k-eles in English, β, (from Table 7.4), $U\beta=\{A,\text{--}\}_{\beta}$,
d. a set OEβ of o-exps in English (from Table 7.2), OEβ={(*a*, *f*1, *e*), (*a*, *f*1, *e*, *e*2), (*a*, *f*1, *a*2, *e*), (*a*, *f*1, *e*, *e*2, ya), (*a*), (*fad*, *f*1)}.

To achieve the objective:

1. Find the o-exps in α that yields the given k-exp in α. (shrill sounds) $\in R1 \subseteq A_{\alpha}$, (from (a) and (b)). Then, the o-exp that yields the k-exp is $(a)_{\alpha}$.
2. Select an o-exp as an element of a set in β from OEβ of (d). A random selection of an o-exp that satisfies all Reqs 1,--,4 is $(a)_{\beta}$.
3. Give values, *ui*β, to the variables, *vi*β, of the o-exps of the established set of (2). And the given k-exp, $(\text{shrill sounds})_{\alpha}$, is changed to the k-exp, $(\text{bird predators})_{\beta}$. The English sentence is: "There are bird predators."

7.6 SUMMARY AND CONCLUSIONS

Important needs exist for electronic translation, e.g., the need for the translation of messages in the World-Wide Web (Bulkeley 1998, Immergut 1996).

Defining Terms

1. **Form and style (fs):** fs refers to how vocal sounds, gestures, signs, signals, or the written symbols for them, occur as expressions of knowledge for communication. An example for a fs is any of the languages of humans.
2. **Changing of expressions from one fs, α, to another fs, β:** requires that, after change, the expressions in β communicate the same (or nearly the same) knowledge as the expressions of α.
3. **Knowledge sentence (k-sent):** a k-sent is an accepted way of communicating knowledge in a particular fs.
4. **Knowledge element (k-ele):** a k-ele is either a basic k-ele or a support k-ele, and, in the fss of the languages of man, k-eles are in the form of words; support k-eles support basic k-eles.

5. **Filler element (f-ele):** a f-ele is inserted between k-eles of a k-sent to give support to the k-sent; an example of a f-ele is the word (an) in the k-sent, "He bought **an** umbrella."
6. **A family of sets of k-eles:** the member sets of a family categorize the k-eles by their role or function in conveying knowledge in a fs; example member sets of families are,

   A1={we, He, teacher, Mr. Brown, ---} in the fs of English,
   A2={wer, Er, Lehrer, Herr Brown, ---} in the fs of German.
7. **Knowledge expression (k-exp):** a k-exp is a sequence of k-eles as they appear in a k-sent, i.e., the f-eles are excluded in the sequence; the k-exp describes the structure of the k-sent. The k-exp, (He, buys, umbrella) is a sequence of k-eles of the k-sent, "He buys an umbrella."
8. **Open-expression (o-exp):** an o-exp is obtained from a k-exp by replacing each k-ele of the k-exp with a variable; each k-ele is considered a possible future value for the variable that replaced it, as well as any other k-ele as an element of the same member set of a family of sets in the same fs. For the above example k-exp, the o-exp that results is, say, (a, $f1$, e), where, for example, **He** as a k-ele is replaced with the variable a, and **He** is considered a possible future value for the variable a that replaced it, as well as any other k-ele as an element of a same member set of a family of sets in the same fs, A1={we, He, teacher, Mr. Brown, ---}.
9. **Open sentence (o-sent):** an o-sent is obtained from a k-sent by replacing each k-ele of the k-sent with a variable under the same conditions described in the definition of an o-exp; the f-eles appearing in the resulting o-sent are subject to change in value depending on values given the variables when reverting to k-sents. For the example k-sent, the o-sent that results is, say, a $f1$ (an) e, where (an) is the f-ele value.
10. **Three-part list (3-part list):** each entry of a 3-part list includes an o-exp and a corresponding o-sent; each list is for a particular fs and is required in selecting and establishing acceptable sentence structures for communicating knowledge.

Procedures for Changing Expressions of Knowledge from fs, α, to fs, β

Given:

a. a k-exp in α obtained from a given k-sent in α, e.g., k-sent, "He buys an umbrella." k-exp, (He, buys, umbrella),

b. a family of sets of k-eles in α, e.g., member set, $A\alpha$={we, He, teacher, Mr. Brown, ---}$_\alpha$,

c. a family of sets of k-eles in β and the equivalence of meanings of the k-eles of α and β, e.g., member set, $A\beta$={wer, Er, Lehrer, Herr Brown, ---}$_\beta$,

d. a 3-part list in β with 3-part entries, product set, o-exp, o-sent, e.g., $A \times F1 \times E$, $(a, f1, e)$, *a f1* (einen) *e*.

To achieve the objective:

1. Find the o-exp in α that yields the given k-exp in α, and, if the o-exp is not already in the list in α, establish a new entry in the List with that o-exp, e.g., the o-exp for the k-exp of (a), (He, buys, umbrella), is established as $(a, f1, e)$ which is checked for its presence in the list.
2. Select a set of o-exps from the 3-part list in β of (d) that meets the following four requirements:
   A selected o-exp in β is to yield a k-exp that contains only knowledge contained in the given k-exp in α.
   A currently selected o-exp in β is to provide for additional knowledge beyond what is provided by previously selected o-exps in β.
   A currently selected o-exp in β is to provide knowledge linked to knowledge provided by previously selected o-exps in β.
   A complete set of selected o-exps in β is to yield a set of k-exps that communicate the same (or nearly the same) knowledge that is communicated by the k-exp of the given k-sent.
3. From (c), give values to the variables of the o-exps of the selected set of (2). And the changing of an expression of knowledge for communication from α to a set of expressions in β is complete.

Concluding Features of the Procedures

1. The procedures are flexible in application to a wide range of knowledge-based systems.
2. The procedures are compatible with procedures for achieving other facets of knowledge-based systems when the systems perform under the principles of general automata. These other facets, as previously established, include the following:
   a. extracting and storing the knowledge of sentences (Sections 1.2, 4.2),
   b. associating knowledge (Sections 1.2, 4.3),
   c. communicating stored knowledge (generating sentences) (Koenig 1997a),
   d. obtaining effective operation paths in knowledge structures (for finding and sequencing knowledge) (Appendices A, B),
   e. establishing intelligent-like features (by reasoning, generating inferences) (Sections 4.4, 4.5).
3. The procedures allow for growth and expansion of a knowledge-based system.

4. The procedures allow the k-eles in one fs to be combined to obtain equivalence of meaning with one or more k-eles in another fs. For example, **rings** and **bell** could be combined to be equivalent in meaning with the single k-ele **klingelt** in German.
5. The procedures are flexible for expanding from a one-way mode of communication to a conversational mode of communication.
6. The procedures allow for adding lingual fss to existing lingual systems.
7. A procedure for randomly selecting a set of o-exps in β from a list means the selected set may be different than a previously selected set in an effort to communicate the same knowledge. This is considered to parallel a human's experiences in selecting a set of sentence structures from memory for communicating knowledge; and the set will likely be different than a previously selected set for communicating the same knowledge.
8. A selection of a set of o-exps by the procedures can be made partial to o-exps that yield k-sents for the individual's understanding. This is considered to parallel a human's experiences in selecting sentence structures and word values to match an individual's understanding capabilities.
9. The procedures recognize opportunities to add new entries to a 3-part list in α; thus the capabilities for communicating knowledge may continually increase. This is considered to parallel a human's experiences in adding to his list of sentence structures in memory thus increasing an individual's choices of structures for communicating knowledge.

EXERCISES

7.1 Instead of an English to German translation for the given Example 7.1, substitute some other language for German and follow with the same translation procedure.

7.2 Same as Exercise 7.1 except use the English sentence of Example 7.2.

7.3 Choose a sentence whose meaning to be understood requires the level of a graduate student. Using the translation procedures presented in this text, obtain a set of sentences in the same language whose meaning is the same and whose meaning can be understood by a student of the mid-primary grade level.

CHAPTER 8

Electronic Security Through Pseudo Languages

8.1 INTRODUCTION

The development of a scheme for establishing security systems through pseudo languages is presented. The scheme relates to symmetric key encryption in that the sender and receiver both use the same key to scramble and unscramble messages (Sikorski and Peters 1999). Maintaining security when sender and receiver exchange private key information need not be a problem for the development presented.

A theorem in the form of an algorithm is developed for establishing an operating system. The system accepts private keys as data for processing a message into a more secure message (encryption operation) or for processing a message into a less secure message (decryption operation). Examples demonstrate the theorem.

Users create their own encryption systems by designing and establishing the private keys. The keys are in the form of families of sets of words of languages. Four principal variables of design are determined. The choices of values for the variables provide great flexibility in the design of encryption systems and can be made on an arbitrary basis or on some mathematical basis. By comparison, some of the other encryption schemes, such as RSA encryption, are based on well-defined mathematical techniques (Chen 1999, Comerford 2000, Hellemans 1999, Sikorski and Peters 1999).

E-signing is accomplished by altering the encryption uses of the private key. Public-key cryptography has often been the basis for e-signing (Chen 1999, Comerford 2000).

A password using the development presented is treated as a message to be made secure, i.e., it is encrypted like any other message and is then decrypted by the receiver, such as a server.

Knowledge Structures for Communications in Human-Computer Systems: General Automata-Based, by Eldo C. Koenig

8.2 DEFINITIONS, SETS, AND RELATIONS

1. *Form and style of a language*: refers to how vocal sounds, gestures, signs, signals, or the written symbols for them, occur as expressions of knowledge for communication.
2. γs, $s=0, 1, 2, -, n$: identifies a language by its form and style; an encryption or decryption changes the form and style of the language of a message.
3. *message in* γs, ($M\gamma s$): a message to be encrypted or decrypted in the language of γs.
4. *secured message*, ($M\gamma s$, $s\neq 0$): a message that can communicate knowledge only to a select few.
5. *unsecured message*, ($M\gamma s$, $s=0$): a message that can communicate knowledge to more than a select few.
6. *pseudo language*: the language that communicates a $M\gamma s$, $s\neq 0$; i.e., a language understood only by a select few.
7. *encryption step*, ($M\gamma s \rightarrow M\gamma c$), $s\neq n$; $c=s+1$: a message in γs becomes more secure in γc.
8. *decryption step*, ($M\gamma s \rightarrow M\gamma c$), $s\neq 0$, $c=s-1$: a message in γs becomes less secure in γc.
9. *word of a message*, ($wi\gamma s$): a word in position i of a message, $M\gamma s=(wi\gamma s, i=1, 2, -, m)$, such as, a letter or symbol or a group of letters or symbols, not necessarily written.
10. *A family of sets of words of a language* γs

$$\mathrm{U}\gamma \mathrm{s}=\{Wj\gamma s, j=1, 2, \text{--}, f\}$$

$$W\gamma s=\{Wej\gamma s, e=1, 2, \text{--}, y\}$$

11. *Relation RW*: With $Wj\gamma s \in \mathrm{U}\gamma s$ and $Wj\gamma c \in \mathrm{U}\gamma c$, for $(Wj\gamma s, Wj\gamma c) \in RW \subseteq \mathrm{U}\gamma s \times \mathrm{U}\gamma c$, ($Wj\gamma s$, $Wj\gamma c$) is an element of RW whenever $Wj\gamma c$ is known to be equivalent in meaning to $Wj\gamma s$ by only a select few

$$\text{for encryption } s\neq n, c=s+1$$

$$\text{for decryption } s\neq 0, c=s-1$$

12. *Relation Rw*: With $wej\gamma s \in Wj\gamma s$ and $wej\gamma c \in Wj\gamma c$, for $(wej\gamma s, wej\gamma c) \in Rw \subseteq Wj\gamma s \times Wj\gamma c$, ($wej\gamma s$, $wej\gamma c$) is an element of Rw whenever $wej\gamma c$ is known to be equivalent in meaning to $wej\gamma s$ by only a select few

$$\text{for encryption } s\neq n, c=s+1$$

$$\text{for decryption } s\neq 0, c=s-1$$

8.3 ANALYSIS FOR E-SECURITY THROUGH PSEUDO LANGUAGES

8.3.1 A Basic E-Security System

Encryption

Given:

An unsecured message, $M\gamma s = (wi\gamma s, i = 1, 2, \text{--}, m), s = 0$
A family of sets of words for the language $\gamma s, s = 0$

$$U\gamma 0 = \{Wj\gamma 0, j = 1, 2, \text{--}, f\}$$
$$Wj\gamma 0 = \{wej\gamma 0, e = 1, 2, \text{--}, y\}$$

A family of sets of words for a pseudo language $\gamma s, s = n = 1$

$$U\gamma 1 = \{Wj\gamma 1, j = 1, 2, \text{--}, f\}$$
$$Wj\gamma 1 = \{wej\gamma 1, e = 1, 2, \text{--}, y\}$$

Example 8.1. For an example to demonstrate the encryption system, the vocabularies of the pseudo language $\gamma 1$ and the language $\gamma 0$ of the given message are chosen to be the same. Language distinction is accomplished by positioning the elements differently in the sets. The words are categorized as either knowledge words, $K\gamma s$, or functional words $F\gamma s$. A different choice might have been made in the categorization of words.

Consider an example message $M\gamma 0$ to pass through an encryption system, $M\gamma 0 \rightarrow M\gamma 1 \rightarrow M\gamma 0$

$$M\gamma 0 = (wi\gamma 0, i = 1, 2, \text{--}, m) = (\textit{John buys an umbrella}), m = 4$$

Example family of sets of words in the language of $M\gamma 0$

$$U\gamma 0 = \{K\gamma 0, F\gamma 0\}, f = 2$$
$$K\gamma 0 = \{\textit{John, Jane, Joe, umbrella, rocket, car, buys, takes, steals}\}, y = 9$$
$$F\gamma 0 = \{\textit{a, an, the, or, and, also}, ., ;, :\}, y = 9$$

Example family of sets of words in the pseudo language. (Make the member sets $K\gamma 1$ and $F\gamma 1$ the same as $K\gamma 0$ and $F\gamma 0$ but make the positions of the elements in $U\gamma 0$ and $U\gamma 1$ different.)

$$U\gamma 1 = \{K\gamma 1, F\gamma 1\}, f = 2$$
$$F\gamma 1 = \{\textit{a, an, the, or, and, also}, ., ;, :\}, y = 9$$
$$K\gamma 1 = \{\textit{John, Jane, Joe, umbrella, rocket, car, buys, takes, steals}\}, y = 9$$

$U\gamma 0$ and $U\gamma 1$ function as a private key to be shared in the operations of encryption and decryption. Changes to be made in $U\gamma 0$ and $U\gamma 1$ as a private key in the operation of encryption must be communicated to the operation of decryption. Suppose, for example, $F\gamma 0$ and $F\gamma 1$ are to be modified in the operation of encryption by adding the question mark (?) as an element in these two sets. The modified sets can be considered a message $M\gamma 0$ and sent through the existing secure system in order to maintain a private identical key.

Encryption Operations, $M\gamma 0 \rightarrow M\gamma 1$

a. $\boldsymbol{wej\gamma 0 = wi\gamma 0}$: $wej\gamma 0 \in Wj\gamma 0$ is found to be the same as word $wi\gamma 0$ of $M\gamma 0$.

   For the example, with $i=1$

   The word $wi\gamma 0 = w1\gamma 0 = John$, in position $i=1$ of the given message $M\gamma 0$, is found to be the same as the word, $mej\gamma 0 = w11\gamma 0 = John$, in position $e=1$ of set $K\gamma 0$, i.e.,

   $w11\gamma 0 = w1\gamma 0$, $w11\gamma 0 \in K\gamma 0$.

b. $\boldsymbol{(Wj\gamma 0,\ Wj\gamma 1) \in RW}$: From (a), $wi\gamma 0 = wej\gamma 0 \in Wj\gamma 0$; $Wj\gamma 1 \in U\gamma 1$ and $Wj\gamma 0 \in U\gamma 0$ are considered equivalent in meaning by a select few and the position of $Wj\gamma 1$ in set $U\gamma 1$ is the same as the position of $Wj\gamma 0$ in the set $U\gamma 0$.

   For the example, with $i=1$

   The position of $F\gamma 1$ in set $U\gamma 1$ is found to be the same as the position of $K\gamma 0$ in the set $U\gamma 0$ where $w1\gamma 0 = John$ was found to be located, i.e., $(K\gamma 0, F\gamma 1) \in RW$.

c. $\boldsymbol{(wej\gamma 0,\ wej\gamma 1) \in Rw}$: From (b), $wej\gamma 0 \in Wj\gamma 0$; $wej\gamma 1 \in Wj\gamma 1$ and $wej\gamma 0 \in Wj\gamma 0$ are considered equivalent in meaning by a select few, and the position of $wej\gamma 1$ in set $Wj\gamma 1$ is the same as the position of $wej\gamma 0$ in set $Wj\gamma 0$.

   For the example, with $i=1$

   The position of the word $wej\gamma 1 = w11\gamma 1 = \boldsymbol{a}$ in set $F\gamma 1$ is found to be the same as the position of the word $wej\gamma 0 = w11\gamma 0 = John$ in the set $K\gamma 0$, i.e., $(John, \boldsymbol{a}) \in Rw$.

d. $\boldsymbol{wi\gamma 1 = wej\gamma 1}$: The position of i of the message $M\gamma 1$ that is being established is given the word $wej\gamma 1$ which was determined in (c); for $i=m$, $M\gamma 1$ is complete, $M\gamma 1 = (wi\gamma 1, i=1, 2, -, m)$.

   For the example, with $i=1$

   The position $i=1$, of the message $M\gamma 1$ being established, receives the word $wej\gamma 1 = w11\gamma 1 = \boldsymbol{a}$ which was determined in (c); then $M\gamma 1 =$ ($\boldsymbol{a}$ __ __ __).

For the example (continued), with $i=2, 3, 4$

$i=2$

a. $wi\gamma0=w2\gamma0=buys$, $wej\gamma0=w71\gamma0=buys \in K\gamma0$

b. $(K\gamma0, F\gamma1) \in RW$

c. $(buys, .) \in Rw$

d. position $i=2$ of $M\gamma1$ receives the word $wej\gamma1=w71\gamma1=$**.**

$i=3$

a. $wi\gamma0=w3\gamma0=an$, $wej\gamma0=w22\gamma0=an \in F\gamma0$

b. $(F\gamma0, K\gamma1) \in RW$

c. $(an, Jane) \in Rw$

d. position $i=3$ of $M\gamma1$ receives the word $wej\gamma1=w22\gamma1=Jane$

$i=4$

a. $wi\gamma0=w4\gamma0=umbrella$, $wej\gamma0=w41\gamma0=umbrella \in K\gamma0$

b. $(K\gamma0, F\gamma1) \in RW$

c. $(umbrella, or) \in Rw$

d. position $i=4$ of $M\gamma1$, receives the word $wej\gamma1=w41\gamma1=or$

$m=4$

$M\gamma1=(wi\gamma1, i=1, 2, \text{-}, m)=(wi\gamma1, i=1, 2, 3, 4)=($***a* . *Jane or***$)$

Encryption complete, $M\gamma0 \rightarrow M\gamma1$.

Decryption

Given:
The secured message, $M\gamma1=(wi\gamma1, i=1, 2, \text{--}, m)$
For the example, $M\gamma1=($***a*** . *Jane or*$)$, $m=4$
Families of sets of words of $M\gamma0$ and the pseudo language $\gamma1$; i.e., the private key is shared with the operation of Encryption.

Decryption Operation, $M\gamma1 \rightarrow M\gamma0$:

a. $\boldsymbol{wej\gamma1=wi\gamma1}$: $wej\gamma1 \in Wj\gamma1$ is found to be the same as word $wi\gamma1$ of $M\gamma1$

For the example with $i=1$

$wi\gamma1=w1\gamma1=$***a***, $wej\gamma1=w11\gamma1=$***a***$\in F\gamma1$

b. $\boldsymbol{(Wj\gamma1, Wj\gamma0) \in RW}$: The position of $Wj\gamma0$ in set $U\gamma0$ is the same as the position of $wj\gamma1$ in the set $U\gamma1$

For the example with $i=1$

$(F\gamma1, K\gamma0) \in RW$

c. $\boldsymbol{(wej\gamma1, wej\gamma0) \in Rw}$: The position of $wej\gamma0$ in set $Wj\gamma0$ is the same as the position of $wej\gamma1$ in set $Wj\gamma1$

For the example with $i=1$

$(w11\gamma1, w11\gamma0)=($***a***, *John*$) \in Rw$

d. ***wiγ0 = wejγ0***: The position *i* of *M*γ0 is given the word *wej*γ0 which was determined in (c), *M*γ0 = (*wi*γ0, *i* = 1, 2, --, *m*)
For the example with i = 1
*wej*γ0 = w11γ0 = *John* determined in (c), thus, *M*γ0 = (*John* __ __ __)

For the example (continued) with *i* = 2, 3, 4

i = 2

a. *wi*γ1 = *w*2γ1 = ., *wej*γ1 = *w*71γ1 = . ∈ *F*γ1
b. (*F*γ1, *K*γ0) ∈ *RW*
c. (., *buys*) ∈ *Rw*
d. position *i* = 2 of *M*γ0 receives the word *w*71γ0 = *buys*, thus, *M*γ0 = (*John buys* __ __)

i = 3

a. *wi*γ1 = *w*3γ1 = *Jane*, *wej*γ1 = *w*22γ1 = *Jane* ∈ *K*γ1
b. (*K*γ1, *F*γ0) ∈ *RW*
c. (*Jane*, *an*) ∈ *Rw*
d. position *i* = 3 of *M*γ0 receives the word *w*22γ0 = *an*, thus, *M*γ0 = (*John buys an* __)

i = 4

a. *wi*γ1 = *w*4γ1 = *or*, *wej*γ1 = *w*41γ1 = *or* ∈ *F*γ1
b. (*F*γ1, *K*γ0) ∈ *RW*
c. (*or*, *umbrella*) ∈ *Rw*
d. position *i* = 4 of *M*γ0 receives the word *w*41γ0 = *umbrella* thus, *M*γ0 = (*John buys an umbrella*)

m = 4
*M*γ0 = (*wi*γ0, *i* = 1, 2, --, *m*) = (*John buys an umbrella*)

Decryption complete; the message is the original unsecured message.

8.3.2 A Two-Step Encryption System

Consider an example encryption system of two steps. A message *M*γ0 is to pass through a two-step encryption system as indicated.

$$M\gamma 0 \rightarrow M\gamma 1 \rightarrow M\gamma 2 \rightarrow M\gamma 1 \rightarrow M\gamma 0,\ s = 0, 1, 2, \text{--}, n,\ n = 2$$

Four principal variables of design were considered:

a. vocabulary
b. categories of words

c. ordering of words
d. number of encryption steps, i.e., the number of pseudo languages in the encryption system

Encryption

Given:

An unsecured message, $M\gamma 0=(wi\gamma 0, i=1, 2, -, m)$
For the Example 8.1, $M\gamma 0=$(*John buys an umbrella*), $m=4$
Families of sets of words of three languages, γs, $s=0, 1, 2$, i.e., the private key consists of three families instead of two.
Family of sets of words of language of $M\gamma s$, $s=0$

$$U\gamma 0=\{K\gamma 0, F\gamma 0\}$$

$K\gamma 0=\{$*John, Jane, Joe, umbrella, rocket, car, buys, takes, steals*$\}$

$F\gamma 0=\{$***a**, an, the, or, and, also*, . , ; , : $\}$

Family of sets of words of a pseudo language in γs, $s=1$

$$U\gamma 1=\{F\gamma 1, K\gamma 1\}$$

$F\gamma 1=\{$***a**, an, the, or, and, also*, . , ; , : $\}$

$K\gamma 1=\{$*John, Jane, Joe, umbrella, rocket, car, buys, takes, steals*$\}$

Family of sets of words of a pseudo language in γs, $s=n=2$

$$U\gamma 2=\{F\gamma 2, K\gamma 2\}$$

$F\gamma 2=\{$: , ; , . , *the, an,* ***a**, also, and, or*$\}$

$K\gamma 2=\{$*steals, takes, buys, Joe, Jane, John, car, rocket, umbrella*$\}$

Encryption Operation, $M\gamma 0 \rightarrow M\gamma 1 \rightarrow M\gamma 2$

Summary of the results of an operating system for the two-step encryption system:
From the given data

$s=0, 1, 2, --, n; n=2$

$j=1, 2, --, f; f=2$

$e=1, 2, --, y; y=9$

$i=1, 2, --, m; m=4$

Encryption operation results

$s=0$
$M\gamma 0=$(*John buys an umbrella*)

$s=1$
$M\gamma 1=$(***a*** **.** *Jane or*)

$s=n=2$
$M\gamma 2=$(**:** *also takes the*)

Encryption complete, message secured.

Decryption

Given:

The secured message, $M\gamma 2=(wi\gamma 2, i=1, 2, \text{--}, m)$
For the example, $M\gamma 2=$(**:** *also takes the*)
Families of sets of words of three languages, γs, $s=0, 1, 2$, i.e., the private key determined in the Encryption.

Decryption Operation, $M\gamma 2 \rightarrow M\gamma 1 \rightarrow M\gamma 0$

Summary of the results for the two-step encryption system:
From the given data

$s=0, 1, 2, \text{--}, n; n=2$

$j=1, 2, \text{--}, f; f=2$

$e=1, 2, \text{--}, y; y=9$

$i=1, 2, \text{--}, m; m=4$

Decryption Operation Results

$s=n=2$
$M\gamma 2=$(**:** *also takes the*)

$s=1$
$M\gamma 1=$(***a*** **.** *Jane or*)

$s=0$
$M\gamma 0=$(*John buys an umbrella*)

Decryption complete. The message is the original unsecured message.

The above discussions support the following theorem in the form of an algorithm.

Theorem 8.1 (Algorithm): *Electronic Security through Pseudo Languages; An n-Step Encryption System:*

$$M\gamma 0 \rightarrow M\gamma 1 \rightarrow M\gamma 2 \rightarrow \text{--} \rightarrow M\gamma n \rightarrow \text{--} \rightarrow M\gamma 2 \rightarrow M\gamma 1 \rightarrow M\gamma 0$$

Given data for an n-Step Operating System Algorithm:

1. *Messages:*
   For encryption operation, unsecured message

$$M\gamma 0 = (wi\gamma 0, i = 1, 2, \text{--}, m)$$

   For decryption operation, secured message

$$M\gamma n = (wi\gamma n, i = 1, 2, \text{--}, m)$$

2. *Families of sets of words for n+1 languages, for both encryption and decryption operations*

$$\mathrm{U}\gamma s = \{Wj\gamma s, j = 1, 2, \text{--}, f\}, s = 0, 1, 2, \text{--}, n; n \neq 0$$

$$Wj\gamma s = \{wej\gamma s, e = 1, 2, \text{--}, y\}$$

   where $\mathrm{U}\gamma 0$ *is the family of sets of words of the language of* $M\gamma 0$, *and* $\mathrm{U}\gamma s$, $s \neq 0$, *are families of sets of words of n pseudo languages.*

The sets $\mathrm{U}\gamma s$ and $Wj\gamma s$ *for a given encryption system must be made mathematically equivalent to sets* $\mathrm{U}\gamma c$ *and* $Wj\gamma c$, *respectively*

$$\mathrm{U}\gamma s \sim \mathrm{U}\gamma c, c = s + 1$$

$$Wj\gamma s \sim Wj\gamma c$$

The families of sets of words function as private keys that are shared in the operations of encryption and decryption. Changes made in them in the operation of encryption must be communicated to the operation of decryption. The modified keys are communicated in security by treating them as messages to be sent through the existing secure system.

Operating System Algorithm for an n-step Encryption System

If the operation is encryption,

$s = 0, 1, 2, \text{--}, n-1; n \neq 0$

$c = s + 1$

If the operation is decryption,

$s=n, n-1, --, 2, 1$

$c=s-1$

a. ***wejγs = wiγs***;

wejγs ∈ Wjγs is found to be the same as word wiγs of the message,

$$(Wi\gamma s, i=1, 2, --, m);\ j=1, 2, --, f;\ e=1, 2, --, y$$

b. **(Wjγs, Wjγc) ∈ *RW***;

Wjγc ∈ Uγc and Wjγs ∈ Uγs are considered equivalent in meaning by a select few, and the position of Wjγc in set Uγc is the same as the position of Wjγs in the set Uγs.

c. **(*wejγs*, *wejγc*) ∈ *Rw***;

wejγc ∈ Wjγc and *wejγs ∈ Wjγs are considered equivalent in meaning by a select few and the position of wejγc* in set *Wjγc is the same as the position of wejγs in set Wjγs.*

d. ***wiγc = wejγc***;

*The position **i** of the message Mγc that is being established is given the Word wejγc, which was determined in* (c), *and for i=m, the message Mγc=(wiγc, i=*1, 2, --, *m) is complete.*

The secured message following encryption:

$$M\gamma n=(wi\gamma n, i=1, 2, --, m),\ c=n$$

The unsecured message following decryption:

$$M\gamma 0=(wi\gamma 0, i=1, 2, --, m),\ c=0$$

8.3.3 E-Signing

The operating software described by the Theorem 8.1 also accommodates e-signing. The e-signing occurs in the following manner when a signature is carried through the operation as a message:

*Mγ*0: Signature of person *A* in a language γ0.

Mγn: A secured signature in a pseudo language *γn*

*Mγ*0→*Mγn*: With the operating software in the encryption mode, *Mγn* is obtained from *Mγ*0 by *A*'s private key in n-steps.

Mγv: Signature in a pseudo language *γv* for identification and verification.

Mγn→*Mγv*: With the operating software in the decryption mode, *Mγv* is obtained from *Mγn* by any person *B* by *A*'s public key.

Since the decryption key is a public key, anyone can verify that the signature $M\gamma\nu$ is that of the person who has the private key, and all receive an identical signature.

8.4 SUMMARY AND CONCLUSIONS

The development in Section 8.3 concludes with a Theorem 8.1 in the form of an algorithm for obtaining electronic security through pseudo languages which have the following features:

1. The security systems perform as symmetric key systems.
2. Four principal variables of design give users many choices of pseudo languages in creating encryption systems:
   a. vocabulary
   b. categories of words
   c. ordering of words
   d. number of encryption steps, i.e., the number of pseudo languages in the encryption system
3. The operating software accommodates the wide variety of encryption systems created by users and presented as private keys.
4. The encryption systems need not be based on fixed rules or disciplines that can be discovered by eavesdroppers and used to break security.
5. The development allows the creation of multi-step security systems operating as either a fixed security system or as a system that can be periodically expanded by adding steps to maintain security confidence.
6. Encryption systems can be intentionally created to produce misleading messages in that the messages can lead eavesdroppers into believing the messages are authentic.
7. Encryption systems can be created ranging from low security to high security; thus systems can be customized to meet specific security requirements.
8. Users of the scheme can create encryption systems that are very secure based on the prêmise that it can be difficult for eavesdroppers to determine when success is achieved during a decoding process.
9. The operating software also accommodates e-signing and e-passwords.

EXERCISES

8.1 Establish a pseudo English language to the extent required to translate the given English sentence of Exercise 7.2 to a set of pseudo English sentences as a secure message. Follow the translation procedure of Exercise 7.2 to establish the secure message.

8.2 Establish a secure message in the form of a single sentence in a pseudo English language, and translate this secure message to an unsecured message in the English language. Follow the translation procedure of Exercise 7.2.

8.3 Establish a secure message from the example unsecured sentence of Exercise 7.2 by performing a two-step encryption. Establish a pseudo English language for each of the two encryptions. Follow the translation procedure of Exercise 7.2 to establish the secure message.

APPENDIX A

Analysis for an Effective Operation of a General Automaton

A.1 INTRODUCTION

A natural extension of the analysis for a general automaton (Section 2.1) is the development and application of recursive methods for determining the effective operation of a general automaton through select properties of these established graphs. The recursive methods, in a practical sense, provide a means for modeling and studying a general automaton by a computer. The material follows the work of Frederick, Koenig (1971a).

Classical recursive theory deals with number-theoretic functions or relations and has grown out of induction or recursion definition (Kleene 1967, Minsky 1967, Peter 1967, Rogers 1967). Various functions involving the fundamental automaton and graph functions for the purpose of determining an effective operation of the general automaton through the graph model are perceived as recursion-like but are not necessarily number-theoretic by nature. It is important to introduce a nonnumeric recursion concept for such functions that provides an algorithm for carrying out the function calculations.

A basic part of this effort is the development of recursive methods that incorporate a concept of nonnumeric recursion for showing certain general automaton functions are computable, and a concept of time-variant adjacency for points in graphs derived from a physical system. The latter concept enables recursive graph functions to be established for the graphs of the model paralleling those for the general automaton.

First, a review is given of recursive function theory for number-theoretic recursion along with existing approaches for including broader nonnumerical procedures within the formal theory of recursive functions. This is followed by a concept of recursion that brings certain nonnumeric functions into the class of general recursive functions by providing a computable algorithm for carrying out the calculations of those functions. The nonnumeric recursion

---

Knowledge Structures for Communications in Human-Computer Systems: General Automata-Based, by Eldo C. Koenig

concept is used to define a set of moment functions, which formally describes a fundamental general automaton principle as it relates to automaton sets and functions characterizing the physical system, and to show that they belong to the class of general recursive functions.

Second, the concept of time-variant adjacency for points of the model for a general automaton will be defined. Using the recursion concepts, effective adjacency functions will be defined for the graphs in terms of the moment functions. The effective adjacency functions describe the unique pair of points and connecting line for each pair of consecutive moments throughout the moment base used in the formal analysis for the automaton. The concluding part of this analysis will be the application of the aforementioned methods to determine an effective operation function for each graph defined by a point, line, point sequence that determines a unique, effective operation path in each graph. Since the effective operation functions will be defined in terms of the effective adjacency functions, they too are included in the class of general recursive functions and are computable.

A.2 RECURSIVE METHODS

The purpose of this section is to develop recursive methods that can be applied to graphs of the model for a general automaton to determine select properties peculiar to an effective operation. The methods are categorized into four subsections called nonnumeric recursion, moment functions, adjacency concept, and effective adjacency function. As results are detailed, the abstract example presented in 2.1 is interwoven to illustrate the results. Nonnumeric recursion is a concept that enables certain functions interrelating fundamental automaton principles, functions, and sets, called moment functions, to be included in the class of general recursive functions. The adjacency concept for graphs in the model enables the nature of the moment functions to be paralleled by effective adjacency functions that also belong to the class of general recursive functions.

Nonnumeric Recursion. Number-theoretic recursion is important in its relation to computing automata of the digital class. Since the work is concerned with general automata, a concept of nonnumeric recursion is important as well. Both types are reviewed and comparisons made to establish the nonnumeric concept needed in relation to general automata.

The notion of classic function theory of recursion deals with number-theoretic functions and has grown out of a fundamental concept of induction (recursion definition). Recurrent sequences, such as the Fibonacci sequence, are considered the forerunners of the simplest class of recursive functions called the primitive recursive functions. These functions are formally characterized by the smallest of functions based on the zero, successor, identity, and those functions defined in terms of the above using composition and the primitive-recursion definition (Kleene 1967, Minsky 1967, Rogers 1967). The term general recursive functions is used to denote a broad class of recursive functions and represents all functions computable in the most general sense. By the most general sense of computability is meant that there is an effective procedure for calculation that is independent of the arbitrariness of the computing agent and that is repeatable at any time. It is well known that the functions mechanically calculable by computing automata are identical to the class of general recursive functions (Peter 1967).

In many mathematically disciplined areas, there are functions that are perceived as "recursion-like" yet are not number-theoretic by nature. Some approaches exist for including such broader nonnumerical procedures within the formal theory of recursive functions. One approach is based on the concept that a nonnumerical mapping is defined to be recursive if an algorithm exists thereby enabling a computing automaton to carry out the calculations of the mapping (Rogers 1967). Another involves extending the theory of recursive functions to include those defined on sets which can be constructed like numbers (Peter 1967). Such sets have been termed holomorph sets for their nonnegative integer-like structure. Primitive recursive functions have been defined on holomorph sets that are analogous to some of the important functions in recursive number theory. Also, definitions for holomorph sets have been introduced that are analogous to different number-theoretic sorts of recursion, such as, course-of-values recursion, simultaneous, nested, and partial and general recursion.

In the formal analysis for a general automaton of Section 2.1, there are functions that are perceived as recursion-like but are not necessarily number-theoretic by nature. These functions are brought into the class of general recursive functions by a concept of nonnumeric recursion that describes an algorithm for carrying out the calculations of the functions. This concept of nonnumeric recursion closely parallels the basic notion behind primitive-recursion for number-theoretic functions.

The concept of nonnumeric recursion for certain functions pertaining to a general automaton as it parallels primitive recursion is as follows: First, the domain set is generally nonnumeric, and the co-domain is nonnumeric and different from the domain. Secondly, an ordering for the domain elements is established by showing the domain set is the simplest kind of holomorph set in that there is a one-to-one correspondence between it and a subset of the nonnegative integers. Thirdly, a condition exists for a function mapping the domain into the codomain, such that the initial domain element has its image specified, and a defined intermediate function parallels the inductive step function of primitive recursion for numbers. This function described the image of an arbitrary domain element in terms of the predecessor of that domain element and its image under the original function. Any function defined by nonnumeric recursion as detailed above belongs to the class of general recursive functions because the algorithmic nature of the definition enables a computing automaton to carry out the calculations.

Moment Functions. Several functions were introduced in 2.1 for the purpose of formally describing the fundamental automaton principle as it relates to the general automaton sets and functions. The objective of the following discussion is to show the moment functions are included in the class of general recursive functions based on the aforementioned concept of nonnumeric recursion. In order to accomplish this, it is necessary to assume knowledge of a minimal set of performance descriptors of the automaton from the formal analysis.

The following set of performance descriptors of the physical system obtained from the formal analysis for a general automaton constitutes a minimal set of information from which to establish an analysis for an effective operation of the automaton. First, the moment base M is specified along with the set Q of states, set S of stimuli, and the set $V \subseteq Q \times S$ of state-stimulus pairs as determined by the nature of the automaton. Also, the environment space is given as the real n-dimensional vector space R^n for some n along with initial conditions (conditions at the initial moment m) of stimulus location $E(m)$, time $T(m)$, state $Q(m)$ and stimulus $S(m)$. The minimal amount of information needed is complete with the knowledge of several functions and their image sets. These include the function ϕ_0 mapping a subset of the environment onto set S describing the stimulus associated with each stimulus location at the initial moment, the automaton functions σ, δ, τ, and ω along with the image sets $\sigma(V)=F$, $\delta(V)=D$, and $\tau(V)=C$ of responses; and the function α which associates a stimulus from S with each transformation response.

Moment functions 2.1M to 2.8M, introduced in Section 2.1, formally describe the General automaton principle as it relates to the automaton sets and functions as expressed by the automaton equations. These include functions associating the proper stimulus location, time, and state with each moment throughout the moment base; the partial functions that describe the proper stimulus at each moment other than the final moment if one exists; and the partial functions that denote the proper transformation, spatial change, and time change response for each moment other than the initial one.

The main purpose of the discussion of the moment functions is to establish a defining algorithm for them within the framework of the nonnumeric recursion concept that is computable. The domain sets for the moment functions are ordered since they have already been shown to be the simplest of holomorph sets by the correspondence $m+i \rightarrow i \rightarrow m+i$. Assuming the minimal set of performance descriptors is given, the images of the initial domain elements are known. It remains to establish the images of the successive domain elements. Fundamental to this is the knowledge of the automaton sets and functions given in the minimal set of performance descriptors, the set of automaton equations as detailed in Section 2.1, and the set ϕ of environment functions of which ϕ_0 is also given in the minimal set of performance descriptors.

Let the initial stimulus location(environment) be denoted by $E(m)=e_0$, the initial time by $T(m)=t_0$, the initial state by $Q(m)=q_0$, and the initial stimulus by $S(m)=\phi_0(e_0)$. The first transformation response is $F(m+1)=\sigma(Q(m),$

$S(m))=\sigma(q_0, \phi_0(e_0))$ denoted by σ_1. Similarly, the first spatial change response is $D(m+1)=\delta(Q(m), S(m))=\delta(q_0, \phi_0(e_0))$ denoted by δ_1, and the first time change response is $C(m+1)=\tau(Q(m), S(m))=\tau(q_0, \phi_0(e_0))$ denoted by τ_1. Consider the set of moment functions $\{\theta_E, \theta_T, \theta_Q, \theta_S, \theta_F, \theta_D, \theta_C\}$ which formally describes the automaton principle as it relates to the automaton sets and functions (see 2.1M–2.8M). On the basis of the above remarks, the images of the initial domain elements under the moment functions are

$$\theta_E(m)=e_0$$

$$\theta_T(m)=t_0$$

$$\theta_Q(m)=q_0$$

$$\theta_S(m)=\phi_0\,(e_0)$$

denoting the initial environment, time, state, and stimulus. Owing to the partial nature of the functions θ_F, θ_D, and θ_C, then

$$\theta_F(m+1)=\sigma(\theta_Q(m), \theta_S(m))=\sigma_1$$

$$\theta_D(m+1)=\delta(\theta_Q(m), \theta_S(m))=\delta_1$$

$$\theta_C(m+1)=\tau(\theta_Q(m), \theta_S(m))=\tau_1$$

denote the initial transformation, spatial change, and time change responses.

From the definition of the set of environment functions ϕ, namely $\phi=\{\phi_i|\phi_i: E' \subseteq \bar{E} \rightarrow \phi_i(E')=S$, ϕ is a function, i corresponds to the moment $m+i\}$, the function associated with the moment $m+1$ is denoted by ϕ_1. By definition, ϕ_1 is defined on the same set as ϕ_0, and the range of ϕ_1 is the set S. At moment $m+1$, the automaton affects the stimulus location $\theta_E(m)$ by the response $\theta_F(m+1)$. Therefore, the image of $\theta_E(m)$ under the function ϕ_1 is $\phi_1(\theta_E(m))=\alpha(\theta_F(m+1))$, where α is the automaton function which maps the set of transformation responses into the set of stimuli. Since the automaton affects no other locations in E' at moment $m+1$, the images at all the other locations in E' under ϕ_1 for moment $m+1$ are the same as under ϕ_0 for moment m. More precisely, the environment function ϕ_1 is related to ϕ_0 by

$$\phi_1(\theta_E(m))=\alpha(\theta_F(m+1))$$

$$\phi_1(e_k)=\phi_0(e_k), \quad \forall e_k \in E', \quad e_k \neq \theta_E(m)$$

The next stimulus location to be accessed is $E(m+1)=\gamma'(E(m), D(m+1))=\gamma'(E(m), \delta(Q(m), S(m)))=\gamma'(e_0, \delta(q_0, \phi_0(e_0)))=\gamma'(e_0, \delta_1)$, and the next time is $T(m+1)=\lambda'(t(m), C(m+1))=\lambda'(t(m), \tau(Q(m), S(m)))=\lambda'(t_0, \tau(q_0, \phi_0(e_0)))=\lambda'(t_0, \tau_1)$. Thus, the images of the second domain element $m+1$ under the functions θ_E and θ_T are defined as

$$\theta_E(m+1)=\gamma'(e_0,\delta_1)=\gamma'(\theta_E(m),\delta_1)$$

$$\theta_T(m+1)=\lambda'(t_0,\tau_1)=\lambda'(\theta_T(m),\tau_1)$$

Note that these images $\theta_E(m+1)$ and $\theta_T(m+1)$ are functions of the preceding domain element's images, $\theta_E(m)$ and $\theta_T(m)$. The next state is $Q(m+1)=\omega(Q(m), S(m))=\omega(q_0, \phi_0(e_0))=\omega_1$. By putting $\theta_Q(m+1)=Q(m+1)$, then the image of the second domain element

$$\theta_Q(m+1)=\omega(q_0,\phi_0(e_0))=\omega(\theta_Q(m),\phi_0(e_0))$$

is a function of the preceding domain element's image. By definition, the stimulus received at the moment $m+1$ is the one associated with the current environment $E(m+1)$. Therefore, $S(m+1)=\phi_1(E(m+1))=\phi_1(\gamma'(e_0, \delta_1))$. Now if $\delta_1=\delta(q_0, \phi_0(e_0))$ is the zero vector, then $\gamma'(e_0, \delta_1)=e_0$, and $\phi_1(\gamma'(e_0, \delta_1))=\alpha(\sigma_1)=\alpha(\sigma(q_0, \phi_0(e)))$. If not, then $E(m+1)\neq E(m)$, hence $\phi_1(E(m+1))=\phi_1(\gamma'(e_0, \delta_1))$. Put $\theta_S(m+1)=S(m+1)$, so

$$\theta_S(m+1)=\alpha(\sigma(q_0,\theta_S(m)))\text{if } \delta_1=0$$

$$\phi_1(\gamma'(e_0,\delta(q_0,\theta_S(m))))\text{otherwise}$$

Note here, also, that the image $\theta_S(m+1)$ is a function of $\theta_S(m)$. Consider next $i=1$, and the second transformation response is $F(m+2)=\sigma(Q(m+1), S(m+1))=\sigma(\omega(q_0, \phi_0(e_0)), \phi_1(\gamma'(e_0, \delta(q_0, \phi_0(e_0)))))$, the second spatial change response is $D(m+2)=\delta(Q(m+1), S(m+1))=\delta(\omega(q_0, \phi_0(e_0)), \phi_1(\gamma'(e_0, \delta(q_0, \phi_0(e_0)))))$, and the next time change response is $C(m+2)=\tau(Q(m+1), S(m+1))=\tau(\omega(q_0, \phi_0(e_0)), \phi_1(\gamma'(e_0, \delta(q_0, \phi_0(e_0)))))$. Following as before, put $\theta_F(m+2)=F(m+2)$, $\theta_D(m+2)=D(m+2)$, and $\theta_C(m+2)=C(m+2)$ so that

$$\theta_F(m+2)=\sigma(\theta_Q(m+1),\theta_S(m+1))$$

$$\theta_D(m+2)=\delta(\theta_Q(m+1),\theta_S(m+1))$$

$$\theta_C(m+2)=\tau(\theta_Q(m+1),\theta_S(m+1))$$

Note that, for each of these three moment functions involving responses, the images of the second domain element are functions of the moment functions θ_Q and θ_S already discussed.

Continuing in this manner for all values of i corresponding to the moments $m+i$ in the moment base, the images of an arbitrary domain element under the various functions are similarly derived. Referring to earlier remarks that the automaton affects one and only one stimulus location with a transformation response at each moment, then the function ϕ_{i+1} is defined by $\phi_{i+1}(E(m+i))=\alpha(\sigma_{i+1})$, and $\phi_{i+1}(e_k)=\phi_i(e_k)$, $\forall e_k\in E'$, $e_k\neq E(m+i)$. Next, the $i+1^{st}$ environment is $E(m+(i+1))=\gamma'(\gamma'(\text{---}(\gamma'(\gamma'(e_0, \delta_1), \delta_2), \text{---}), \delta_1), \delta_{i+1})$, so define $\theta_E(m+(i+1))=E(m+(i+1))$ and

$$\theta_E(m+(i+1))=\gamma'(\theta_E(m+i), \delta_{i+1})$$

Thus, the image of an arbitrary domain element $m+(i+1)$ under θ_E, namely $\theta_E(m+(i+1))$, is a function of the preceding domain element's image under θ_E. Similarly, the $i+1^{st}$ time is $T(m+(i+1))=\lambda'(\lambda'(\text{---}(\lambda'(\lambda'(t_0,\tau_1), \tau_2), \text{---}), \tau_i), \tau_{i+1})$. Define $\theta_T(m+(i+1))=T(m+(i+1))$, so

$$\theta_T(m+(i+1))=\lambda'(\theta_T(m+i), \tau_{i+1})$$

As above, the image is a function of the preceding image. The $i+1^{st}$ state is given by $Q(m+(i+1))=\omega(Q(m+i), S(m+i))=\omega(\omega(\text{---}(\omega(\omega(q_0, \phi_0(e_0)), \phi_1(\gamma'(e_0, \delta_1))), \text{---}\ \phi_{i-1}(\gamma'(\gamma'(\text{---}(\gamma'(\gamma'(e_0, \delta_1), \delta_2), \text{---}), \delta_{i-2}), \delta_{i-1}))), \phi_i(\gamma'(\gamma'(\text{---}(\gamma'(\gamma'(e_0, \delta_1), \delta_2), \text{---}), \delta_{i-1}), \delta_i))))$.(The ω's correspond to the ϕ_j's, $j=0, 1, \text{---}, i$). Therefore, define the image of the moment $m+(i+1)$ under θ_Q as

$$\theta_Q(m+(i+1))=\omega(\theta_Q(m+i), \theta_S(m+i))$$

and it is clear that the image of an arbitrary domain element is a function of the preceding domain element's image. The $i+1^{st}$ stimulus $S(m+(i+1))= \phi_{i+1}(E(m+(i+1)))$, so $S(m+(i+1))=\phi_{i+1}(\gamma'(\gamma'(\text{---}(\gamma'(\gamma'(e_0, \delta_1), \delta_2), \text{---}), \delta_1), \delta_{i+1}))=\alpha(\sigma_{i+1})$ if $\delta_{i+1}=0$ and $\phi_i(E(m+(i+1)))$ otherwise. Define the image $\theta_S(m+(i+1))$ by

$$\theta_S(m+(i+1))=\alpha(\sigma(\theta_Q(m+i), \theta_S(m+i)))\ \text{if}$$
$$\delta(\theta_Q(m+i), \theta_S(m+i))=0$$
$$\phi_i(\theta_E(m+(i+1)))\ \text{otherwise}$$

Since $\theta_E(m+(i+1))$ is a function of $\theta_S(m+i)$ as already shown, it follows that $\theta_S(m+(i+1))$ is a function of $\theta_S(m+i)$. Finally, the response equations apply as before, hence the images of the moment functions relating to the components of response for the domain element $m+(i+1)$ are

$$\theta_F(m+(i+1))=\sigma(\theta_Q(m+i), \theta_S(m+i))$$
$$\theta_D(m+(i+1))=\sigma(\theta_Q(m+i), \theta_S(m+i))$$
$$\theta_C(m+(i+1))=\tau(\theta_Q(m+i), \theta_S(m+ssi))$$

Again, the images of an arbitrary domain element under the response-moment functions are functions of the moment functions θ_Q and θ_S already defined.

The preceding remarks pertaining to an algorithmic definition for the moment functions that interrelate the fundamental automaton principle with the automaton sets and functions are now summarized in the following theorem.

Theorem A1. *Let A be a general automaton, and suppose the minimal set of performance descriptors for A as previously outlined are known. Let* $\sigma_{i+1}=F(m+(i+1))$, $\delta_{i+1}=D(m+(i+1))$, *and* $\tau_{i+1}=C(m+(i+1))$ *denote the transformation, spatial change, and time change responses, respectively, at each moment* $m+(i+1), i=0,1,2,$ --- *in the moment base M used in the formal analysis for A. Then the aforementioned moment functions are*

a. $\theta_E: M \to \theta_E(M)=E \subseteq \bar{E}$ *defined by:*
$\theta_E(m)=e_0$
$\theta_E(m+(i+1))=\gamma'(\gamma'(---(\gamma'(\gamma'(e_0,\delta_1),\delta_2),---),\delta_1),\delta_{i+1})$
$=\gamma'(\theta_E(m+i),\delta_{i+1})$

b. $\theta_T: M \to \theta_T(M)=T \subseteq \bar{T}$ *defined by:*
$\theta_T(m)=t_0$
$\theta_T(m+(i+1))=\lambda'(\lambda'(---(\lambda'(\lambda'(t_0,\tau_1),\tau_2),---),\tau_i),\tau_{i+1})$
$=\lambda'(\theta_T(m+i),\tau_{i+1})$

c. $\theta_S: M-\{m+h\} \to \theta_S(M-\{m+h\})=S \subseteq \bar{S}$
($m+h$ is the final moment if one exists) defined by:
$\theta_S(m)=\phi_0(e_0)$
$\theta_S(m+(i+1))=\phi_{i+1}(\theta_E(m+(i+1)))$
$=\alpha(\sigma(\theta_Q(m+i),\theta_S(m+i)))$ *if*
$\delta(\theta_Q(m+i),\theta_S(m+i))=0$
$\phi_i(\theta_E(m+(i+1)))$ *otherwise*

d. $\theta_Q: M \to \theta_Q(M)=Q \subseteq \bar{Q}$ *defined by:*
$\theta_Q(m)=q_0$
$\theta_Q(m+(i+1))=\omega(\omega(---(\omega(\omega(q_0,\phi_0(e_0)),\phi_1(\gamma'(e_0,\delta_1))),$
$---,\phi_{i-1}(\gamma'(\gamma'(---(\gamma'(\gamma'(e_0,\delta_1),\delta_2),---),$
$\delta_{i-2}),\delta_{i-1}))),\phi_i(\gamma'(\gamma'(---(\gamma'(\gamma'(e_0,\delta_1),\delta_2),$
$---),\delta_{i-1}),\delta_i))))$
$=\omega(\theta_Q(m+i),\theta_S(m+i))$

e. $\theta_F: M-\{m\} \to \theta_F(M-\{m\})=F \subseteq \bar{F}$ *defined by:*
$\theta_F(m+1)=\sigma(\theta_Q(m),\theta_S(m))=\sigma_1$
$\theta_F(m+(i+2))=\sigma(\theta_Q(m+(i+1)),\theta_S(m+(i+1)))=\sigma_{i+2}$

f. $\theta_D: M-\{m\} \to \theta_D(M-\{m\})=D \subseteq \bar{D}$ *defined by:*
$\theta_D(m+1)=\delta(\theta_Q(m),\theta_S(m))=\delta_1$
$\theta_D(m+(i+2))=\delta(\theta_Q(m+(i+1)),\theta_S(m+(i+1)))=\delta_{i+2}$

g. $\theta_C: M-\{m\} \to \theta_C(M-\{m\})=C \subseteq \bar{C}$ *defined by:*
$\theta_C(m+1)=\tau(\theta_Q(m),\theta_S(m))=\tau_1$
$\theta_C(m+(i+2))=\tau(\theta_Q(m+(i+1)),\theta_S(m+(i+1)))=\tau_{i+2}$

To illustrate the theorem, consider the example of the abstract general automaton used in 2.1. The following information obtained from the formal analysis for A is the minimal set of performance descriptors of the system:

moment base—$M=\{m, m+1, ---, m+12\}$
environment space—R^3(three dimensional)
initial conditions—$E(m)=e_{0,0,0}=(0, 0, 0)\in R^3$
$T(m)=t_0=0\in R^1$
$Q(m)=q_0$
$S(m)=S_3=\phi_0(e_{0,0,0})$
set of states—$Q=\{q_0, q_1, ---, q_5\}$
set of stimuli—$S=\{s_0, s_1, ---, s_4\}$

Included in the contents of Table A1 is the initial environment function ϕ_0, the set $V\subseteq Q\times S$ determined by the nature of the automaton, the functions σ, δ, τ, and ω along with the defined response sets, and the function α which associates a stimulus with each transformation response.

The results of applying Theorem A1 to the example automaton A are listed in Table A2. The initial domain element's images under the moment functions θ_E, θ_T, θ_Q, θ_S are obtainable from the minimal set of performance descriptors. These are

$$\theta_E(m)=e_{0,0,0}$$

$$\theta_T(m)=t_0$$

$$\theta_Q(m)=q_0$$

$$\theta_S(m)=\phi_0\,(e_{0,0,0})=s_3$$

Also given in the minimal set of performance descriptors are response functions σ, δ, τ. The initial domain element's images under the moment functions θ_F, θ_D, θ_C are

$$\theta_F(m+1)=\sigma(\theta_Q(m),\theta_S(m))=\sigma(q_0,s_3)=f_0$$

$$\theta_D(m+1)=\delta(\theta_Q(m),\theta_S(m))=\delta(q_0,s_3)=d_1=(1,0,0)$$

$$\theta_C(m+1)=\tau(\theta_Q(m),\theta_S(m))=\tau(q_0,s_3)=c_1=(1)$$

The environment function ϕ_1 is defined by $\phi_1(\theta_E(m))=\phi_1(e_{0,0,0})=\alpha(f_0)=s_0$, and $\phi_1(e_k)=\phi_0(e_k)$, $\forall e_k\in E'$, $e_k\neq e_{0,0,0}$. The image of $m+1$ under θ_E is $\gamma'(\theta_E(m), \delta_1)=\gamma'(e_{0,0,0}, d_1)=e_{1,0,0}$; i.e.,

TABLE A1. Sets and functions for an example automaton that belong in the minimal set of performance descriptors

E'	$\phi_0(E')=S$	V	$\sigma(V)=F$	$\delta(V)=D$	$\tau(V)=C$	$\omega(V)$	$\alpha(\sigma(V))$
$e_{0,0,0}=(0, 0, 0)$	s_3	(q_0, s_3)	f_0	$d_1=(1, 0, 0)$	$c_1=1$	q_1	s_0
$e_{1,0,0}=(1, 0, 0)$	s_2	(q_1, s_0)	f_2	$-d_2=(0, -1, 0)$	$c_2=2$	q_3	s_2
$e_{1,-1,0}=(1, -1, 0)$	s_2	(q_1, s_1)	f_2	$d_3=(0, 0, 1)$	$c_2=2$	q_2	s_2
$e_{1,0,1}=(1, 0, 1)$	s_1	(q_1, s_2)	f_0	$d_0=(0, 0, 0)$	$c_0=0$	q_1	s_0
$e_{0,0,1}=(0, 0, 1)$	s_2	(q_2, s_0)	f_0	$d_0=(0, 0, 0)$	$c_0=0$	q_3	s_0
$e_{0,1,0}=(0, 1, 0)$	s_1	(q_2, s_2)	f_2	$-d_3=(0, 0, -1)$	$c_2=2$	q_3	s_2
$e_{-1,1,0}=(-1, 1, 0)$	s_4	(q_2, s_4)	f_0	$d_0=(0, 0, 0)$	$c_0=0$	q_5	s_0
e_k	s_0	(q_3, s_0)	f_1	$d_2=(0, 1, 0)$	$c_2=2$	q_4	s_1
		(q_3, s_2)	f_1	$d_0=(0, 0, 0)$	$c_0=0$	q_1	s_1
		(q_4, s_1)	f_1	$-d_1=(-1, 0, 0)$	$c_1=1$	q_2	s_1
e_k=all others							

TABLE A2. Results of the application of Theorem A1 to the performance descriptors of the abstract example automaton of 2.1

M	$\theta_E(M)$	$\theta_T(M)$	$\theta_Q^*(M)$	$\theta_S(M)$	$\theta_F(M)$	$\theta_D(M)$	$\theta_C(M)$	ϕ_i changes
m	$e_{0,0,0}$	t_0	q_0	s_3	—	—	—	—
$m+1$	$e_{1,0,0}$	t_1	q_1	s_2	f_0	d_1	c_1	$\phi_1(e_{0,0,0})=s_0$
$m+2$	$e_{1,0,0}$	t_1	q_1	s_0	f_0	d_0	c_0	$\phi_2(e_{1,0,0})=s_0$
$m+3$	$e_{1,-1,0}$	t_3	q_3	s_2	f_2	$-d_2$	c_2	$\phi_3(e_{1,0,0})=s_2$, $\phi_3=\phi_1$
$m+4$	$e_{1,-1,0}$	t_3	q_1	s_1	f_1	d_0	c_0	$\phi_4(e_{1,-1,0})=s_1$
$m+5$	$e_{1,-1,1}$	t_5	q_2	s_0	f_2	d_3	c_2	$\phi_5(e_{1,-1,0})=s_2$, $\phi_5=\phi_3$
$m+6$	$e_{1,-1,1}$	t_5	q_3	s_0	f_0	d_0	c_0	$\phi_6(e_{1,-1,1})=s_0$, $\phi_6=\phi_5$
$m+7$	$e_{1,0,1}$	t_7	q_4	s_1	f_1	d_2	c_2	$\phi_7(e_{1,-1,1})=s_1$
$m+8$	$e_{0,0,1}$	t_8	q_2	s_2	f_1	$-d_1$	c_1	$\phi_8(e_{1,0,1})=s_1$, $\phi_8=\phi_7$
$m+9$	$e_{0,0,0}$	t_{10}	q_3	s_0	f_2	$-d_3$	c_2	$\phi_9(e_{0,0,1})=s_2$, $\phi_9=\phi_8$
$m+10$	$e_{0,1,0}$	t_{12}	q_4	s_1	f_1	d_2	c_2	$\phi_{10}(e_{0,0,0})=s_1$
$m+11$	$e_{-1,1,0}$	t_{13}	q_2	s_4	f_1	$-d_1$	c_1	$\phi_{11}(e_{0,1,0})=s_1$, $\phi_{11}=\phi_{10}$
$m+12$	$e_{-1,1,0}$	t_{13}	q_5		f_0	d_0	c_0	$\phi_{12}(e_{-1,1,0})=s_0$

$$\theta_E(m+1)=e_{0,0,0}$$

Similarly, the image of $m+1$ under θ_T is $\lambda'(\theta_T(m), \tau_1)=\lambda'(t_0, c_1)=t_1$, and

$$\theta_T(m+1)=t_1$$

Next, the image of the second domain element under the moment function θ_S is given by $\theta_S(m+1)=\phi_1(\theta_E(m+1))=\phi_1(e_{1,0,0})=\phi_0(e_{1,0,0})$, and

$$\theta_S(m+1)=s_2$$

Also, for the function θ_Q, $\theta_Q(m+1)=\omega(\theta_Q(m), \theta_S(m))=\omega(q_0, s_3)$ and with ω given, then

$$\theta_Q(m+1)=q_1$$

is the image of the next domain element under θ_Q. Finally, the images of the second domain element $m+2$ under the response moment functions are

$$\theta_F(m+2)=\sigma(\theta_Q(m+1), \theta_S(m+1))=\sigma(q_1, s_2)=f_0$$
$$\theta_D(m+2)=\delta(\theta_Q(m+1), \theta_S(m+1))=\delta(q_1, s_2)=d_0$$
$$\theta_C(m+2)=\tau(\theta_Q(m+1), \theta_S(m+1))=\tau(q_1, s_2)=c_0$$

Continuing in the same manner for all choices of $i=0, 1, 2, \text{---}, 11$, Table A2 is established.

Several results describing the fundamental automaton principle as it relates to other automaton sets and functions follow from Theorem A1. Recall the set $V \subseteq Q \times S$ of state-stimulus pairs over the moment base determined by the nature of the automaton. Clearly, there is one and only one state-stimulus pair for each moment. Corollary A1.1 is an obvious result of Theorem A1 in light of the above remarks.

Corollary A1.1. *There is a moment function θ_V describing the state-stimulus pair associated with each moment. That is, θ_V: $M-\{m+h\} \rightarrow \theta_V(M-\{m+h\})=V \subseteq Q \times S$ defined by*

$$\theta_V(m)=(\theta_Q(m), \theta_S(m))$$
$$\begin{aligned}\theta_V(m+(i+1)) &= (\theta_Q(m+(i+1)), \theta_S(m+(i+1)))\\ &= (\omega(\theta_V(m+i)), \phi_{i+1}(\gamma'(\theta_E(m+i),\\ &\quad \delta(\theta_V(m+i)))\end{aligned}$$

where $m+h$ denotes the final moment if one exists.

As shown in 2.1, the dependence of the function γ' on the set $D=\delta(V)$ interrelates the automaton function δ with γ'. By composition, there is a function $\gamma=\gamma' \circ \gamma''$ with γ: $W \subseteq E \times V \rightarrow \gamma(W) \subseteq E \subseteq \bar{E}$ that describes the image of an environment-state-stimulus as the next environment. Again, from the fundamental automaton principle, one and only one environment-state-stimulus triple is associated with each moment. Applying Theorem A1 along with Corollary A1.1 to the above remarks, the following result is obtained:

Corollary A1.2. *There is a moment function θ_W describing the environment-state-stimulus triple associated with each moment. That is, $\theta_W: M-\{m+h\} \rightarrow \theta_W(M-\{m+h\})=W \subseteq E \times V$, and θ_W is defined by*

$$\theta_W(m)=(\theta_E(m), \theta_Q(m), \theta_S(m))$$
$$\begin{aligned}\theta_W(m+(i+1)) &= (\theta_E(m+(i+1)), \theta_Q(m+(i+1)), \theta_S(m+(i+1)))\\ &= (\gamma(\theta_W(m+i)), \theta_V(m+(i+1)))\end{aligned}$$

where $m+h$ denotes the final moment if one exists.

In the same manner, the dependence of the function λ' on the set $C=\tau(V)$ interrelates the automaton function τ with λ'. Again, by composition, there is a function $\lambda=\lambda'\circ\lambda''$ with $\lambda: N\subseteq T\times V\rightarrow\lambda(N)\subseteq T\subseteq\bar{T}$ that describes the image of the time-state-stimulus triple as the next time. Since there is one and only one such triple, a Corollary A1.3 parallels the preceding corollary.

Corollary A1.3. *There is a function θ_N describing the time-state-stimulus triple associated with each moment. That is, $\theta_N: M-\{m+h\}\rightarrow\theta_N(M-\{m+h\})=N\subseteq T\times V$, and θ_N is defined by*

$$\theta_N(m)=(\theta_T(m),\theta_Q(m),\theta_S(m))$$

$$\begin{aligned}\theta_N(m+(i+1))&=(\theta_T(m+(i+1)),\theta_Q(m+(i+1)),\theta_S(m+(i+1)))\\&=(\lambda(\theta_N(m+i)),\theta_V(m+(i+1)))\end{aligned}$$

where $m+h$ denotes the final moment $m+h$ if one exists.

Finally, let $Y\subseteq T\times W$ denote the set of time-environment-state-stimulus 4-tuples over the moment base used in the formal analysis for a General automaton. Defined in 2.1 was the function $\pi: Y\subseteq T\times W\rightarrow\pi(Y)\subseteq\gamma(W)\times\lambda(N)\subseteq E\times T$. Also defined was the parallel function μ such that $\mu: Y'\subseteq E\times N\rightarrow\mu(Y')\subseteq\lambda(N)\times\gamma(W)\subseteq T\times E$, where $Y'\subseteq E\times N$ denotes the set of environment-time-state-stimulus 4-tuples over the moment base. Applying Theorem A1 along with Corollary A1.1 and the aforementioned functions π and μ, Corollary A1.4 is obtained.

Corollary A1.4. *There is a moment function θ_Y describing the time-environment-state-stimulus 4-tuple associated with each moment. That is, $\theta_Y: M-\{m+h\}\rightarrow\theta_Y(M-\{m+h\})=Y\subseteq T\times W$, where θ_Y is defined by*

$$\theta_Y(m)=(\theta_T(m),\theta_E(m),\theta_Q(m),\theta_S(m))$$

$$\begin{aligned}\theta_Y(m+1)&=(\theta_T(m+1),\theta_E(m+1),\theta_Q(m+1),\theta_S(m+1)),\ \theta_Y(m+(i+1))\\&=(\theta_T(m+(i+1)),\theta_E(m+(i+1)),\theta_Q(m+(i+1)),\theta_S(m+(i+1)))\\&=(\mu(\pi(\theta_Y(m+(i-1)))),\theta_V(m+i)),\theta_V(m+(i+1))))\end{aligned}$$

Similarly, there is a moment function $\theta_{Y'}$ describing the environment-time-state-stimulus 4-tuple corresponding to each moment. That is, $\theta_{Y'}: M-\{m+h\}\rightarrow\theta_{Y'}(M-\{m+h\})=Y'\subseteq E\times N$, where $\theta_{Y'}$ is defined by

$$\theta_{Y'}(m)=(\theta_E(m),\theta_T(m),\theta_Q(m),\theta_S(m))$$

$$\theta_{Y'}(m+1)=(\theta_E(m+1),\theta_T(m+1),\theta_Q(m+1),\theta_S(m+1))$$

$$\begin{aligned}\theta_{Y'}(m+(i+1))&=(\theta_E(m+(i+1)),\theta_T(m+(i+1)),\theta_Q(m+(i+1)),\ \theta_S(m+(i+1)))\\&=(\pi(\mu(\theta_Y(m+(i-1)))),\theta_V(m+i)),\theta_V(m+(i+1)))\end{aligned}$$

where $m+h$ denotes the final moment if one exists. Furthermore, $\theta_{Y'}(m+(i+1)) = (\rho_2(\theta_Y(m+(i+1))),\ \rho_1(\theta_Y(m+(i+1))),\ \rho_3(\theta_Y(m+(i+1))),\ \rho_4(\theta_Y(m+(i+1))))$, $\forall i$, *with* $m+i \in M$, *where* ρ_j *denotes the jth projection(identity) function.*

The intention was to keep the defining algorithms for the functions interrelating the fundamental automaton principle with the automaton sets and functions (as presented in Theorem A1 and associated corollaries) within the framework of the nonnumeric recursion concept. The moment functions θ_E, θ_T, θ_Q, θ_S as defined in Theorem A1 fit in this framework. First, both the domain and codomain sets consist of different elements. However, the domain set is the simplest holomorph set in that the domain elements are in a one-to-one correspondence with a naturally ordered subset of the nonnegative integers; i.e., the correspondence is given by $m+i \rightarrow i \rightarrow m+i$. Furthermore, the image of an arbitrary domain element is a function of the image of the preceding domain element (determined by the nature of the domain set). The intermediate functions in the algorithms are compositions of functions given in the minimal set of performance descriptors for a general automaton. Since the aforementioned moment functions are described by a nonnumeric recursion concept paralleling primitive-recursion for number-theoretic functions, the algorithms for carrying out the calculations of the functions are computable. It follows that the moment functions θ_E, θ_T, θ_Q, and θ_S belong to the class of general recursive functions. Since the remaining moment functions θ_F, θ_D, θ_C of Theorem A1 are defined in terms of θ_Q and θ_S using the automaton functions σ, δ, τ given in the minimal set of performance descriptors, the algorithms for carrying out their calculations are also computable. The moment functions θ_V, θ_W, θ_N, θ_Y, $\theta_{Y'}$ defined in Corollaries A1.1–A1.4 are based on the functions defined in Theorem A1, hence the algorithms for carrying out their calculations are also computable. Furthermore, in the case of θ_V, θ_W, θ_N, the image of an arbitrary domain element is a function of preceding domain elements' images under the moment functions. However, the image of an arbitrary element under the moment functions θ_Y and $\theta_{Y'}$ are functions of the images of the predecessor to preceding domain element under θ_Y and $\theta_{Y'}$. In summary

Theorem A2. *The moment functions in the set* $\{\theta_T, \theta_E, \theta_Q, \theta_S, \theta_F, \theta_D, \theta_C, \theta_V, \theta_W, \theta_N, \theta_Y, \theta_{Y'}\}$ *belong to the class of general recursive functions.*

Effective Adjacency Functions. The formal analysis of 2.1 concluded with a graph model for a general automaton based on the physical system. The purpose of the ensuing discussion is to define a concept of adjacency for points of the graph that will enable the interrelation of the fundamental automaton principle and the automaton set and functions to be translated in graph terms.

Utilizing the basic automaton sets and functions, a graph model was defined for a general automaton in Section 2.1. The model consists of a set of general directed graphs established by sets and functions based on those describing the physical system. There are three types of graphs in the model called processor, environment, and time graphs, and each reflects the operation on the automaton as it performs internally and externally in space and time. The graphs are general directed graphs because the lines joining distinct points are directed, and because parallel (multiple) lines, loops, and parallel loops are permitted.

The concept of adjacency for points of graphs as defined in the literature for graph theory is that any two points joined by a line are adjacent (Harary, Norman, Cartwright 1965). The word adjacent is used in the sense that the two points are not separated by another point or set of points. However, in graphs defined by sets and functions formally describing a physical system, such as found in the graph model for a general automaton, adjacency is a time-variant concept. Consider the following definition:

Definition A1. In a graph of the graph model for a general automaton, any two points connected by a directed line (loop) are adjacent only at the moment(s) whenever the automaton's current status with respect to the type of graph is represented by the origin point, and the automaton operates according to the conditions described by the connecting line to achieve the status represented by the terminal point. This concept of adjacency is called time-variant adjacency.

In dealing with graphs of the graph model, the term conditional adjacency is used to denote the usual graph-theoretic meaning of the word adjacency. As a result, any two points of a graph in the model connected by a line are conditionally adjacency points. If the points are distinct, the origin is conditionally adjacent to the terminal point, and the terminal point is conditionally adjacent from the origin point. Hereafter, any use of the word adjacency without the adjective *conditional* refers to the defined concept of time-variant adjacency.

The fundamental automaton principle translated in terms of the graphs of the model states that at each moment one and only one point represents a current status of the automaton with respect to the type of graph. Moreover, the operating condition of an automaton in achieving a change of status from a given moment to the next is described by one and only one line directed from a given point to the next. A fundamental graph principle for graphs of the model for a general automaton parallels the fundamental automaton principle.

General Automaton Graph Principle. There is one and only one pair of adjacent points connected by one and only one line in the graphs of the model for a general automaton corresponding to each pair of consecutive moments from the moment base used in the formal analysis.

The automaton graph principle translates the fundamental automaton principle in terms of the graph by means of the concept of time-variant adjacency.

Inherent in the results of Theorems A1, A2, and corollaries is the recursive nature of the moment functions for formally describing the automaton principle in terms of the physical system. The purpose of the following remarks is to establish effective adjacency functions formally describing the graph principle in terms of the graphs in the model. Since the graph principle is based on the automaton principle, the effective adjacency functions are defined in terms of the moment functions. The effective adjacency functions are general recursive and form the basis for developing other functions for determining the effective operation of the automaton through select properties of the graphs.

Given any one of the graphs in the model, and an arbitrary pair of consecutive moments from the moment base, the graph principle states that there is one and only one pair of adjacent points and only one line joining these points. These points and line are determined by the effective adjacency function. Consider the following definition:

Definition A2. Given a graph of the model for a general automaton, and given an arbitrary pair of consecutive moments from the moment base, an effective adjacency function is a function that determines the pair of adjacent points and joining line as described in the general automaton graph principle.

The main objective of the discussion on effective adjacency functions is to define such functions by means of an algorithm within the concept of nonnumeric recursion that is computable. Based on the previous remarks, effective adjacency functions by concept are the means of formally describing the graph principle in terms of the graphs in the model. It follows from the graph principle that the domain set for the effective adjacency functions is a set of pairs of consecutive moments from the moment base. This set is a simple holomorph set using the correspondence $(m+i, m+(i+1)) \rightarrow i \rightarrow (m+i, m+\rho_s(i))$, where ρ_s denotes the successor function. Moreover, it follows from the graph principle that the range elements should denote the pair of points and proper line joining them. Relying on the graph primitives, a natural codomain is the set of all possible n-tuples as described in Primitives P3, E_p3, T_p3. Assuming knowledge of the moment functions as described in Theorems A1, A2, and corollaries and knowledge of the graph functions as defined using the automaton function given in the minimal set of performance descriptors, the images of successive domain elements under the effective adjacency functions are functions of specific moment and graph functions. For reasons of simplicity, the effective adjacency functions are pursued by type of graph starting with the processor graph.

Suppose $\overline{M} \subseteq M \times M$ denotes the set of pairs of consecutive moments from the moment base used to analyze a general automaton; i.e., $\overline{M} = \{(m, m+1), (m+1, m+2), ---, (m+i, m+(i+1)), ---\}, \forall i = 0, 1, ---$, such that $m+i \in M$. Let $G_M(P)$ denote the processor graph defined by Primitives P1 through P4 and the graph function Ω (from 2.1). Let ψ_p denote a function mapping $\overline{M}$ into a subset of the product space $Q\times S\times F\times D\times C\times Q$ corresponding to the sextuples described in Primitive P3; i.e., $\psi_p:\overline{M}\rightarrow\psi_p(\overline{M})\subseteq Q\times S\times F\times D\times C\times Q$. Define $\psi_p(m, m+1)=(\theta_Q(m), \theta_S(m), \theta_F(m+1), \theta_D(m+1), \theta_C(m+1), \theta_Q(m+1))$. Then $\psi_p(m, m+1)$ is the initial sextuple outlined in Primitive P3. Applying the graph function Ω and Corollary A1.1 then $\psi_p(m, m+1)=(\theta_V(m), \Omega(\theta_V(m)))$. The adjacent points are labeled by the first and last entries of the sextuple, namely $\theta_Q(m)$ and $\theta_Q(m+1)$; and the joining line is denoted by the remaining entries, namely $\theta_S(m)$, $\theta_F(m+1)$, $\theta_D(m+1)$, and $\theta_C(m+1)$. Next, define $\psi_p(m+1, m+2)=(\theta_Q(m+1), \theta_S(m+1), \theta_F(m+2), \theta_D(m+2), \theta_C(m+2), \theta_Q(m+2))=(\theta_V(m+1), \Omega(\theta_V(m+1)))$. To establish the relationship of the second image under ψ_p to the preceding domain element's image, let ρ_j denote the jth projection (identity) function. With $\psi_p(m, m+1)=(\theta_Q(m), \theta_S(m), \theta_F(m+1), \theta_D(m+1), \theta_C(m+1), \theta_Q(m+1))$, then $\rho_6(\psi_p(m, m+1))=\theta_Q(m+1)$, and $\rho_4(\psi_p(m, m+1))=\theta_D(m+1)$. By Theorem A1, $\theta_S(m+1)=\phi_1(\gamma'(\theta_E(m), \theta_D(m+1)))=\phi_1(\gamma'(\theta_E(m), \rho_4(\psi_p(m, m+1))))$. Applying the automaton equation describing the transformation response, $\theta_F(m+2)=\sigma(\theta_Q(m+1), \theta_S(m+1))$. Substituting from the above remarks, then $\theta_F(m+2)=\sigma(\rho_6(\psi_p(m, m+1)), \phi_1(\gamma'(\theta_E(m), \rho_4(\psi_p(m, m+1)))))$. Similarly, $\theta_D(m+2)$ and $\theta_C(m+2)$ parallel $\theta_F(m+2)$. Finally, substituting for all parts in the expression for $\psi_p(m+1, m+2)$ paralleling that for $\psi_p(m, m+1)$ given above, $\psi_p(m+1, m+2)=(\rho_6(\psi_p(m, m+1)), \phi_1(\gamma'(\theta_E(m), \rho_4(\psi_p(m, m+1)))), \sigma(\rho_6(\psi_p(m, m+1)), \phi_1(\gamma'(\theta_E(m), \rho_4(\psi_p(m, m+1))))), \delta(\rho_6(\psi_p(m, m+1)), \phi_1(\gamma'(\theta_E(m), \rho_4(\psi_p(m, m+1))))), \tau(\rho_6(\psi_p(m, m+1)), \phi_1(\gamma'(\theta_E(m), \rho_4(\psi_p(m, m+1))))), \omega(\rho_6(\psi_p(m, m+1)), \phi_1(\gamma'(\theta_E(m), \rho_4(\psi_p(m, m+1))))))$. Continuing in this manner $\forall i=0, 1, \text{---}$ with $m+(i+2)\in M$, then $\psi_p(m+(i+1), m+(i+2))=(\theta_Q(m+(i+1)), \theta_S(m+(i+1)), \theta_F(m+(i+2)), \theta_D(m+(i+2)), \theta_C(m+(i+2)), \theta_Q(m+(i+2)))=(\theta_V(m+(i+1)), \Omega(\theta_V(m+(i+1))))$. To establish the relationship of an arbitrary domain element's image under ψ_p to the preceding domain element's image, as before, let ρ_j denote the jth projection (identity function). With $\psi_p(m+i, m+(i+1))=(\theta_Q(m+i), \theta_S(m+i), \theta_F(m+(i+1)), \theta_D(m+(i+1)), \theta_C(m+(i+1)), \theta_Q(m+(i+1)))$, then $\rho_6(\psi_p(m+i, m+(i+1)))=\theta_Q(m+(i+1))$, and $\rho_4(\psi_p(m+i, m+(i+1)))=\theta_D(m+(i+1))$. By Theorem A1, $\theta_S(m+(i+1))=\phi_{i+1}(\gamma'(\theta_E(m+i), \theta_D(m+(i+1))))=\phi_{i+1}(\gamma'(\theta_E(m+i), \rho_4(\psi_p(m+i, m+(i+1)))))$. Again, applying the automaton equation describing the transformation response, $\theta_F(m+(i+2))=\sigma(\theta_Q(m+(i+1)), \theta_S(m+(i+1)))$. Substituting from the above remarks, then $\theta_F(m+(i+2))=\sigma(\rho_6(\psi_p(m+i, m+(i+1))), \phi_{i+1}(\gamma'(\theta_E(m+1), \rho_4(\psi_p(m+i, m+(i+1))))))$. Similarly, $\theta_D(m+(i+2))$ and $\theta_C(m+(i+2))$ parallel $\theta_F(m+(i+2))$. Finally, substituting for all parts in the original expression $\psi_p(m+(i+1), m+(i+2))=(\rho_6(\psi_p(m+i, m+(i+1)))$

$$\begin{aligned}&\phi_{i+1}(\gamma'(\theta_E(m+i),\rho_4(\psi_p(m+i,m+(i+1))))),\sigma(\rho_6(\psi_p(m+i,m+(i+1))),\\&\quad\phi_{i+1}(\gamma'(\theta_E(m+i),\rho_4(\psi_p(m+i,m+(i+1)))))),\delta(\rho_6(\psi_p(m+i,m+(i+1))),\\&\quad\phi_{i+1}(\gamma'(\theta_E(m+i),\rho_4(\psi_p(m+i,m+(i+1)))))),\tau(\rho_6(\psi_p(m+i,m+(i+1))),\\&\quad\phi_{i+1}(\gamma'(\theta_E(m+i),\rho_4(\psi_p(m+i,m+(i+1)))))),\omega(\rho_6(\psi_p(m+i,m+(i+1))),\\&\quad\phi_{i+1}(\gamma'(\theta_E(m+i),\rho_4(\psi_p(m+i,m+(i+1)))))))\end{aligned}$$

Thus, for arbitrary choices of pairs of consecutive moments in the moment base, the image of the choice under the function ψ_p is a sextuple of the processor graph $G_M(P)$ which denotes the proper pair of adjacent points and joining line. Note that both the domain and codomain sets for the function ψ_p are nonnumeric although the domain set $\overline{M}$ is isomorphic to a naturally ordered subset of the nonnegative integers under the mapping $(m+i, m+(i+1)) \to i \to (m+i, m+\rho_s(i))$, where ρ_s is the successor function. As observed from the definition, the image of an arbitrary domain element is a function of the preceding domain element's image. The intermediate function involves either defined functions or recursive functions, hence there is a computable algorithm for carrying out the calculations of ψ_p. The foregoing remarks are summarized in the following theorem:

Theorem A3. *Let $\overline{M} \subseteq M \times M$ denote the set of pairs of consecutive moments from the moment base used in the formal analysis for a general automaton, and let $G_M(P)$ denote the processor graph defined by the Primitives* P1 *through* P4 *and the graph function Ω. Let ρ_j denote the jth projection (identity) function. Then the function $\psi_p : \overline{M} \to \psi_p(\overline{M}) \subseteq Q \times S \times F \times D \times C \times Q$ defined by*

$$\begin{aligned}\psi_p(m, m+1) &= (\theta_Q(m), \theta_S(m), \theta_F(m+1), \theta_D(m+1), \theta_C(m+1), \theta_Q(m+1)) \\ &= (\theta_V(m), \Omega(\theta_V(m)))\end{aligned}$$

$$\begin{aligned}\psi_p(m+(i+1), m+(i+2)) &= (\theta_V((m+(i+1)), \Omega(\theta_V(m+(i+1))))) \\ &= (\rho_6(\psi_p(m+i, m+(i+1))), \\ &\quad \phi_{i+1}(\gamma'(\theta_E(m+i), \rho_4(\psi_p(m+i, m+(i+1))))), \\ &\quad \Omega(\rho_6(\psi_p(m+i, m+(i+c1))), \\ &\quad \phi_{i+1}(\gamma'(\theta_E(m+i), \rho_4(\psi_p(m+i, m+(i+)))))))\end{aligned}$$

$\forall i = 0, 1, ---,$ *such that $m+(i+2) \in M$, is an effective adjacency function for $G_M(P)$ and belongs in the class of general recursive functions.*

Consider again the example automaton. The graph function Ω is defined in Table 2.1b, and the corresponding processor graph $G_M(P)$ is shown in Figure 2.2a. With $\theta_Q(m) = q_0$, $\theta_S(m) = s_3$, $\theta_F(m+1) = f_0$, $\theta_D(m+1) = d_1$, and $\theta_C(m+1) = c_1$ as initial conditions, and applying Theorem A1, then $\theta_Q(m+1) = \omega(q_0, s_3) = q_1$ (see Tables A1, A2). So $\psi_p(m, m+1) = (\theta_Q(m), \theta_S(m), \theta_F(m+1), \theta_D(m+1), \theta_C(m+1), \theta_Q(m+1)) = (q_0, s_3, f_0, d_1, c_1, q_1)$, and the adjacent points and connecting line in the graph at moments m, $m+1$ is q_0, q_1, and (s_3, f_0, d_1, c_1). Next, applying the recursion, $\rho_6(\psi_p(m, m+1)) = q_1$, $\phi_1(\gamma'(\theta_E(m), \rho_4(\psi_p(m, m+1)))) = \phi_1(\gamma'(e_{0,0,0}, d_1)) = \phi_1(e_{1,0,0}) = \phi_0(e_{1,0,0}) = s_2$. Therefore, $\psi(m+1, m+2) = (q_1, s_2, \Omega(q_1, s_2)) = (q_1, s_2, f_0, d_0, c_0, q_1)$, and the pair of adjacent points corresponding to moment $m+1$, $m+2$ are q_1 and q_1 with the joining line (loop) as(s_2, f_0, d_0, c_0). For $i=1$, $\rho_6(\psi_p(m+1, m+2)) = q_1$, and $\phi_2(\gamma'(\psi_E(m+1, \rho_4(\psi_p(m+1, m+2))))) = \phi_2(\gamma'(e_{1,0,0}, d_0)) = \phi_2(e_{1,0,0}) = \alpha(f_0)$. $\alpha(f_0) = s_0$, hence $\psi_p(m+2, m+3) = (q_1, s_0, \Omega(q_1, s_0)) = (q_1, s_0, f_2, -d_2, c_2, q_3)$. Thus, the adjacent points at moments $m+2$, $m+3$

are q_1 and q_3 joined by the line $(s_0, f_1, -d_2, c_2)$. Continuing, the function ψ_p is defined as it appears in Table A3. Note that Table A3 is obtainable from a proper use of Table A2 based on Theorem A3.

Next, consider the principal environment graph $G_M(E_p)$ as defined by the Primitives E_p1 through E_p4 and the graph function Γ (see 2.1). Let ψ_E denote a function mapping $\overline{M}$ into a subset of the product space $E \times Q \times S \times F \times D \times C \times Q \times E$ corresponding to the octuples described in Primitive E_p3; i.e., $\psi_E : \overline{M} \to \psi_E(\overline{M}) \subseteq E \times Q \times S \times F \times D \times C \times Q \times E$. Using arguments similar to those preceding Theorem A3, ψ_E is defined in the same manner as ψ_p allowing for the function θ_E. Moreover, replacing the function θ_V with θ_W and the graph function Ω with the graph function Γ, ψ_E parallels ψ_p in that $\psi_E(m+i, m+(i+1)) = (\theta_W(m+i), \Gamma(\theta_W(m+i)))$. Theorem A4 is a natural result for the function ψ_E paralleling Theorem A3.

Theorem A4. *Let $\overline{M} \subseteq M \times M$ denote the set of pairs of consecutive moments from the moment base, and let $G_M(E_p)$ denote the principal environment graph defined by the Primitives E_p1 through E_p4 and the graph function Γ. If p_j denotes the jth projection (identity) function, then the function $\psi_E : \overline{M} \to \psi_E(\overline{M}) \subseteq E \times \psi_p(\overline{M}) \times E \subseteq E \times Q \times S \times F \times D \times C \times Q \times E$ defined by*

$$\psi_E(m, m+1) = (\theta_E(m), \theta_Q(m), \theta_S(m), \theta_F(m+1), \theta_D(m+1), \theta_C(m+1), \theta_Q(m+1), \theta_E(m+1)) = (\theta_W(m), \Gamma(\theta_W(m)))$$

$$\begin{aligned}\psi_E(m+(i+1), m+(i+2)) &= (\theta_W(m+(i+1)), \Gamma(\theta_W(m+(i+1))))\\ &= (\rho_8(\psi_E(m+i, m+(i+1))), \rho_7(\psi_E(m+i, m+(i+1))),\\ &\quad \phi_{i+1}(\gamma'(\theta_E(m+i), \rho_5(\psi_E(m+i, m+(i+1))))),\\ &\quad \Gamma(\rho_8(\psi_E(m+i, m+(i+1))), \rho_7(\psi_E(m+i, m+(i+1))),\\ &\quad \phi_{i+1}(\gamma'(\theta_E(m+i), \rho_5(\psi_E(m+i, m+(i+1)))))))\end{aligned}$$

TABLE A3. Effective adjacency function ψ_p for the processor graph G_M (*P*) of the example automaton

$\psi_p : \overline{M} \subseteq M \times M \to \psi_p(\overline{M}) \subseteq V \times \Omega(V) \subseteq Q \times S \times F \times D \times C \times Q$	
$(m, m+1)$	$\to (q_0, s_3, f_0, d_1, c_1, q_1)$
$(m+1, m+2)$	$\to (q_1, s_2, f_0, d_0, c_0, q_1)$
$(m+2, m+3)$	$\to (q_1, s_0, f_2, -d_2, c_2, q_3)$
$(m+3, m+4)$	$\to (q_3, s_2, f_1, d_0, c_0, q_1)$
$(m+4, m+5)$	$\to (q_1, s_1, f_2, d_3, c_2, q_2)$
$(m+5, m+6)$	$\to (q_2, s_0, f_0, d_0, c_0, q_3)$
$(m+6, m+7)$	$\to (q_3, s_0, f_1, d_2, c_2, q_4)$
$(m+7, m+8)$	$\to (q_4, s_1, f_1, -d_1, c_1, q_2)$
$(m+8, m+9)$	$\to (q_2, s_2, f_2, -d_3, c_2, q_3)$
$(m+9, m+10)$	$\to (q_3, s_0, f_1, d_2, c_2, q_4)$
$(m+10, m+11)$	$\to (q_4, s_1, f_1, -d_1, c_1, q_2)$
$(m+11, m+12)$	$\to (q_2, s_4, f_0, d_0, c_0, q_5)$

$\forall i=0, 1,$ ---, *such that* $m+(i+2)\in M$, *is an effective adjacency function for* $G_M(E_p)$ *and belongs in the class of general recursive functions.*

Based on the relationship of the set W to the set V, namely, $W\subseteq E\times V$, the graph function Γ is related to the graph function Ω. More precisely, the range set under the environment graph function is $\Gamma(W)\subseteq\Omega(V)\times E$. Therefore, it is natural for the effective adjacency function ψ_E for the environment graph to reflect this relationship with the effective adjacency function ψ_p for the processor graph for the initial moment pair $(m, m+1)$, $\psi_E(m, m+1)=(\theta_E(m), \psi_p(m, m+1), \theta_E(m+1))$. It follows from Theorems A3, A4, that for arbitrary i, $\psi_E(m+i, m+(i+1))=(\theta_E(m+i), \psi_p(m+i, m+(i+1)), \theta_E(m+(i+1)))$; hence the result is Corollary A4.1.

Corollary A4.1. *The effective adjacency function* ψ_E *for the environment graph* $G_M(E_p)$ *is defined in terms of the effective adjacency function* ψ_p *for the processor graph* $G_M(P)$ *by*

$$\psi_E(m+i, m+(i+1))=(\theta_E(m+i), \psi_P(m+i, m+(i+1)), \theta_E(m+(i+1)))$$

$\forall i=0, 1,$ ---, *such that* $m+(i+1)\in M$.

The effective adjacency function ψ_E for the principal environment graph $G_M(E_p)$ of the example automaton may be determined directly by applying Theorem A4 or indirectly by applying Corollary A4.1 along with Theorem A1 and the definition of ψ_p in Table A3. The principal environment function Γ and the environment graph $G_M(E_p)$ for the example automaton as established in 2.1 are presented in Table 2.2(b) and Figure 2.3(a). To determine the effective adjacency function ψ_E directly, consider the initial pair of consecutive moments $(m, m+1)$ and $\psi_E(m, m+1)=(e_{0,0,0}, q_0, s_3, f_0, d_1, c_1, q_1, e_{1,0,0})$. Applying the recursive form of ψ_E, $\rho_8(\psi_E(m, m+1))=e_{1,0,0}$, $\rho_7(\psi_E(m, m+1))=q_1$, and $\phi_1(\gamma'(\theta_E(m), \rho_5(\psi_E(m, m+1))))=\phi_1(\gamma'(e_{0,0,0}, d_1))=\phi_1(e_{1,0,0})=s_2$. Thus, the next image is $\psi_E(m+1, m+2)=(e_{1,0,0}, q_1, s_2, \Gamma(e_{1,0,0}, q_1, s_2))=(e_{1,0,0}, q_1, s_2, f_0, d_0, c_0, q_1, e_{1,0,0})$. Continuing in the same manner for all necessary choices of i, the effective adjacency function ψ_E defined directly by Theorem A4 is presented in Table A4. Note that the Table is obtainable from a proper use of Table 2.10 based on Theorem A4. In addition, ψ_E is determined indirectly based on Corollary A4.1. For the initial domain element $(m, m+1)$, $\psi_E(m, m+1)=(\theta_E(m), \psi_p(m, m+1), \theta_E(m+1))=(e_{0,0,0}, q_0, s_3, f_0, d_1, c_1, q_1, e_{1,0,0})$. With $\theta_E(m+1)=e_{1,0,0}$, $\psi_p(m+1, m+2)=(q_1, s_2, f_0, d_0, c_0, q_1)$, and $\theta_E(m+2)=e_{1,0,0}$, then $\psi_E(m+1, m+2)=(e_{1,0,0}, q_1, s_2, f_0, d_0, c_0, q_1, e_{1,0,0})$, which agrees with the result using Theorem A4 directly. Again, continuing for all necessary choices of i, ψ_E is defined as given in Table A4.

Finally, consider the principal time graph $G_M(T_p)$ as defined by the Primitives T_p1 through T_p4 and the graph function Λ (see 2.1). Suppose ψ_T denotes a function mapping $\overline{M}$ into a subset of the product space $T\times E\times Q\times S\times F\times D\times C\times Q\times E\times T$ corresponding to the 10-tuples described in Primitive T_p3; i.e., ψ_T: $M\to\psi_T(M)\subseteq T\times E\times Q\times S\times F\times D\times C\times Q\times E\times T$. Paralleling the discussion of

TABLE A4. Effective adjacency function ψ_E for the environment graph of the example automaton

$\psi_E: M \rightarrow \psi_E(m) \subseteq E\times\psi_P(M)\times E \subseteq E\times Q\times S\times F\times D\times C\times Q\times E$	
$(m, m+1)$	$\rightarrow(e_{0,0,0},\ q_0, s_3, f_0,\ d_1, c_1, q_1, e_{1,0,0})$
$(m+1, m+2)$	$\rightarrow(e_{1,0,0},\ q_1, s_2, f_0,\ d_0, c_0, q_1, e_{1,0,0})$
$(m+2, m+3)$	$\rightarrow(e_{1,0,0},\ q_1, s_0, f_2, -d_2, c_2, q_3, e_{1,-1,0})$
$(m+3, m+4)$	$\rightarrow(e_{1,-1,0},\ q_3, s_2, f_1,\ d_0, c_0, q_1, e_{1,-1,0})$
$(m+4, m+5)$	$\rightarrow(e_{1,-1,0},\ q_1, s_1, f_2,\ d_3, c_2, q_2, e_{1,-1,1})$
$(m+5, m+6)$	$\rightarrow(e_{1,-1,1},\ q_2, s_0, f_0,\ d_0, e_0, q_3, e_{1,-1,1})$
$(m+6, m+7)$	$\rightarrow(e_{1,-1,1},\ q_3, s_0, f_1,\ d_2, c_2, q_4, e_{1,0,1})$
$(m+7, m+8)$	$\rightarrow(e_{1,0,1},\ q_4, s_1, f_1, -d_1, c_1, q_2, e_{0,0,1})$
$(m+8, m+9)$	$\rightarrow(e_{0,0,1},\ q_2, s_2, f_2, -d_3, c_2, q_3, e_{0,0,0})$
$(m+9, m+10)$	$\rightarrow(e_{0,0,0},\ q_3, s_0, f_1,\ d_2, c_2, q_4, e_{0,1,0})$
$(m+10, m+11)$	$\rightarrow(e_{0,1,0},\ q_4, s_1, f_1, -d_1, c_1, q_2, e_{-1,1,0})$
$(m+11, m+12)$	$\rightarrow(e_{-1,1,0},\ q_2, s_4, f_0,\ d_0, c_0, q_5, e_{-1,1,0})$

preceding Theorems A3, A4, ψ_T is defined in the same manner as ψ_E allowing for the moment function θ_T. By replacing the moment function θ_W with θ_Y and the graph function Γ with Λ, $\psi_T(m+i, m+(i+1))=(\theta_Y(m+i), \Lambda(\theta_Y(m+i)))$. Theorem A5 is a natural result for the function ψ_T.

Theorem A5. *Let $\overline{M}\subseteq M\times M$ denote the set of pairs of consecutive moments from the moment base, and let $G_M(T_p)$ denote the Time graph defined by Primitives* T_p1 *through* T_p4 *and the graph function Λ. Denoting the jth projection function by ρ_j, then the function ψ_T: $\overline{M}\rightarrow\psi_T(\overline{M})\subseteq T\times\psi_E(\overline{M})\times T\subseteq T\times E\times Q\times S\times F\times D\times C\times Q\times E\times T$ defined by*

$$\begin{aligned}\psi_T(m, m+1) &= (\theta_T(m), \theta_E(m), \theta_Q(m), \theta_S(m), \theta_F(m+1), \theta_D(m+1), \theta_C(m+1),\\ &\quad \theta_Q(m+1), \theta_E(m+1), \theta_T(m+1))\\ &= (\theta_Y(m), \Lambda(\theta_Y(m)))\end{aligned}$$

$$\begin{aligned}&\psi_T(m+(i+1), m+(i+2))+(\theta_Y(m+1)), \Lambda(\theta_Y(m+(i+1)))\\ &=(\rho_{10}(\psi_T(m+i, m+(i+1))), \rho_9(\psi_T(m+i, m+(i+1))), \rho_8(\psi_T(m+i, m+(i+1))),\\ &\quad \phi_{i+1}(\gamma'(\theta_E(m+i), \rho_6(\psi_T(m+i, m+(i+1))))),\\ &\quad \Lambda(\rho_{10}(\psi_T(m+i, m+(i+1))), (\rho_9(\psi_T(m+i, m+(i+1))),\\ &\quad \rho_8(\psi_T(m+i, m+(i+1))), \phi_{i+1}(\gamma'(\theta_E(m+i), \rho_6(\psi_T(m+i, m+(i+1)))))))\end{aligned}$$

$\forall i = 0, 1, ---,$ *such that $m + (i + 2) \in M$ is an effective adjacency function for $G_M(T_p)$ and belongs in the class of general recursive functions.*

Recall that $Y\subseteq T\times W$, and the graph function Λ is based on the graph function Γ in that $\Lambda(Y)\subseteq\Gamma(W)\times T$. Thus, the effective adjacency function ψ_T for the time graph is related to its counterpart ψ_E for the environment graph. In particular, $\psi_T(m, m+1)=(\theta_T(m), \psi_E(m, m+1), \theta_T(m+1))$, and by applying

Theorems A4, A5, $\psi_T(m+i,\ m+(i+1))=(\theta_T(m+i),\ \psi_E(m+i,\ m+(i+1)),\ \theta_Y(m+(i+1)))$ as stated in Corollary A5.1.

Corollary A5.1. *The effective adjacency function ψ_T for the time graph $G_M(T_p)$ is defined in terms of the effective adjacency function ψ_E for the environment graph by*

$$\psi_T(m+i, m+(i+1))=(\theta_T(m+i), \psi_E(m+i, m+(i+1))), \theta_T(m+(i+1)))$$
$$\forall i=0, 1, \text{---}, \textit{ such that } m+(i+1)\in M.$$

In the same manner as for the preceding graph, the effective adjacency function ψ_T for the Time graph may be determined directly by Theorem A5 or indirectly by Corollary A5.1.

A.3 EFFECTIVE OPERATION ANALYSIS

A concept of nonnumeric recursion was used to define a set of moment functions that formally describe the fundamental automaton principle as it relates to the automaton sets and functions characterizing the physical system. In addition, the concept was used to show that the moment functions belong to the class of general recursive functions. Next, a concept of time-variant adjacency for points of the graphs of the model was used to determine an automaton graph principle that translates the fundamental automaton principle in terms of the graphs. Then, an effective adjacency function was defined for each graph in terms of the moment functions for the purpose of describing the unique pair of points and connecting line for each pair of consecutive moments (graph principle). The recursive methods were completed by showing that the effective adjacency functions were also included in the class of general recursive functions.

The effective adjacency functions form the basis for developing other functions for determining properties of the graphs. Moreover, the properties are interpretable in terms of properties of the automaton. This section presents a significant result for defining an effective operation function that is based on the respective effective adjacency function and that determines a unique effective operation path in the graph. This path reflects the effective operation of the general automaton throughout the moment base used in the formal analysis.

Effective Operation Function. The concept of time-variant adjacency for points of the automaton graphs is equivalent to the concept of element operation for the physical system. Recall the definition of time-variant adjacency; namely, any two (distinct) points connected by a (directed) line are adjacent only at the moment(s) whenever the automaton's current status with respect to the type of graph is represented by the origin point, and the automaton operates according to the conditions described by the connecting line to achieve the status represented by the terminal point. The latter part of this definition corresponds directly to the physical system, and describes the notion of an element operation of the automaton. More specifically, an internal element operation is the current state, stimulus received and the triadic response produced, and the next state associated with any given pair of consecutive moments. The space element operation is the current environment, the internal element operation, and the next environment associated with any given pair of consecutive moments. Lastly, the time element operation is the current time, space element operation, and the next time corresponding to any given pair of consecutive moments. In as much as the effective adjacency function is a means of calculating the adjacent points of a graph in the model, given an arbitrary pair of consecutive moments, the function could also be termed an element operation function in that it determines the element operation for the arbitrary pair of consecutive moments.

The term *effective operation path* is introduced and defined as the following concept:

Definition A3. An effective operation path for a graph of the model for a general automaton is an alternating sequence of points and lines beginning with the point representing the respective initial status and including all remaining points in proper order representing the appropriate statuses, and all lines in proper order representing the appropriate operating conditions. Equivalently, the effective operation path is a sequence of element operations associated with an ordered set of pairs of consecutive moments spanning a moment base.

Note that for a moment base of cardinality two, the effective operation path, element operation, and adjacent points are one and the same. In particular, an element operation, or adjacent points, represent the simplest effective operation path.

Recall that the effective adjacency functions (element operation functions) introduced in Theorems A3, A4, A5 determine at each pair of consecutive moments, a point, connecting line, and adjacent point of the graph in the form of the appropriate n-tuple. In a natural way, these functions are used to define effective operation functions for the graphs that determine effective operation paths. The following remarks are directed toward defining the effective operation function in terms of a sequence describing the effective operation path, and showing the function belongs to the class of general recursive functions.

The procedure for defining an effective operation function involves building the aforementioned sequence using the proper effective adjacency function. Once the sequence is completed, the function defined by the sequence will be the desired effective operation function. The following discussion is restricted to the processor graph, but those involving the principal environment graph or the principal time graph are identical in the arguments used.

Let M denote the moment base used in the formal analysis for a general automaton, and let $G_M(P)$ denote the processor graph. Thus, ψ_p denotes the effective adjacency function for $G_M(P)$, and by Theorem A3, $\psi_p(m, m+1)=(\theta_Q(m), \theta_S(m), \theta_F(m+1), \theta_D(m+1), \theta_C(m+1), \theta_Q(m+1))$. Using ρ_j to denote the jth projection (identity) function, the point representing the initial state is described by $\theta_Q(m)=\rho_1(\psi_p(m, m+1))$. Therefore, let $\rho_1(\psi_p(m, m+1))$ be the term of a single term sequence. By the nature of the graph function Ω and the sextuple $\psi_p(m, m+1)$, the line associated with the sextuple is given by $(\theta_S(m), \theta_F(m+1), \theta_D(m+1), \theta_C(m+1))$. But this is precisely $(\rho_2(\psi_p(m, m+1)), \rho_3(\psi_p(m, m+1)), \rho_4(\psi_p(m, m+1)), \rho_5(\psi_p(m, m+1)))=\rho_{2,5}(\psi_p(m, m+1))$, so put $\rho_{2,5}(\psi_p(m, m+1))$ equal to the second term of a sequence beginning with the term $\rho_1(\psi_p(m, m+1))$. Using the time-variant concept of adjacency and the nature of the effective adjacency function, the point adjacent to $\rho_1(\psi_p(m, m+1))$ is the one denoted by $\rho_6(\psi_p(m, m+1))=\theta_Q(m+1)$. However, it follows that $\rho_6(\psi_p(m, m+1))=\theta_Q(m+1)=\rho_1(\psi_p(m+1, m+2))$. Continuing as before, let $\rho_1(\psi_p(m+1, m+2))$ be the third term of the sequence just started. By the same line of argument used previously, put $\rho_{2,5}(\psi_p(m+1, m+2))$ equal to the fourth term of the sequence since it denotes the second line. Continuing in this manner for all pairs of consecutive moments $(m+i, m+(i+1))$, $i=0, 1, ---$, the sequence given by $\rho_1(\psi_p(m, m+1))$, $\rho_{2,5}(\psi_p(m, m+1))$, ---, $\rho_1(\psi_p(m+i, m+(i+1)))$, $\rho_{2,5}(\psi_p(m+i, m+(i+1)))$, --- is an alternating sequence of points and lines describing the adjacent points (element operation) for every pair of consecutive moments spanning the moment base. By definition, this sequence represents an effective operation path in the processor graph $G_M(P)$, or in terms of the automaton, represents an effective operation of the automaton. Furthermore, this sequence is unique with respect to the moment base. If any different sequence described an effective operation, there would be, at some given moment, two or more states entered, or stimuli received, or responses produced that contradicts the fundamental automaton principle.

In the case where M is finite, say $M=\{m, m+1, ---, m+h\}$, the terminating point of the sequence representing the effective operation path in the graph is $\rho_6(\psi_p(m+(h-1), m+h))$. Since no moment $m+(h+1) \in M$ exists, there is no $\rho_1(\psi_p(m+h, m+(h+1)))$. Thus, the sequence representing the effective operation function is given by $\rho_1(\psi_p(m, m+1))$, $\rho_{2,5}(\psi_p(m, m+1))$, ---, $\rho_1(\psi_p(m+i, m+(i+1)))$, $\rho_{2,5}(\psi_p(m+i, m+(i+1)))$, ---, $\rho_6(\psi_p(m+(h-1), m+h))$, where $i=0, 1, ---, h-2$.

In either case, whether M is finite or infinite, the sequence describes a unique effective operation path in $G_M(P)$ and defines an effective operation function for the processor graph. The recursive nature of the function follows directly from the fact that the effective adjacency function ψ_p belongs to the class of general recursive functions.

The building of sequences that represent effective operation paths in the principal environment graph $G_M(E_p)$ and in the principal time graph $G_M(T_p)$ take the format used in the above remarks. The appropriate effective adjacency functions ψ_E and ψ_T are used along with different values for j in the jth projection functions to account for the octuple and 10-tuple differences. The following theorem summarizes these significant remarks.

Theorem A6. *Let ψ_p, ψ_E, ψ_T denote the effective adjacency functions for the processor graph $G_M(P)$, the principal environment graph $G_M(E_p)$ and the principal time graph $G_M(T_p)$, respectively. Let ρ_j denote the jth projection function.*

a. *The sequence given by $\rho_1(\psi_p(m, m+1))$, $\rho_{2,5}(\psi_p(m, m+1))$, ---, $\rho_1(\psi_p(m+i, m+(i+1)))$, $\rho_{2,5}(\psi_p(m+i, m+(i+1)))$, --- for $i=0, 1, 2$, ---, defines an effective operation function that describes a unique effective operation path in the processor graph $G_M(P)$ and belongs to the class of general recursive functions.*
b. *The sequence given by $\rho_1(\psi_E(m, m+1))$, $\rho_{2,7}(\psi_E(m, m+1))$, ---, $\rho_1(\psi_E(m+i, m+(i+1)))$, $\rho_{2,7}(\psi_E(m+i, m+(i+1)))$, --- for $i=0, 1, 2$, ---, defines an effective operation function that describes a unique effective operation path in the principal environment graph $G_M(E_p)$ and belongs to the class of general recursive functions.*
c. *The sequence given by $\rho_1(\psi_T(m, m+1))$, $\rho_{2,9}(\psi_T(m, m+1))$, ---, $\rho_1(\psi_T(m+i, m+(i+1)))$, $\rho_{2,9}(\psi_T(m+i, m+(i+1)))$, --- for $i=0, 1, 2$, ---, defines an effective operation function that describes a unique effective operation path in the principal time graph $G_M(T_p)$ and belongs to the class of general recursive functions.*

If the moment M is finite, say $M=\{m, m+1, ---, m+h\}$, the last term of the above sequences of (a), (b), (c) is given by $\rho_6(\psi_p(m+(h-1), m+h))$, and $\rho_8(\psi_E(m+(h-1), m+h))$, and $\rho_{10}(\psi_T(m+(h-1), m+h))$, respectively.

Owing to the hierarchical nature of the principal environment graph over the processor graph and the principal time graph over the principal environment graph as pointed out in previous remarks, the sequences describing the effective operation paths are related. In particular, due to the relationship of the respective n-tuples described in Primitives P3, E_p3, T_p3 in 2.1, $\rho_{2,7}(\psi_E(m+i, m+(i+1)))=\psi_p(m+i, m+(i+1))$ and $\rho_{2,9}(\psi_T(m+i, m+(i+1)))=\psi_E(m+i, m+(i+1))$. Hence, the following corollary reflects the hierarchical nature of the graphs through the effective operation functions.

Corollary A6.1. *The sequence given by $\rho_1(\psi_E(m, m+1))$, $\psi_p(m, m+1)$, ---, $\rho_1(\psi_E(m+i, m+(i+1)))$, $\psi_p(m+i, m+(i+1))$, ---, $\forall i=0, 1, 2$, ---, describes the effective operation path for the principal environment graph. Similarly, the sequence given by $\rho_1(\psi_T(m, m+1))$, $\psi_E(m, m+1)$, ---, $\rho_1(\psi_T(m+i, m+(i+1)))$, $\psi_E(m+i, m+(i+1))$, ---, $\forall i=0, 1, 2$, ---, describes the effective operation path for the principal time graph.*

To illustrate the significant result, consider again the previous abstract example. Given the processor graph $G_M(P)$ as represented in Figure 2.2a and the results of applying Theorem A3 to the example as given in Table A3, the sequence of Theorem A6a for the example is easily determined. Moreover, as the sequence is established, the corresponding effective operation path may be visually traced in the graph (Figure 2.2a).

Since $M=\{m, m+1, ---, m+12\}$, the sequence given by $\rho_1(\psi_p(m, m+1))$, $\rho_{2,5}(\psi_p(m, m+1))$, ---, $\rho_1(\psi_p(m+i, m+(i+1)))$, $\rho_{2,5}(\psi_p(m+i, m+(i+1)))$, ---, $\rho_6(\psi_p(m+(h-1), m+h)))$ for $i=0, 1, 2, ---, 11$, $h=12$ defines the effective operation function and describes the unique effective operation path in $G_M(P)$. Put $i=0$, and by Theorem A3, $\psi_p(m, m+1)=(q_0, s_3, f_0, d_1, c_1, q_1)$ (see Table A3). Thus, $\rho_1(\psi_p(m, m+1))=q_0$ and $\rho_{2,5}(\psi_p(m, m+1))=(s_3, f_0, d_1, e_1)$. With $i=1$, $\psi_p(m+1, m+2)=(q_1, s_2, f_0, d_0, c_0, q_1)$, so $\rho_1(\psi_p(m+1, m+2))=q_1$ and $\rho_{2,5}(\psi_p(m+1, m+2))=(s_2, f_0, d_0, c_0)$. Thus, the effective operation path is the sequence beginning as q_0, (s_3, f_0, d_1, c_1), q_1, (s_2, f_0, d_0, c_0), ---. Continuing in this manner for $i=0$, 1, ---, 11 and $h=12$, the complete sequence is established and presented in Table A5. Note that the Table A5 is also readily obtainable from Table A2.

In a similar manner, the sequence describing the effective operation paths in the principal environment and time graphs for the abstract example automaton are established. That is, given the principal environment graph $G_M(E_p)$ (Figure 2.3a) and the results of applying Theorem A4 to the example (Table A4), put $i=0$, then $\psi_E(m, m+1)=(e_{0,0,0}, q_0, s_3, f_0, d_1, c_1, q_1, e_{1,0,0})$. So $\rho_1(\psi_E(m, m+1))=e_{0,0,0}$, and $\rho_{2,7}(\psi_E(m, m+1))=(q_0, s_3, f_0, d_1, c_1, q_1)$. Thus, the sequence begins $e_{0,0,0}, (q_0, s_3, f_0, d_1, c_1, q_1)$, ---. Continuing for $i=1, 2, ---, 11$, $h=12$, the sequence describing the effective operation path for the principal environment graph is established and given in Table A6.

TABLE A5. The sequence that describes the effective operation path in the processor graph G_M (*P*) for the example abstract automaton

q_0, (s_3, f_0, d_1, c_1), q_1, (s_2, f_0, d_0, c_0), q_1, $(s_0, f_2, -d_2, c_2)$,
q_3, (s_2, f_1, d_0, c_0), q_1, (s_1, f_2, d_3, c_2), q_2, (s_0, f_0, d_0, c_0),
q_3, (s_0, f_1, d_2, c_2), q_4, $(s_1, f_1, -d_1, c_1)$, q_2, $(s_2, f_2, -d_3, c_2)$,
q_3, (s_0, f_1, d_2, c_2), q_4, $(s_1, f_1, -d_1, c_1)$, q_2, (s_4, f_0, d_0, c_0),
q_5

TABLE A6. The sequence that describes the effective operation path in the principal environment graph $G_M(E_p)$ for the example abstract automaton

$e_{0,0,0}$, $(q_0,s_3,f_0,d_1,c_1,q_1)$, $e_{1,0,0}$, $(q_1,s_2,f_0,d_0,c_0,q_1)$,
$e_{1,0,0}$, $(q_1,s_0,f_2,-d_2,c_2,q_3)$, $e_{1,-1,0}$, $(q_3,s_2,f_1,d_0,c_0,q_1)$,
$e_{1,-1,0}$, $(q_1,s_1,f_2,d_3,c_2,q_2)$, $e_{1,-1,1}$, $(q_2,s_0,f_0,d_0,c_0,q_3)$,
$e_{1,-1,1}$, $(q_3,s_0,f_1,d_2,c_2,q_4)$, $e_{1,0,1}$, $(q_4,s_1,f_1,-d_1,c_1,q_2)$,
$e_{0,0,1}$, $(q_2,s_2,f_2,-d_3,c_2,q_3)$, $e_{0,0,0}$, $(q_3,s_0,f_1,d_2,c_2,q_4)$,
$e_{0,1,0}$, $(q_4,s_1,f_1,-d_1,c_1,q_2)$, $e_{-1,0,0}$, $(q_2,s_4,f_0,d_0,c_0,q_5)$,
$e_{-1,1,0}$

Extending Theorem A6 to include the alternate environment graph $G_M(E_a)$ and alternate time graph $G_M(T_a)$ is left to an exercise.

EXERCISES

A.1 Write the Corollary A1.3 for the moment function θ_N that parallels Corollary A1.2.

A.2 Establish a theorem for defining ψ_E for the alternate environment graph (paralleling Theorem A4 for the principal environment graph).

A.3 Determine the effective adjacency function ψ_T for the principal time graph $G_M(T_p)$ of the example automaton (Figure 2.4) directly by applying Theorem A5.

A.4 Extend Theorem A6 to include a) the alternate environment graph $G_M(E_a)$, and b) the alternate time graph $G_M(T_a)$.

A.5 Detail algorithms from Theorem A6 for calculating effective operations for a) the processor graph, b) alternate environment graph, and c) principal time graph.

A.6 Write programs from the algorithms established in Exercise A5 in the language of your choice for a) the processor graph, b) the alternate environment graph, and c) the principal Time graph. You may use the abstract example of an automaton of Section 2.1 for checking your program.

APPENDIX B

Analysis for an Effective Operation of a General System of Interactive Automata

B.1 INTRODUCTION

The work presented here covers the development and application of recursive methods and graph-theoretic principles for determining the effective operation of the general system of interactive automata through select properties of the graphs in the model (Frederick, Koenig 1971b). A system of k-automata, denoted by $A=\{A_j | j \in J=\{1, 2, ---, k\}\}$, was analyzed in Section 5.1 over a system moment base $M=\{m+i | i \in I=\{0, 1, ---, h\}\}$ in an environment space R^n. Functions based on those describing the general system in conjunction with primitives utilizing sets of the general system were developed to establish a graph model containing two classes of graphs called macro and microsystem graphs. Each class consists of three types, called processor, environment, and time graphs, which reflect the system internally and externally in a hierarchical sense. The latter two types are subdivided into principal and alternate kinds in the hierarchy. Recall, the macrosystem graphs are of special interest when a general system is viewed as a single automaton. Such graphs consist of points in the form of k-tuples representing the particular system sets and lines depicting macrosystem operating conditions. The microsystem graphs are of special interest in viewing an interactive system on the component automaton level. Such graphs consist of k-subgraphs representing the k-component automata, and each subgraph is similar to that described in Section 2.1 for the general automaton.

Starting with the graph model for a general system of interactive automata as developed, the work of this Appendix B applies recursive methods and graph principles to determine the effective operation of the system. The k-subgraph nature of the microsystem graphs lends itself to the analysis

---

Knowledge Structures for Communications in Human-Computer Systems: General Automata-Based, by Eldo C. Koenig

developed in Appendix A when the concept of interaction is brought into account to keep track of what stimuli are received by each component automaton at each moment. Furthermore, the macrosystem graphs lend themselves to the analysis of Appendix A since they represent the system when it is viewed as a single automaton.

The key result of Appendix A was determining the effective operation of the general automaton through graphs in its model. The development of an effective operation path for each graph in the form of an alternating sequence of points and lines was based on the recursive nature of the established moment functions and effective adjacency functions. These latter functions were developed from a basic foundation consisting of a concept of nonnumeric recursion, knowledge of a minimal set of performance descriptors of the physical system, and a set of environment functions for describing the stimuli associated with the environments at each moment.

Before proceeding to determine the effective operation of a general system of interactive automata, some basic concepts defining the system are reviewed. Following this, the set of performance descriptors of the system constituting a minimal set of information from which the effective operation of the system can be determined through the graphs is presented. Finally, the environment functions are defined to account for the interaction in the system and any possible environmental media effects.

Basic System Concept. Each component automaton A_j of a general system of interactive automata is analyzed over its own moment base $M_j=\{m_j+i \mid i\in I\}$. Using this, a system moment base is defined as a set of k-tuples composed of corresponding moments from each component automaton; i.e., $M=\{m+i \mid i\in I, m+i=(m_j+i \mid i\in I)\}$. An arbitrary element $m+i\in M$ corresponds to the time slice necessary to analyze the k-automata at their respective moments m_j+i. In any given system moment, the absolute times associated with the moments corresponding to each component automaton are considered distinct. That is, if $T_n(m_n+i)$ denotes the time associated with an arbitrary component automaton A_n at moment m_n+i, then for all $n, k\in J$, $T_n(m_n+i)\neq T_k(m_k+i)$ where m_n+i, $m_k+i, \in m+i=(m_j+i \mid j\in J)$. The intervals of time associated with consecutive system moments, as time slices, are considered disjoint. That is, if $[min_{j\in J}T_j(m_j+i), max_{j\in J}T_j(m_j+i)]$ defines the time slice associated with an arbitrary system moment $m+i$, then $\{[min_{j\in J}T_j(m_j+i), max_{j\in J}T_j(m_j+i)]\}\cap\{[min_{j\in J}T_j(m_j+(i+1)), max_{j\in J}T_j(m_j+(i+1))]\}=\phi$ and $max_{j\in J}T_j(m_j+i)<min_{j\in J}T_j(m_j+(i+1))$.

Recall, there is a fundamental system principle associated with the moment base analogous to that for the general automaton. Such a principle, in addition to the timing already outlined, establishes an operating framework for the system as given in Section 5.1.1. Given a discrete system moment $m+i=(m_j+i \mid j\in J)$, each component automaton is accessing one and only one environment from a subset $E_j\subseteq R^n$ of the environment space, receiving one and only one stimulus from a set S_j of potential stimuli, is in one and only one state from a set of possible states Q_j, and has one and only one absolute time associated with m_j+i from a set T_j of times. Furthermore, during the next system moment $m+(i+1)=(m_j+(i+1)\,|\,j\in J)$, each component automaton A_j produces one and only one response based on the aforementioned conditions at moment m_j+i from a set R_j of possible responses. As presented in Section 2.1, the response by each component automaton is a triadic response consisting of an ordered triple from a set F_j of transformation responses, a set D_j of spatial change responses, and a set C_j of time change responses. The transformation response by A_j at moment $m_j+(i+1)$ modifies the stimulus received from the environment location accessed at the previous moment m_j+i.

Supplementing the system principle in detailing the operational framework of the general system of interactive automata are pertinent sets, functions, and equations describing the physical system. These fundamentals presented in Section 5.1 are repeated here in their equation form.

$$Q_j(m_j+(i+1))=\omega_j(Q_j(m_j+i), S_j(m_j+i)) \qquad 5.1E$$

$$R_j(m_j+(i+1))=\beta_j(Q_j(m_j+i), S_j(m_j+i)) \qquad 5.2E$$

$$F_j(m_j+(i+1))=\sigma_j(Q_j(m_j+i), S_j(m_j+i)) \qquad 5.3E$$

$$D_j(m_j+(i+1))=\delta_j(Q_j(m_j+i), S_j(m_j+i)) \qquad 5.4E$$

$$C_j(m_j+(i+1))=\tau_j(Q_j(m_j+i), S_j(m_j+i)) \qquad 5.5E$$

$$E_j(m_j+(i+1))=\gamma_j'(E_j(m_j+i), D_j(m_j+(i+1))) \qquad 5.6E$$

$$T_j(m_j+(i+1))=\lambda_j'(T_j(m_j+i), C_j(m_j+(i+1))) \qquad 5.7E$$

$$B_j(m_j+(i+1))=\eta_j'(E_j(m_j+i), F_j(m_j+(i+1))) \qquad 5.8E$$

$$S_n(m_n+(i+p))=\alpha_{nj}(B_j(m_j+i)) \qquad 5.9E$$

where Equation 5.9E is restricted in time to $((T_j(m_j+i)<T_n(m_n+(i+p))\,|\,i\in I-\{0, h\})\,|\,p\in P)$, $I=\{0, 1, 2, ---, h\}$, $p=\{0, 1, 2, ---, h-(i+1)\}$, $n\in J$.

Equations 5.1E and 5.2E describe the basic action of the jth automaton, A_j, in the system. That is, the state of A_j at an arbitrary moment is a function ω_j of the state and stimulus of the previous moment. The response at an arbitrary moment is a function β_j of the state and stimulus of the previous moment. Equations 5.3E, 5.4E, 5.5E describe the components of the triadic response of the jth automaton, while Equations 5.6E and 5.7E describe the environment and time for A_j. From these fundamental equations, sets, and functions for the component automata, analogous equations, sets, and functions were established for the system. Equations 5.8E and 5.9E will be reviewed in greater detail.

Recall that A is said to be a *general system of interactive automata* over a system moment base if the environment sets are not disjoint under any pairwise partition; i.e., there does exist a proper subset $N\subset 2^J$ in the power set of J such that $(\cup_{n\in N}E_n)\cap(\cup_{j\in J/N}E_j)=\phi$. It is clear that each component automaton accesses all the locations in its environment set during the moment base, therefore, the responses of some component automata are being interpreted as stimuli by others.

The effect of the environmental media on the transformation responses is described by a set B_j, called recorded responses, resulting in Equation 5.8E, $B_j(m_j+(i+1))=\eta_j'(E_j(m_i+i), F_j(m_j+(i+1)))$. That is, a recorded response is the modification of the transformation response as might be induced by environmental media. Given that the environmental media is such that no effect is made on the transformation response, the set of recorded responses B_j is assumed to be the same as the set of transformation responses F_j.

Since the system is an interactive system, any arbitrarily recorded response $B_j(m_j+i)$ by automaton A_j has the potential of being interpreted as a stimulus by an arbitrary automaton A_k at any time on or after $T_j(m_j+i)$ assuming common environments (Equation 5.9E). This potential is described by a set of functions $\{\alpha_{kj} | \alpha_{kj}:B_j \rightarrow S_k, \forall k, j \in J\}$ that details how each component automaton A_k interprets the recorded response B_j as a stimulus. Note that if $k=j$, then α_{kj} is the same as the function introduced in Section 2.1. Given that an arbitrary component automaton A_k has produced a transformation response recorded as b_{km} at some environment $e_{kp} \in E_k$, the set of possible stimuli that may be received by the component automata in the system at the next accessing of e_{kp} is the set $\{\alpha_{kj}(b_{km}) | j \in J\} \subseteq \Pi_j S_j$. In particular, if automaton A_n accesses e_{km} next, the stimulus received is $\alpha_{kn}(b_{km}) \in S_n$.

Minimal Set of Performance Descriptors. The increased complexity brought about by dealing with a system of k-general automata together with the interaction concept requires that the minimal set of performance descriptors for the system reflect this added complexity. First of all, the set described in Appendix A can be enlarged to include information on all component automata. In addition, the recorded response and set of functions $\{\alpha_{kj} | j \in J\}$ for the interpretation of responses as stimuli must be incorporated in it.

The following set of performance descriptors of the physical system established from the formal analysis for a general system of Interactive automata of Section 5.1 constitutes a minimal set of information fundamental to determining the effective operation of the system. First, the environment space R^n for some n for the system is given, and the moment base M_k for each component automaton A_k is specified. Also, the sets of states Q_k, stimuli S_k, and state-stimulus pairs $V_k \subseteq Q_k \times S_k$ are established. In addition, initial conditions of environment $E_k(m_k)$, time $T_k(m_k)$, state $Q_k(m_k)$, and stimulus $S_k(m_k)$ representing conditions at the initial moment m_k are needed for each component automaton A_k. Finally, the knowledge of several functions and their image sets complete the amount of information needed. These include the functions ϕ_{k0} mapping the environments E_k onto the sets of stimuli S_k at the initial moment m_k; the automaton functions σ_k, δ_k, τ_k, and ω_k along with the image sets $\sigma_k(V_k)=F_k$, $\delta_k(V_k)=D_k$, and $\tau_k(V_k)=C_k$ of transformation, spatial change, and time change responses, respectively; the functions γ_k' and λ_k' that determine the next environment and associated time; the functions η_k and the associated image sets $\eta_k(W_k)=B_k$ of recorded responses that reflect the effect of the environmental media on the transformation responses; and the set of functions $\{\alpha_{kj} | j \in J\}$ that determines how the recorded responses of all component automata A_j are interpreted as stimuli by A_k.

Environment Functions. The environment functions for the general automaton, as introduced in Appendix A for the purpose of keeping track of what stimuli were associated with the environment at various moments, are inadequately defined for the component automata of a general system of interactive automata. The inadequacy lies in the previous definition's inability to account for interaction. The following remarks directed toward defining environment functions for a general system of interactive automata are considered fundamental to the subsequent analysis:

The environment functions for each component automaton at the initial moment are specified in the minimal set of performance descriptors. That is, if $A=\{A_j | j \in J=\{1, 2, ---, k\}\}$ denotes a general system of interactive automata, the functions $\phi_{10}, \phi_{20}, ---, \phi_{k0}$ with $\phi_{j0}: E'_j \rightarrow S_j$ have their images known. Let $A_n \in A$ be an arbitrary component automaton and suppose the environment function ϕ_{n1} for automaton A_n at moment m_n+1 is to be defined. First, let H_n denote the subset of $J/\{n\}$ for which $h \in H_n$, $A_h \in A$, and $E_n \cap E_h \neq \phi$. Then, the set of common environments of A_n with other component automata is denoted by $\cup_{h \in s_j}\{E_n \cap E_h\}$. Note that it is possible that for more than one $h \in H_n$, $E_n \cap E_h$ is the same set. Next, recall that the system moment $m+i$ is the k-tuple $(m_1+i, m_2+i, ---, m_k+i)$ and represents the time slice necessary to view all automata A_j at m_j+i. In particular, there is a time associated with each moment, namely $T_j(m_j+i)$. Let $T_n(m_n+1)$ denote the time associated with moment m_n+1 for automaton A_n. Denote by P_{n1} the subset of $J/\{n\}$ for which $p \in P_{n1}$ and $T_p(m_p+1) < T_n(m_n+1)$. Then the set of component automata $\{A_x | x \in H_n \cap P_{n1}\}$ have the potential for affecting the environment function ϕ_{n1} by interaction. That is, component automata A_x share common environments with A_n and have the proper timing whereby their responses could serve as stimuli for A_n at moment m_n+1. Let $G_{n1} \subseteq H_n \cap P_{n1}$ such that $g \in G_{n1}$ and $E_g(m_g) \in \cup_{h \in \bar{E}_j}\{E_n \cap E_h\}$. Then $\{E_g(m_g) | g \in G_{n1}\}$ denotes the subset of the environment of A_n for which interaction is taking place with the proper timing. Note that if there is more than one $g \in G_{n1}$ such that automata A_g is accessing the same environment location, the choice of g for which $T_g(m_g+1)$ is maximum will be the one whose interaction contributes to ϕ_{n1}. If there are any other component automata accessing the same environment as A_n and having times associated with the second system moment less than that for A_n, their responses at the location would occur prior to that of A_n and thus would not affect the environment function ϕ_{n1} unless the response by A_n is the null recorded response. The environment function ϕ_{n1} for component automaton A_n at moment m_n+1 can be defined as follows: $\phi_{n1}(E_n(m_n))=\alpha_{nn}(B_n(m_n+1))$ for $B_n(m_n+1)$ not the null recorded response, $\phi_{n1}(E_g(m_g))=\alpha_{ng}(B_g(m_g+1))$, $\forall g \in G_{n1}$, with the maximum condition stated above, and $\phi_{n1}=\phi_{n0}$ for all other domain elements. If $B_n(m_n+1)$

is the null recorded response, the image $\phi_{n1}(E_n(m_n))$ is covered by one of the other parts of the definition depending upon $E_n(m_n)$.

It is assumed that for all $j \in J$, the environment functions ϕ_{j1} for each automaton A_j are defined at moment m_j+1 by the procedure outlined above. Next, suppose ϕ_{n2} is to be defined. First, consider those automata interacting with automaton A_n during the system moment $m+1$ for which their timing is too late to affect the environment function ϕ_{n1}; i.e., $\{A_y \mid y \in H_n, E_y(m_y) \in \cup_{h \in H_n}\{E_n \cap E_h\}$ and $T_n(m_n+1) < T_y(m_y+1)\}$. Suppose that the environments $E_y(m_y)$ are not involved in any properly timed interaction during the system moment $m+2$. Then the effect of this delayed interaction with A_n must be specifically accounted for in the environment function ϕ_{n2}. Let $G_{n2} \subseteq H_n \cap P_{n2}$ be determined as before. Suppose $L_{n1} = \{y \in H_n \mid E_y(m_y) \in \cup_{h \in H_n}\{E_n \cap E_h\}$, $T_n(m_n+1) < T_y(m_y+1)$, and $y \notin G_{n2}\}$. This latter requirement eliminates those automata whose interaction at moment $m+2$ would override the delayed interaction at moment $m+1$. Again, note that if, for any y, $E_y(m_y) = E_n(\mathrm{m}_n+1)$, it follows that $T_n(m_n+2) > T_y(m_y+1)$ and the response of A_n will be reflected in ϕ_{n2} instead of the delayed interaction unless the response by A_n is the null recorded response. Define $\phi_{n2}(E_y(m_y)) = \alpha_{ny}(B_y(m_y+1))$ for all $y \in L_{n1}$, $\phi_{n2}(E_g(m_g+1)) = \alpha_{ng}(B_g(m_g+2))$ with the maximum condition and $g \in G_{n2}$, $\phi_{n2}(E_n(m_n+1)) = \alpha_{nn}(B_n(m_n+2))$ for $B_n(m_n+2)$ not the null recorded response, and $\phi_{n2} = \phi_{n1}$ otherwise.

Generalizing, the environment function $\phi_{n(n+1)}$ for an arbitrary component automaton A_n at moment $m_n+(i+1)$ is defined as follows: $\phi_{n(i+1)}(E_n(m_n+i)) = \alpha_{nn}(B_n(m_n+(i+1)))$ for $B_n(m_n+(i+1))$ not the null recorded response, $\phi_{n(i+1)}(E_g(m_g+i)) = \alpha_{ng}(B_g(m_g+(i+1)))$ with the maximum condition and $g \in G_{n(i+1)}$, $\phi_{n(i+1)}(E_y(m_y+(i-1))) = \alpha_{ny}(B_y(m_y+i))$ with the maximum condition and $y \in L_{ni}$, $\phi_{n(i+1)} = \phi_{ni}$ otherwise. If $B_n(m_n+(i+1))$ is the null recorded response, the image $\phi_{n(i+1)}(E_n(m_n+i))$ is covered by one of the other parts of the definition depending upon $E_n(m_n+i)$.

The result of the preceding discussion is a family of sets of environment functions for a general system of interactive automata $\phi = \{\phi_1, \phi_2, ---, \phi_k\}$ where $\phi_j = \{\phi_{j0}, \phi_{j1}, ---, \phi_{jh}\}$. Each set ϕ_j consists of the environment functions for the component automaton A_j at moments $m_j, m_j+1, ---, m_j+h$. The functions serve to describe the stimulus at each location in the environment as it would be interpreted by the component automaton at each moment in its formal analysis.

B.2 MICROSYSTEM GRAPHS

The purpose of this section is to apply recursive methods and graph principles to determine the effective operation of a general system of interactive automata through the microsystem graphs in the model. As stated in the introduction, the microsystem graphs are of special interest in viewing an interactive system on the component automaton level. Since each microsystem graph is composed of k subgraphs representing the k-component automata, the analysis developed in Appendix A can be applied to each subgraph. As a result, the effective operation of a general system of interactive automata can be defined in terms of the microsystem graphs by k-paths, one in each subgraph occurring simultaneously with respect to the graph.

The material presented in this section is a proper paralleling of the methods developed in Appendix A. First, the moment functions associating the proper status condition with each moment are presented. Next, the graph principles related to the microsystem graphs are determined, and the effective adjacency functions that formally describe the graph principles in terms of the graphs in the model are established. Finally, the effective operation of a general system of interactive automata is developed by means of effective operation functions.

Moment Functions. Several functions relating the fundamental automaton principle to sets and functions describing the physical situation called moment functions were defined and shown to belong to the class of general recursive functions in Appendix A. These functions associated the proper status condition (e.g., environment, state, stimulus, response, etc.) with each moment. Corresponding moment functions for each component automaton introduced in Section 5.1 are stated here to establish notation for later use. The differences over that for a single automaton occur in the changed environment functions and response interpreting functions which necessarily account for the interaction. The statement is in the form of a theorem, and since the proof follows that for the corresponding theorem of Appendix A, it is omitted here.

Theorem B1. *Let $A=\{A_j \mid j\in J=\{1, 2, ---, h\}\}$ be a general system of interactive automata analyzed over a moment base $M=\{m+i \mid i\in I=\{0, 1, ---, h\}\}$. Suppose the minimal set of performance descriptors for A as previously outlined are known. Then the moment functions for each component automaton A_j include*

a. θ_{E_j}: $M_j \rightarrow \theta_{E_j}(M_j)=E_j\subseteq\bar{E}_j$ *defined by*
$\theta_{E_j}(m_j)=E_j(m_j)$
$\theta_{E_j}(m_j+(i+1))=\gamma_j'(\theta_{E_j}(m_j+i), \theta_{D_j}(m_j+(i+1)))$

b. θ_{T_j}: $M_j \rightarrow \theta_{T_j}(M_j)=T_j\subseteq\bar{T}_j$ *defined by*
$\theta_{T_j}(m_j)=T_j(m_j)$
$\theta_{T_j}(m_j+(i+1))=\lambda_j'(\theta_{T_j}(m_j+i), \theta_{C_j}(m_j+(i+1)))$

c. θ_{S_j}: $M_j-\{m_j+h\} \rightarrow \theta_{S_j}(M_j-\{m_j+h\})=S_j\subseteq\bar{S}_j$ *defined by*
$\theta_{S_j}(m_j)=\phi_{j0}(\theta_{E_j}(m_j))$
$\theta_{S_j}(m_j+(i+1))=\phi_{j(i+1)}(\theta_{E_j}(m_j+(i+1)))$

d. $\theta_{Q_j}: M_j \to \theta_{Q_j}(M_j) = Q_j \subseteq \bar{Q}_j$ *defined by*
$\theta_{Q_j}(m_j) = Q_j(m_j)$
$\theta_{Q_j}(m_j + (i+1)) = \omega_j(\theta_{Q_j}(m_j + i), \theta_{S_j}(m_j + i))$

e. $\theta_{F_j}: M_j - \{m_j\} \to \theta_{F_j}(M_j - \{m_j\}) = F_j \subseteq \bar{F}_j$ *defined by*
$\theta_{F_j}(m_j + 1) = \sigma_j(\theta_{Q_j}(m_j), \theta_{S_j}(m_j))$
$\theta_{F_j}(m_j + (i+2)) = \sigma_j(\theta_{Q_j}(m_j + (i+1)), \theta_{S_j}(m_j + (i+1)))$

f. $\theta_{D_j}: M_j - \{m_j\} \to \theta_{D_j}(M_j - \{m_j\}) = D_j \subseteq \bar{D}_j$ *defined by*
$\theta_{D_j}(m_j + 1) = \delta_j(\theta_{Q_j}(m_j), \theta_{S_j}(m_j))$
$\theta_{D_j}(m_j + (i+2)) = \delta_j(\theta_{Q_j}(m_j + (i+1)), \theta_{S_j}(m_j + (i+1)))$

g. $\theta_{C_j}: M_j - \{m_j\} \to \theta_{C_j}(M_j - \{m_j\}) = C_j \subseteq \bar{C}_j$ *defined by*
$\theta_{C_j}(m_j + 1) = \tau_j(\theta_{Q_j}(m_j + 1), \theta_{S_j}(m_j + 1))$
$\theta_{C_j}(m_j + (i+2)) = \tau_j(\theta_{Q_j}(m_j + (i+1)), \theta_{S_j}(m_j + (i+1)))$

h. $\theta_{B_j}: M_j - \{m_j\} \to \theta_{B_j}(M_j - \{m_j\}) = B_j \subseteq \bar{B}_j$ *defined by*
$\theta_{B_j}(m_j + 1) = \eta_j'(\theta_{E_j}(m_j), \theta_{F_j}(m_j + 1))$
$\theta_{B_j}(m_j + (i+2)) = \eta_j'(\theta_{E_j}(m_j + (i+1)), \theta_{F_j}(m_j + (i+2)))$

Furthermore, the defined functions belong to the class of general recursive functions (based on the recursion concepts established in Appendix A).

Recall, there is the set $V_j \subseteq Q_j \times S_j$ of state-stimulus pairs over the moment base determined by the nature of the automaton A_j. Clearly, there is one and only one state-stimulus pair for each moment. Corollary B1a is an obvious result of Theorem B1.

Corollary B1a. *There is a moment function θ_{V_j} describing the state-stimulus pair associated with each moment from a set V_j of state-stimulus pairs for each component automaton A_j. That is,*

$$\theta_{V_j}: M_j - \{m_j + h\} \to \theta_{V_j}(M_j - \{m_j + h\}) = V_j \subseteq Q_j \times S_j \text{ defined by}$$
$$\theta_{V_j}(m_j) = (\theta_{Q_j}(m_j), \theta_{S_j}(m_j))$$
$$\theta_{V_j}(m_j + (i+1)) = (\omega_j(\theta_{V_j}(m_j + i)), \phi_{j(i+1)}(\gamma_j'(\theta_{E_j}(m_j + i), \delta_j(\theta_{V_j}(m_j + i)))))$$

belongs to the class of general recursive functions.

There is a function $\gamma_j: W_j \subseteq E_j \times V_j \to \gamma_j(W_j) \subseteq E_j \subseteq \bar{E}_j$ that describes the image of an environment-state-stimulus as the next environment. Again, from the fundamental system principle, one and only one environment-state-stimulus triple is associated with each moment for each component automaton. Applying Theorem B1 along with Corollary B1a to the above remarks, the following result is obtained:

Corollary B1b. *There is a moment function θ_{W_j} describing the environment-state-stimulus triple associated with each moment from a set W_j of such triples* *for each component automaton. That is,* θ_{W_j}: $M_j - \{m_j + h\} \rightarrow \theta_{W_j}(M_j - \{m_j + h\}) = W_j \subseteq E_j \times V_j$ *and* θ_{W_j} *is defined by*

$$\theta_{W_j}(m_j) = (\theta_{E_j}(m_j), \theta_{V_j}(m_j))$$

$$\theta_{W_j}(m_j + (i+1)) = (\gamma_j(\theta_{W_j}(m_j + i)), \theta_{V_j}(m_j + (i+1)))$$

belongs to the class of general recursive functions.

There is a function λ_j: $N_j \subseteq T_j \times V_j \rightarrow \lambda_j(N_j) \subseteq T_j \subseteq \overline{T}_j$ that describes the image of the time-state-stimulus triple as the next time. Since there is one and only one such triple for each moment, Corollary B1c parallels the preceding corollary.

Corollary B1c. *There is a moment function* θ_{N_j} *describing the time-state-stimulus triple associated with each moment from a set* N_j *of such triples for each component automaton. That is,* θ_{N_j}: $M_j - \{m_j + h\} \rightarrow \theta_{N_j}(M_j - \{m_j + h\}) = N_j \subseteq T_j \times V_j$ *where* θ_{N_j} is defined by

$$\theta_{N_j}(m_j) = (\theta_{T_j}(m_j), \theta_{V_j}(m_j))$$

$$\theta_{N_j}(m_j + (i+1)) = (\lambda_j(\theta_{N_j}(m_j + i)), \theta_{V_j}(m_j + (i+1)))$$

belongs to the class of general recursive functions.

Defined in Section 5.1 was the function π_j whose image of an arbitrary four-tuple described above was the next space-time pair. That is

$$\pi_j : Y_j \subseteq T_j \times W_j \rightarrow \pi_j(Y_j) \subseteq \gamma_j(W_j) \times \lambda_j(N_j) \subseteq E_j \times T_j$$

Also defined was a parallel function μ_j such that

$$\mu_j : Y_j' \subseteq E_j \times N_j \rightarrow \mu_j(Y_j') \subseteq \lambda_j(N_j) \times \gamma_j(W_j) \subseteq T_j \times E_j$$

where $Y_j' \subseteq E_j \times N_j$ denotes the set of environment-time-state-stimulus four-tuples over the moment base. Corollary B1d naturally follows.

Corollary B1d. *There is a moment function* θ_{Y_j} *describing the time-environment-state-stimulus four-tuple associated with each moment from a set* Y_j *of four-tuples for each component automaton. That is,* θ_{Y_j}: $M_j - \{m_j + h\} \rightarrow \theta_{Y_j}(M_j - \{m_j + h\}) = Y_j \subseteq T_j \times W_j$ *where* θ_{Y_j} is

$$\theta_{Y_j}(m_j) = (\theta_{T_j}(m_j), \theta_{W_j}(m_j))$$

$$\theta_{Y_j}(m_j + 1) = (\theta_{T_j}(m_j + 1), \theta_{W_j}(m_j + 1))$$

$$\theta_{Y_j}(m_j + (i+1)) = (\mu_j(\pi_j(\theta_{Y_j}(m_j + (i-1))), \theta_{V_j}(m_j + i)), \theta_{V_j}(m_j + (i+1)))$$

belongs to the class of general recursive functions. Similarly, there is a moment function θ'_{Y_j} *describing the environment-time-state-stimulus four-*

tuple. That is, $\theta'_{Y_j}: M_j - \{m_j + h\} \rightarrow \theta'_{Y_j}(M_j - \{m_j + h\}) = Y'_j \subseteq E_j \times N_j$ *where* θ'_{Y_j} *is defined by*

$$\theta_{Y'_j}(m_j) = (\theta_{E_j}(m_j), \theta_{N_j}(m_j))$$

$$\theta_{Y'_j}(m_j + 1) = (\theta_{E_j}(m_j + 1), \theta_{N_j}(m_j + 1))$$

$$\theta_{Y'_j}(m_j + (i+1)) = (\pi_j(\mu_j(\theta_{Y'_j}(m_j + (i-1))), \theta_{V_j}(m_j + 1)), \theta_{V_j}(m_j + (i+1)))$$

belongs to the class of general recursive functions.

Clearly, there is a one-to-one correspondence between the elements of Y and Y' given by a permutation of the first two entries in the four-tuple, hence, there is a corresponding relationship between θ_Y and $\theta_{Y'}$. Note also in Corollary B1d, the initial value of i is taken to be one.

Effective Adjacency Functions. The fundamental system principle translated in terms of the microsystem graphs of the model states that, at each system moment, k points represent the current status of the system with respect to the type of graph (at most k points in the environment type; as here, the points need not be distinct). Moreover, the operating conditions of the system in achieving a change of status from a given system moment to the next are described by k lines, each directed from a given point to a prescribed next.

The concept of time-variant adjacency established in Appendix A is interpreted in terms of the microsystem graphs as follows. Any two points in a subgraph corresponding to a component automaton connected by a directed line (loop) are adjacent only at the moment(s) whenever the component automaton's current status with respect to the type of graph is represented by the origin point, and the automaton operates according to the conditions describing the connecting line to achieve the status represented by the terminal point. Merging this concept of adjacency with the previously mentioned concepts regarding the fundamental system principle gives rise to the following principle for the microsystem graphs:

Microsystem Graph Principle. There are k pairs of adjacent points (not necessarily distinct) pairwise connected by exactly k-distinct lines in the microsystem graphs for a general system of interactive automata $A = \{A_j | j \in J = \{1, 2, ---, k\}\}$ corresponding to each pair of consecutive system moments from the system moment base.

Since the microsystem graphs consist of k subgraphs each reflecting one of the k-component automata in the system, the above principle is a natural extension of the graph principle for the general automaton's model. That is, given each subgraph of the microsystem graph corresponding to the graph for a component automaton and given an arbitrary pair of consecutive moments from its moment base, there is one and only one pair of adjacent points (not necessarily distinct) and only one line joining these points.

The effective adjacency functions introduced in Appendix A were established specifically to determine these points and line when given the consecutive pair of moments of interest. Such functions exist for each subgraph of the microsystem graph corresponding to a component automaton. The effective adjacency functions for these subgraphs are presented here as a theorem. Since the proof follows that for the corresponding theorems of Appendix A, it is omitted here.

Theorem B2. *Let $A=\{A_j | j\in J=\{1, 2, ---, k\}\}$ be a general system of interactive automata analyzed over a system moment base $M=\{m, m+1, ---, m+h\}$. Suppose A_j denotes an arbitrary component automaton and M_j its associated moment base. Suppose $\overline{M}_j \subseteq M_j \times M_j$ denotes the set of pairs of consecutive moments for A_j, and suppose ρ_n denotes the nth projection (identity) function.*

a. *Let $G_{M_j}(P)$ denote the subgraph of the microsystem graph of the processor type G_M (Psi) defined by the Primitives* Psi1 *through* Psi4 *and the graph function Ω_j. Then, the function Ψ_{P_j}: $\overline{M}_j \rightarrow \Psi_{P_j}(\overline{M}_j) \subseteq Q_j \times S_j \times F_j \times D_j \times C_j \times Q_j$ defined by*

$$\psi_{P_j}(m_j, m_j+1)=(\theta_{V_j}(m_j), \Omega_j(\theta_{V_j}(m_j)))$$

$$\begin{aligned}\psi_{P_j}(m_j+(i+1), m_j+(i+2))=(&\rho_6(\psi_{P_j}(m_j+i, m_j+(i+1)))),\\ &\phi_j(i+1)(\gamma_j'(\theta_{E_j}(m_j+i), \rho_4(\psi_{P_j}(m_j+i, m_j+(i+1))))),\\ &\Omega_j(\rho_6(\psi_{P_j}(m_j+i, m_j+(i+1)))),\\ &\phi_{j(i+1)}(\gamma_j'(\theta_{E_j}(m_j+i), \rho_4(\psi_{P_j}(m_j+i, m_j+(i+1)))))\end{aligned}$$

$\forall i=0, 1, ---, h-2$, is an effective adjacency function for the subgraph $G_{M_j}(P)$ and belongs to the class of general recursive functions.

b. *Let $G_{M_j}(E_p)$ denote the subgraph of the microsystem graph of the principle environment type G_M (Es_p i) defined by the respective primitives and the graph function Γ_j. Then the function $\Psi_{E_{pj}}$: $\overline{M}_j \rightarrow \Psi_{E_{pj}}(\overline{M}_j) \subseteq E_j \times Q_j \times S_j \times B_j \times F_j \times D_j \times C_j \times Q_j \times E_j$ defined by*

$$\psi_{E_{P_j}}(m_j, m_j+1)=(\theta_{W_j}(m_j), \Gamma_j(\theta_{W_j}(m_j)))$$

$$\begin{aligned}\psi_{E_{P_j}}(m_j+(i+1), m_j+(i+2))=(&\rho_{9,8}(\psi_{E_{P_j}}(m_j+i, m_j+(i+1))),\\ &\phi_{j(i+1)}(\gamma_j'(\theta_{E_j}(m_j+i), \rho_6(\psi_{E_{P_j}}(m_j+i, m_j+(i+1))))),\\ &\Gamma_j(\rho_{9,8}(\psi_{E_{P_j}}(m_j+i, m_j+(i+1))),\\ &\phi_j(i+1)(\gamma_j'(\theta_{E_j}(m_j+i), \rho_6(\psi_{E_{P_j}}(m_j+i, m_j+(i+1))))))\end{aligned}$$

$\forall i=0, 1, ---, h-2$, is an effective adjacency function for the subgraph $G_{M_j}(E_p)$ and belongs to the class of general recursive functions.

c. *Let $G_{M_j}(T_p)$ denote the subgraph of the microsystem graph of the principal time type G_M (Ts_p i) defined by the Primitives Ts_p i1 through Ts_p i4 and* *the graph function* Λ_j. *Then the function* $\Psi_{T_{pj}}$: $\bar{M}_j \rightarrow \Psi_{T_{pj}}(\bar{M}_j) \subseteq T_j \times E_j \times Q_j \times S_j \times B_j \times F_j \times D_j \times C_j \times Q_j \times E_j \times T_j$ *defined by*

$$\psi_{T_{pj}}(m_j, m_j+1) = (\theta_{Y_j}(m_j), \Lambda_j(\theta_{Y_j}(m_j, m_j+1)))$$

$$\begin{aligned}\psi_{Tp_j}(m_j+(i+1), m_j+(i+2)) = (&\rho_{11,10,9}(\psi_{Tj}(m_j+i, m_j+(i+1))),\\ &\phi_{j(i+1)}(\gamma_j'(\theta_{E_j}(m_j+i), \rho_7(\psi_{Tp_j}(m_j+i, m_j+(i+1))))),\\ &\Lambda_j(\rho_{11,10,9}(\psi_{Tp_j}(m_j+i, m_j+(i+1))),\\ &\phi_j(i+1)(\gamma_j'(\theta_{E_j}(m_j+i), \rho_7(\psi_{Tp_j}(m_j+i, m_j+(i+1))))))\end{aligned}$$

$\forall i = 0, 1, ---, h-2$, *is an effective adjacency function for the subgraph* $G_{M_j}(T_p)$ *and belongs to the class of general recursive functions.*

d. *Let* $G_{M_j}(E_a)$ *denote the subgraph of the microsystem graph of the alternate environment type* $G_M(Es_a\ i)$ *by the Primitives* $Es_a\ i1$ *through* $Es_a\ i4$ *and the graph function* Λ_j'. *Then the function* $\Psi_{E_{aj}}$: $\bar{M}_j \rightarrow \Psi_{E_{aj}}(\bar{M}_j) \subseteq E_j \times T_j \times Q_j \times S_j \times B_j \times F_j \times D_j \times C_j \times Q_j \times T_j \times E_j$ *defined by*

$$\psi_{E_{aj}}(m_j, m_j+1) = (\theta_{Y_j'}(m_j), \Lambda_j'(\theta_{Y_j'}(m_j)))$$

$$\begin{aligned}\psi_{E_{aj}}(m_j+(i+1), m_j+(i+2)) = (&\rho_{11,10,9}(\psi_{E_{aj}}(m_j+i, m_j+(i+1))),\\ &\phi_{j(i+1)}(\gamma_j'(\theta_{E_j}(m_j+i), \rho_7(\psi_{E_{aj}}(m_j+i, m_j+(i+1))))),\\ &\Lambda_j'(\rho_{11,10,9}(\psi_{E_{aj}}(m_j+i, m_j+(i+1))),\\ &\phi_j(i+1)(\gamma_j'(\theta_{E_j}(m_j+i), \rho_7(\psi_{E_{aj}}(m_j+i, m_j+(i+1))))))\end{aligned}$$

$\forall i = 0, 1, ---, h-2$, *is an effective adjacency function for the subgraph* $G_{M_j}(E_a)$ *and belongs to the class of general recursive functions.*

e. *Let* $G_{M_j}(T_a)$ *denote the subgraph of the microsystem graph of the alternate time type* $G_M(Ts_a\ i)$ *defined by the respective primitives and the graph function* Γ_j'. *Then the function* $\Psi_{T_{aj}}$: G_{M_j}: $\bar{M}_j \rightarrow \Psi_{T_{aj}}(\bar{M}_j) \subseteq T_j \times Q_j \times S_j \times F_j \times D_j \times C_j \times Q_j \times T_j$ *defined by*

$$\psi_{T_{aj}}(m_j, m_j+1) = (\theta_{N_j}(m_j), \Gamma_j'(\theta_{N_j}(m_j)))$$

$$\begin{aligned}\psi_{T_{aj}}(m_j+(i+1), m_j+(i+2)) = (&\rho_{8,7}(\psi_{T_{aj}}(m_j+i, m_j+(i+1))),\\ &\phi_{j(i+1)}(\gamma_j'(\theta_{E_j}(m_j+i), \rho_5(\psi_{T_{aj}}(m_j+i, m_j+(i+1))))),\\ &\Gamma_j'(\rho_{8,7}(\psi_{T_{aj}}(m_j+i, m_j+(i+1))),\\ &\phi_j(i+1)(\gamma_j'(\theta_{E_j}(m_j+i), \rho_5(\psi_{T_{aj}}(m_j+i, m_j+(i+1))))))\end{aligned}$$

$\forall i = 0, 1, ---, h-2$, *is an effective adjacency function for the subgraph* $G_{M_j}(T_a)$ *and belongs to the class of general recursive functions.*

The effective adjacency functions presented in Theorem B2 determine at each pair of consecutive moments an appropriate n-tuple for the respective subgraphs corresponding to the component automata in the microsystem graphs. The first entry in the n-tuple names the initial point, the last entry names the adjacent point, and the remaining entries collectively describe the operating conditions associated with the connecting line.

Given an arbitrary pair of system moments and a fixed type of microsystem graph, the effective adjacency functions for all $j \in J$ determine the k pairs of adjacent points and k distinct lines described in the microsystem graph principle in the form of a set of k n-tuples (depending on the type of graph, n may be 6, 8, 9 or 11). Such a set of n-tuples is interpreted in a physical sense as an element operation for the general system of interactive automata. That is, it is the smallest meaningful operation for the system of a kind with respect to the type of graph. An internal element operation for a system is taken to be the set of current states, respective stimuli received and responses produced, and the set of next states as reflected by the points and lines of the processor type graph. The space element operation is defined to be the set of current environments, respective internal element operation, environmental media effects, and the set of next environments as represented by the points and lines of the principal environment graph. Next, the time element operation is considered to be the set of current times, respective internal element operation, and the set of next times, as indicated by the points and lines of the alternate time graph. Finally, there are space-time and time-space element operations reflected by the alternate environment and principal time graphs, respectively. Special importance is placed on the notion of space-time element operation because of the interacting and timing for the system of automata.

Effective Operation. The effective adjacency functions form a basis for developing properties of the graphs, interpretable in terms of the operation of the system. In particular, the notion of effective operation of a given kind for the system is defined by a proper sequence of the previously defined element operations of that kind. As it is used here, a "proper sequence" is taken to mean those sets of n-tuples derived from the effective adjacency functions that are associated with the naturally ordered set of consecutive moments spanning the system moment base; i.e., $\{(m, m+1), (m+1, m+2), ---, (m+(h-1), m+h)\}$. In terms of the microsystem graphs, the effective operation of a given kind is reflected by k paths through the graph of the corresponding type. The k paths consist of one in each subgraph corresponding to each component automaton.

This section presents a significant result for defining a set of effective operation functions, each of which determines a unique path in the subgraph corresponding to the respective component automaton. Furthermore, these functions are based on the effective adjacency functions and also belong to the class of general recursive functions. Before presenting the result, the notion of "a path" in the subgraphs that reflect the effective operation of each component automaton is made precise.

An effective operation path for a subgraph of a microsystem graph corresponding to a component automaton is an alternating sequence of points and lines representing the respective initial status and including all remaining points in proper order representing the appropriate subsequent statuses, and all lines in proper order representing the appropriate subsequent operating conditions. This path reflects the effective operation of the component automaton as a contributing member of the system, and when all k such paths are determined, then the effective operation of the system is known.

The procedure for defining the set of effective operation functions involves building each of the aforementioned sequences using the proper effective adjacency functions. The following discussion is restricted to the alternate environment type of microsystem graph, but those involving other types require identical arguments.

Let $A=\{A_j|j\in J\}$ be a general system of interactive automata and denote an arbitrary component automaton by A_j. Let M_j denote the moment base for A_j and consider the subgraph $G_{M_j}(E_a)$ of the alternate environment type of microsystem graph $G_M(Es_a\, i)$ defined by the Primitives $Es_a\, i1$ through $Es_a\, i4$ and the graph function Λ'_j. From the primitives, the set of points is E_j and the set of lines is $X'_j\subseteq T_j\times Q_j\times S_j\times B_j\times F_j\times D_j\times C_j\times Q_j\times T_j$. Consider the function $\Xi_{E_{aj}}$ mapping a subset $N_{2h+1}=\{1, 2, ---, 2h+1\}$ of the natural numbers into the set $E_j\cup X'_j$ representing the union of the sets of points and lines; i.e., $\Xi_{E_{aj}}: N_{2h+1}\rightarrow \Xi_{E_{aj}}(N_{2h+1})\subseteq E_j\cup X'_j$. The domain set N_{2h+1} is naturally ordered so the initial image is represented by $\Xi_{E_{aj}}(1)$. From Theorem 5.14, $\Psi_{E_{aj}}$ is the defined effective adjacency function for $G_{M_j}(E_a)$. Let ρ_j denote the jth projection function, and follows from the primitives for $G_M(Es_a\, i)$ that $\rho_1(\Psi_{E_{aj}}(m_j+i, m_j+(i+1)))\in E_j$, $\rho_{11}(\Psi_{E_{aj}}(m_j+i, m_j+(i+1)))\in E_j$, and $\rho_{2, ---, 10}(\Psi_{E_{aj}}(m_j+i, m_j+(i+1)))\in X'_j$. Thus, define $\Xi_{E_{aj}}(1)=\rho_1(\Psi_{E_{aj}}(m_j, m_j+1))$, and this represents the initial point for the desired path. Next, define $\Xi_{E_{aj}}(2)=\rho_{2, ---, 10}(\Psi_{E_{aj}}(m_j, m_j+1))$, and this represents the initial line in the path. The adjacent point is given by $\rho_{11}(\Psi_{E_{aj}}(m_j, m_j+1))$ so define $\Xi_{E_{aj}}(3)=\rho_{11}(\Psi_{E_{aj}}(m_j, m_j+1))$. Since the path is an alternating sequence, all the points are represented by images of the odd numbers of N_{2h+1}. So for $n\geq 1$, define $\Xi_{E_{aj}}(2n+1)=\rho_{11}(\Psi_{E_{aj}}(m_j+(n-1), m_j+n))$. Similarly, images of the even numbers of N_{2h+1} represent the lines, so define $\Xi_{E_{aj}}(2n)=\rho_{2, ---, 10}(\Psi_{E_{aj}}(m_j+(n-1), m_j+n))$. The function $\Xi_{E_{aj}}$ as defined above generates a unique path in the subgraph for the component automaton A_j that describes the effective operation of A_j over M_j. The uniqueness follows directly from the automaton principle presented in Appendix A, and the function $\Xi_{E_{aj}}$ is established as an effective operation function for A_j. The recursive nature of $\Xi_{E_{aj}}$ follows directly from that for $\Psi_{E_{aj}}$ and the projection functions. The following theorem summarizes these remarks:

Theorem B3. *Let $A=\{A_j|j\in J=\{1, 2, ---, k\}\}$ be a general system of interactive automata over a system moment base $M=\{m+i|i\in I=\{0, 1, ---, h\}\}$. Denote an arbitrary component automaton by A_j with an associated moment base M_j. Consider the set $N_{2h+1}=\{1, 2, ---, 2h+1\}$ and suppose ρ_j is the jth projection function. Let $G_{M_j}(E_a)$ denote the subgraph of the microsystem graph of the* *alternate environment type* $G_M(Es_a\ i)$. *From the respective primitives,* E_j *represents the set of points in* $G_{M_j}(E_a)$ *and* $X'_j \subseteq T_j \times Q_j \times S_j \times B_j \times F_j \times D_j \times C_j \times Q_j \times T_j$ *the lines. Let* $\Psi_{E_{aj}}$ *denote the effective adjacency function for* $G_{M_j}(E_a)$ *as detailed in Theorem* B2. *Then, the function* $\Xi_{E_{aj}}$: $N_{2h+1} \rightarrow \Xi_{E_{aj}}(N_{2h+1}) \subseteq E_j \cup X'_j$ *defined by*

$$\Xi_{E_{aj}}(1)=\rho_1(\psi_{E_{aj}}(m_j, m_j+1))$$

and for $n \geq 1$

$$\Xi_{E_{aj}}(2n)=\rho_{2,\text{---},10}(\psi_{E_{aj}}(m_j+(n-1), m_j+n))$$

$$\Xi_{E_{aj}}(2n+1)=\rho_{11}(\psi_{E_{aj}}(m_j+(n-1), m_j+n))$$

is an effective operation function for A_j *whose ordered images describe a unique effective operation path with respect to space-time operation in* $G_{M_j}(E_{aj})$. *Furthermore, the function* $\Xi_{E_{aj}}$ *belongs to the class of general recursive functions.*

Recall the effective adjacency functions for the component automata were presented for the processor, principal environment, and principal and alternate time types of microsystem graphs. Effective operation functions for the component automata that determines effective operation paths in the proper subgraphs of the above types are presented in summary form as Theorem B4.

Theorem B4. *Let* $A=\{A_j | j \in J=\{1, 2, \text{---}, k\}\}$ *be a general system of interactive automata over a system moment base* $M=\{m+i | i \in I=\{0, 1, \text{---}, h\}\}$. *Denote an arbitrary component automaton by* A_j *with an associated moment base* M_j. *Consider the set* $N_{2h+1}=\{1, 2, \text{---}, 2h+1\}$ *and suppose* ρ_j *is the jth projection function:*

a. *Let* $G_{M_j}(P)$ *denote the subgraph of the microsystem graph of the processor type* $G_M(Psi)$. *From the respective primitives,* Q_j *represents the points in* $G_{M_j}(P)$ *and* $U_j \subseteq S_j \times F_j \times D_j \times C_j$ *the lines. Let* Ψ_{p_j} *denote the effective adjacency function detailed in Theorem* B2. *Then, the function* Ξ_{p_j}: $N_{2h+1} \rightarrow \Xi_{p_j}(N_{2h+1}) \subseteq Q_j \cup U_j$ *defined by*

$$\Xi_{pj}(1)=\rho_1(\psi_{pj}(m_j, m_j+1))$$

and for $n \geq 1$

$$\Xi_{pj}(2n)=\rho_{2,\text{---},5}(\psi_{pj}(m_j+(n-1), m_j+n))$$

$$\Xi_{pj}(2n+1)=\rho_6(\psi_{pj}(m_j+(n-1), m_j+sn))$$

is an effective operation function for A_j *whose ordered images describe a unique effective operation path in* $G_{M_j}(P)$. *Furthermore,* Ξ_{p_j} *belongs to the class of general recursive functions.*

b. *Let $G_{M_j}(E_p)$ denote the subgraph of the microsystem graph of the principal environment type $G_M(Es_p\ i)$. From the respective primitives, E_j represents the points in $G_{M_j}(E_p)$ and $X_j \subseteq Q_j \times S_j \times B_j \times F_j \times D_j \times C_j \times Q_j$ the lines. Let $\Psi_{E_{pj}}$ denote the effective adjacency function detailed in Theorem* B2. *Then, the function $\Xi_{E_{pj}}: N_{2h+1} \rightarrow \Xi_{E_{pj}}(N_{2h+1}) \subseteq E_j \cup X_j$ defined by*

$$\Xi_{Epj}(1) = \rho_1(\psi_{Epj}(m_j, m_j+1))$$

and for $n \geq 1$

$$\Xi_{Epj}(2n) = \rho_{2,\cdots,8}(\psi_{Epj}(m_j+(n-1), m_j+n))$$

$$\Xi_{Epj}(2n+1) = \rho_9(\psi_{Epj}(m_j+(n-1), m_j+n))$$

is an effective operation function for A_j whose ordered images describe a unique effective operation path in $G_{M_j}(E_p)$. Furthermore, $\Xi_{E_{pj}}$ belongs to the class of general recursive functions.

c. *Let $G_{M_j}(T_p)$ denote the subgraph of the microsystem graph of the principal time type $G_M(Ts_p\ i)$. From the respective primitives, T_j represents the set of points in $G_{M_j}(T_p)$ and $Z_j \subseteq E_j \times Q_j \times S_j \times B_j \times F_j \times D_j \times C_j \times Q_j \times E_j$ the lines. Let $\Psi_{T_{pj}}$ denote the effective adjacency function detailed in Theorem* B2. *Then, the function $\Xi_{T_{pj}}: N_{2h+1} \rightarrow \Xi_{T_{pj}}(N_{2h+1}) \subseteq T_j \cup Z_j$ defined by*

$$\Xi_{Tpj}(1) = \rho_1(\psi_{Tpj}(m_j, m_j+1))$$

and for $n \geq 1$

$$\Xi_{Tpj}(2n) = \rho_{2,\cdots,10}(\psi_{Taj}(m_j+(n-1), m_j+n))$$

$$\Xi_{Tpj}(2n+1) = \rho_{11}(\psi_{Tpj}(m_j+(n-1), m_j+n))$$

is an effective operation function for A_j whose ordered images describe a unique effective operation path in $G_{M_j}(T_p)$. Furthermore, $\Xi_{T_{pj}}$ belongs to the class of general recursive functions.

d. *Let $G_{M_j}(T_a)$ denote the subgraph of the microsystem graph of the alternate time type $G_M(Ts_a\ i)$. From the respective primitives, T_j represents the set of points in $G_{M_j}(T_a)$ and $Z'_j \subseteq Q_j \times S_j \times F_j \times D_j \times C_j \times Q_j$ the lines. Let $\Psi_{T_{aj}}$ denote the effective adjacency function detailed in Theorem* B2. *Then, the function $\Xi_{T_{aj}}: N_{2h+1} \rightarrow \Xi_{T_{aj}}(N_{2h+1}) \subseteq T_j \cup Z'_j$ defined by*

$$\Xi_{Taj}(1) = \rho_1(\psi_{Taj}(m_j, m_j+1))$$

and for $n \geq 1$

$$\Xi_{Taj}(2n) = \rho_{2,\cdots,7}(\psi_{Taj}(m_j+(n-1), m_j+n))$$

$$\Xi_{Taj}(2n+1) = \rho_8(\psi_{Taj}(m_j+(n-1), m_j+n))$$

is an effective operation for A_j whose ordered images describe a unique effective operation path in $G_{M_j}(T_a)$. Furthermore, $\Xi_{T_{aj}}$ belongs to the class of general recursive functions.

Recall that the effective operation of a given kind for the general system of interactive automata was defined as a proper sequence of the element operations of that kind. More specifically, the effective operation for the system is defined as an alternating sequence of sets of status conditions and sets of operating conditions. Given a type of microsystem graph of interest, the k paths determined by the previously defined effective operation functions reflect the effective operation of the proper kind for the system in the following manner. Consider the microsystem graph of the alternate environment type $G_M(Es_a\ i)$ and suppose the effective operation of the space-time kind for the system is desired. From Theorem B3, the set of initial points for the k paths in the graph is given by $\{\Xi_{E_{aj}}(1)\,|\,j \in J\}$. Now each of these points reflect the initial status condition (environment) of a respective component automaton, so taken collectively, they describe the initial status condition of the system. Applying Theorem B3 for $n=1$, the set $\{\Xi_{E_{aj}}(2)\,|\,j \in J\}$ lists exactly k lines in the graph, and these collectively reflect the initial operating conditions of the system. In addition, for $n=1$, the set $\{\Xi_{E_{aj}}(3)\,|\,j \in J\}$ represents the adjacent points, and taken collectively, reflect the next status condition of the system. Continuing in this manner for all choices of n with $1 \leq n \leq h$, the sequence of derived sets describes the effective operation of the space-time kind for the general system of interactive automata. The preceding remarks are summarized as follows:

Corollary B4a. *let $A=\{A_j\,|\,j \in J=\{1, 2, ---, k\}\}$ denote a general system of interactive automata with an associated moment base $M=\{m+i\,|\,i \in I=\{0, 1, 2, ---, h\}\}$. Then the effective operation for A of the space-time kind corresponds to the k-effective operation paths in the microsystem graph of the alternate environment type and is given by the following sequence of sets:*

$$\{\Xi_{E_{aj}}(1)|j\in J\}, \{\Xi_{E_{aj}}(2)|j\in J\}, \{\Xi_{E_{aj}}(3)|j\in J\}, ---, \{\Xi_{E_{aj}}(2h+1)|j\in J\}$$

In the same manner, the effective operation for the system of the internal, space, time, and time-space kind are reflected by the k-effective operation paths in the processor, principal environment, alternate time, and principal time types of microsystem graph, respectively.

Corollary B4b. *Let $A=\{A_j\,|\,j \in J=\{1, 2, ---, k\}\}$ denote a general system of interactive automata with an associated moment base $M=\{m+i\,|\,i \in I=\{0, 1, ---, h\}\}$. Then*

a. *the effective operation for A of the internal kind corresponds to the k-effective operation paths in the microsystem graph of the processor type and is given by the following sequence of sets*:

$$\{\Xi_{pj}(1)|j\in J\},\{\Xi_{pj}(2)|j\in J\},\{\Xi_{pj}(3)|j\in J\},\text{---},\{\Xi_{pj}(2h+1)|j\in J\}$$

b. *the effective operation for A of the space kind corresponds to the k-effective operation paths in the microsystem graph of the principal environment type and is given by the following sequence of sets*:

$$\{\Xi_{Epj}(1)|j\in J\},\{\Xi_{Epj}(2)|j\in J\},\{\Xi_{Epj}(3)|j\in J\},\text{---},\{\Xi_{Epj}(2h+1)|j\in J\}$$

c. *the effective operation for A of the time kind corresponds to the k-effective operation paths in the microsystem graph of the alternate time type and is given by the following sequence of sets*:

$$\{\Xi_{Taj}(1)|j\in J\},\{\Xi_{Taj}(2)|j\in J\},\{\Xi_{Taj}(3)|j\in J\},\text{---},\{\Xi_{Taj}(2h+1)|j\in J\}$$

d. *the effective operation for A of the time-space kind corresponds to the k-effective operation paths in the microsystem graph of the principal time type and is given by the following sequence of sets*:

$$\{\Xi_{Tpj}(1)|j\in J\},\{\Xi_{Tpj}(2)|j\in J\},\{\Xi_{Tpj}(3)|j\in J\},\text{---},\{\Xi_{Tpj}(2h+1)|j\in J\}$$

B.3 MACROSYSTEM GRAPHS

The macrosystem graphs are of special interest when a general system of interactive automata is viewed as a whole. Such graphs contain points in the form of k-tuples describing a particular status condition for the system and contain lines properly labeled to represent the system operating conditions. It is the purpose of this section to determine the effective operation of a general system of interactive automata by means of the macrosystem graphs of the model.

The content presented herein also uses the methods developed in Appendix A and applies some of the results derived in the preceding section. First, system moment functions associating the proper system status condition with each system moment are defined in terms of those for the component automaton. Next, graph principles related to the macrosystem graphs are presented and effective adjacency functions for such graphs are derived. Finally, the effective operation of the system is established by effective operation functions for the macrosystem graphs.

Before proceeding with the analysis of the effective operation of a general system of interactive automata by means of the macrosystem graphs, system status conditions are reviewed. Recall that a system moment is a k-tuple with entries from the corresponding moments for each component automaton, i.e., $m+i=(m_j+i\,|\,j\in J)$. Furthermore, each element in the system sets is a k-tuple consisting of corresponding elements from the component automaton sets. For example, the system state at some arbitrary system moment is a k-tuple of the states of each component automaton at their respective moments, i.e., $Q(m+i)=(Q_j(m_j+i)\,|\,j\in J)$. Even the system functions are k-tuples of component automaton functions. For example, the system function ω, mapping pairs of system states and system stimuli into the set of system states, is defined by $\omega(q, s)=(\omega_j(\rho_j(q_j), \rho_j(s_j))\,|\,j\in J)$ where ρ_j denotes the jth projection function. Moreover, there are system environment functions that are k-tuples corresponding to the environment functions for the component automata. Since the system environment is $E\subseteq\Pi_j E_j$ and the system stimulus set is $S\subseteq\Pi_j S_j$, then at each moment $m+i$, there is a function ϕ_i with ϕ_i: $E\rightarrow S$ defined by $\phi_i(e)=(\phi_{ji}(\rho_j(e))\,|\,j\in J)$. It is natural then that the system moment functions and others pertaining to the subsequent analysis are similar in that they are k-tuples of corresponding functions for the component automaton.

Moment Functions. Since the moment functions for each component automaton are known and since the system status conditions are k-tuples composed of the corresponding status conditions of each component automaton, the procedure for defining the system moment functions follows the k-tuple pattern of the system functions. Suppose $A=\{A_j\,|\,j\in J=\{1, 2, ---, k\}\}$ is a general system of interactive automata with an associated moment base $M=\{m+i\,|\,i\in I=\{0, 1, 2, ---, h\}\}$. Since the moment functions associate the proper status conditions with each moment, the domain set is a subset of M and is naturally ordered by the correspondence $m+i\rightarrow i$. From Theorem B1, the initial environments of the component automata are given by $\theta_{E_j}(m_j)$. Thus, the initial system environment is $\theta_E(m)=\theta_{E_j}(m_j)\,|\,j\in J)$. Similarly, the initial system time, stimulus, and state are given by $\theta_T(m)=(\theta_{T_j}(m_j)\,|\,j\in J)$, $\theta_S(m)=(\theta_{S_j}(m_j)\,|\,j\in J)$, and $\theta_Q(m)=(\theta_{Q_j}(m_j)\,|\,j\in J)$, respectively. Since the initial system response occurs at $m+1$, the initial transformation response of the system is $\theta_F(m+1)=(\sigma_j(\rho_j(\theta_Q(m)), \rho_j(\theta_S(m)))\,|\,j\in J)=(\sigma_Q(m), \theta_S(m))$, the initial recorded response is $\theta_B(m+1)=(\eta_j'(\rho_j(\theta_E(m)), \rho_j(\theta_F(m+1)))\,|\,j\in J)=\eta'(\theta_E(m, \theta_F(m+1)))$, the initial spatial change response is $\theta_D(m+1)=(\delta_j(\rho_j(\theta_Q(m)), \rho_j(\theta_S(m)))\,|\,j\in J)=\delta(\theta_Q(m), \theta_S(m))$, and the initial time change response is $\theta_C(m+1)=(\tau_j(\rho_j(\theta_Q(m)), \rho_j(\theta_S(m)))\,|\,j\in J)=\tau(\theta_Q(m), \theta_S(m))$. Next, the image of the second domain element $m+1$ under the system moment function θ_E is $\theta_E(m+1)=(\theta_{E_j}(m_j+1)\,|\,j\in J)$. But since $\theta_{E_j}(m_j+1)$ is a function of $\theta_{E_j}(m_j)$, namely $\theta_{E_j}(m_j+1)=\gamma_j'(\theta_{E_j}(m_j), \theta_{D_j}(m_j+1))$ then $\theta_E(m+1)$ is also a function of the image of the previous domain element. That is, $\theta_E(m+1)=(\gamma_j'(\rho_j(\theta_E(m)), \rho_j(\theta_D(m+1)))\,|\,j\in J)=\gamma'(\theta_E(m), \theta_D(m+1))$. Similarly, $\theta_T(m+1)=(\lambda_j'(\rho_j(\theta_T(m)), \rho_j(\theta_C(m+1)))\,|\,j\in J)=\lambda'(\theta_T(m), \theta_C(m+1))$, $\theta_S(m+1)=(\phi_{j\ 1}(\rho_j(\theta_E(m+1)))\,|\,j\in J)=\phi_1(\theta_E(m+1))$, and $\theta_Q(m+1)=(\omega_j(\rho_j(\theta_Q(m)), \rho_j(\theta_S(m)))\,|\,j\in J)=\omega(\theta_Q(m), \theta_S(m))$. The system responses of the domain element $m+2$ are given by $\theta_F(m+2)=\sigma(\theta_Q(m+1)$,

$\theta_S(m+1)), \theta_B(m+2)=\eta'(\theta_E(m+1), \theta_F(m+2)), \theta_D(m+2)=\delta(\theta_Q(m+1), \theta_S(m+1)),$ and $\theta_C(m+2)=\tau(\theta_Q(m+1), \theta_S(m+1))$, respectively. Continuing in this manner for all values of i corresponding to the system moments $m+i$, the images of the various functions are similarly derived. Thus $\theta_E(m+(i+1))=(\gamma_j'(\rho_j(\theta_E(m+i)), \rho_j(\theta_D(m+(i+1))))\,|\,j\in J)=\gamma'(\theta_E(m+i), \theta_D(m+(i+1)))$, $\theta_T(m+(i+1))=(\gamma_j'(\rho_j(\theta_T(m+i)), \rho_j(\theta_C(m+(i+1))))\,|\,j\in J)=\lambda'(\theta_T(m+i), \theta_C(m+(i+1)))$, $\theta_S(m+(i+1))=(\phi_{j(i+1)}(\rho_j(\theta_E(m+(i+1))))\,|\,j\in J)=\phi_{i+1}(\theta_E(m+(i+1)))$, and $\theta_Q(m+(i+1))=(\omega_j(\rho_j(\theta_Q(m+i)), \rho_j(\theta_S(m+i)))\,|\,j\in J)=\omega(\theta_Q(m+i), \theta_S(m+i))$. As for the responses, $\theta_F(m+(i+2))=(\sigma_j(\rho_j(\theta_Q(m+(i+1))), \rho_j(\theta_S(m+(i+1))))\,|\,j\in J)=\sigma(\theta_Q(m+(i+1)), \theta_S(m+(i+1)))$, $\theta_B(m+(i+2))=(\eta_j'(\rho_j(\theta_E(m+(i+1))), \rho_j(\theta_F(m+(i+2))))\,|\,j\in J)=\eta'(\theta_E(m+(i+1)), \theta_F(m+(i+2)))$, $\theta_D(m+(i+2))=(\delta_j(\rho_j(\theta_Q(m+(i+1))), \rho_j(\theta_S(m+(i+1))))\,|\,j\in J)=\delta(\theta_Q(m+(i+1)), \theta_S(m+(i+1)))$, and $\theta_C(m+(i+2))=(\tau_j(\rho_j(\theta_Q(m+(i+1))), \rho_j(\theta_S(m+(i+1))))\,|\,j\in J)=\tau(\theta_Q(m+(i+1)), \theta_S(m+(i+1)))$ describes the system transformation, recorded, spatial change, and time change responses, respectively, at arbitrary moments. Since the moment functions for the component automata are recursive, it follows that the system moment functions as defined in the preceding remarks belong to the class of general recursive functions. The following theorem is stated to summarize the results for the system moment functions:

Theorem B5. *Let $A=\{A_j\,|\,j\in J=\{1, 2, ---, k\}\}$ be a general system of interactive automata with an associated moment base $M=\{m+i\,|\,i\in I=\{0, 1, ---, h\}\}$. Suppose the minimal set of performance descriptors for each component automaton are known. Then, the system moment functions for A include*

a. θ_E: $M\to\theta_E(M)=E\subseteq\bar{E}$ *defined by*

$$\theta_E(m)=(\theta_{E_j}(m_j)\,|\,j\in J)$$
$$\theta_E(m+(i+1))=(\gamma_j'(\rho_j(\theta_E(m+i)), \rho_j(\theta_D(m+(i+1))))\,|\,j\in J)$$
$$=\gamma'(\theta_E(m+i), \theta_D(m+(i+1)))$$

b. θ_T: $M\to\theta_T(M)=T\subseteq\bar{T}$ *defined by*

$$\theta_T(m)=(\theta_{T_j}(m_j)\,|\,j\in J)$$
$$\theta_T(m+(i+1))=(\lambda_j'(\rho_j(\theta_T(m+i)), \rho_j(\theta_C(m+(i+1))))\,|\,j\in J)$$
$$=\lambda'(\theta_T(m+i), \theta_C(m+(i+1)))$$

c. θ_S: $M-\{m+h\}\to\theta_S(M-\{m+h\})=S\subseteq\bar{S}$ *defined by*

$$\theta_S(m)=(\theta_{S_j}(m_j)\,|\,j\in J)$$
$$\theta_S(m+(i+1))=(\phi_{j(i+1)}(\rho_j(\theta_E(m+(i+1))))\,|\,j\in J)$$
$$=\phi_{i+1}(\theta_E(m+(i+1)))$$

d. θ_Q: $M\to\theta_Q(M)=Q\subseteq\bar{Q}$ *defined by*

$$\theta_Q(m)=(\theta_{Q_j}(m_j)\,|\,j\in J)$$
$$\theta_Q(m+(i+1))=(\omega_j(\rho_j(\theta_Q(m+i)), \rho_j(\theta_S(m+i)))\,|\,j\in J)$$
$$=\omega(\theta_Q(m+i), \theta_S(m+i))$$

e. θ_F: $M-\{m\}\to\theta_F(M-\{m\})=F\subseteq\bar{F}$ *defined by*

$$\theta_F(m+1)=(\theta_{F_j}(m_j+1)\,|\,j\in J)$$

$$\theta_F(m+(i+2))=(\sigma_j(\rho_j(\theta_Q(m+(i+1))),\ \rho_j(\theta_S(m+(i+1))))\mid j\in J)$$
$$=\sigma(\theta_Q(m+(i+1)),\ \theta_S(m+(i+1)))$$

f. θ_B: $M-\{m\}\rightarrow\theta_B(M-\{m\})=B\subseteq\bar{B}$ *defined by*
$$\theta_B(m+1)=(\theta_{B_j}(m_j+1)\mid j\in J)$$
$$\theta_B(m+(i+2))=(\eta_j'(\rho_j(\theta_E(m+(i+1))),\ \rho_j(\theta_F(m+(i+2))))\mid j\in J)$$
$$=\eta'(\theta_E(m+(i+1)),\ \theta_F(m+(i+2)))$$

g. θ_D: $M-\{m\}\rightarrow\theta_D(M-\{m\})=D\subseteq\bar{D}$ *defined by*
$$\theta_D(m+1)=(\theta_{D_j}(m_j+1)\mid j\in J)$$
$$\theta_D(m+(i+2))=(\theta_j(\rho_j(\theta_Q(m+(i+1))),\ \rho_j(\theta_S(m+(i+1))))\mid j\in J)$$
$$=\delta(\theta_Q(m+(i+1)),\ \theta_S(m+(i+1)))$$

h. θ_C: $M-\{m\}\rightarrow\theta_C(M-\{m\})=C\subseteq\bar{C}$ *defined by*
$$\theta_C(m+1)=(\theta_{C_j}(m_j+1)\mid j\in J)$$
$$\theta_C(m+(i+2))=(\tau_j(\rho_j(\theta_Q(m+(i+1))),\ \rho_j(\theta_S(m+(i+1))))\mid j\in J)$$
$$=\tau(\theta_Q(m+(i+1)),\ \theta_S(m+(i+1)))$$

Furthermore, the defined functions belong to the class of general recursive functions.

The system moment functions for the state-stimulus pairs, environment-state-stimulus triples, etc. are not defined in terms of the corresponding moment functions for the component automata. Instead, they are defined in terms of existing system moment functions. Corollary B5 is an obvious result of Theorem B5.

Corollary B5. *The set of system moment functions include those for determining the state-stimulus pair, environment-state-stimulus triple, time-state-stimulus triple, time-environment-state-stimulus quadruple, and the environment-time-state-stimulus quadruple associated with each moment. These are recursive functions and are defined as follows:*

a. θ_V: $M-\{m+h\}\rightarrow\theta_V(M-\{m+h\})=V\subseteq Q\times S$ *defined by*
$$\theta_V(m)=(\theta_Q(m),\ \theta_S(m))$$
$$\theta_V(m+(i+1))=(\theta_Q(m+(i+1)),\ \theta_S(m+(i+1)))$$

b. θ_W: $M-\{m+h\}\rightarrow\theta_W(M-\{m+h\})=W\subseteq E\times V$ *defined by*
$$\theta_W(m)=(\theta_E(m),\ \theta_V(m))$$
$$\theta_W(m+(i+1))=(\theta_E(m+(i+1)),\ \theta_V(m+(i+1)))$$

c. θ_N: $M-\{m+h\}\rightarrow\theta_N(M-\{m+h\})=N\subseteq T\times V$ *defined by*
$$\theta_N(m)=(\theta_T(m),\ \theta_V(m))$$
$$\theta_N(m+(i+1))=(\theta_T(m+(i+1)),\ \theta_V(m+(i+1)))$$

d. θ_Y: $M-\{m+h\}\rightarrow\theta_Y(M-\{m+h\})=Y\subseteq T\times W$ *defined by*
$$\theta_Y(m)=(\theta_T(m),\ \theta_W(m))$$
$$\theta_Y(m+(i+1))=(\theta_T(m+(i+1)),\ \theta_W(m+(i+1)))$$

e. $\theta_{Y'}: M-\{m+h\} \rightarrow \theta_{Y'}(M-\{m+h\})=Y' \subseteq E \times N$ *defined by*

$\theta_{Y'}(m)=(\theta_E(m), \theta_N(m))$

$\theta_{Y'}(m+(i+1))=(\theta_E(m+(i+1)), \theta_N(m+(i+1)))$

Effective Adjacency Functions. The fundamental system principle, translated in terms of the macrosystem graphs of the model, states that at each system moment one and only one point represents the current status of the system. In addition, the operating conditions of the system in achieving a change of status from a given system moment to the next are described by one and only one line directed from the given point to that for the next status. In this sense, the macrosystem graphs are similar to those for the general automaton, and the resulting principle for the macrosystem graphs naturally parallels that for the general automaton.

Macrosystem Graph Principle. There is one and only one pair of adjacent points (not necessarily distinct) connected by one and only one line in the macrosystem graphs for a general system of interactive automata corresponding to each pair of consecutive system moments from the system moment base.

Effective adjacency functions exist for the macrosystem graphs and determine the pair of points and connecting line for each consecutive pair of moments described in the above principle. These functions are summarized in the form of theorems and since the proofs follow those of the corresponding theorems of Appendix A, they are omitted.

Theorem B6. *Let* $A=\{A_j | j \in J=\{1, 2, ---, k\}\}$ *be a general system of interactive automata analyzed over a system moment base* $M=\{m+i | i \in I=\{0, 1, ---, h\}\}$. *Suppose* $\overline{M} \subseteq M \times M$ *denotes the pairs of consecutive moments for* A, *and* ρ_n *denotes the nth projection (identity) function.*

a. *Let* $G_M(Psa)$ *denote the macrosystem graph of the processor type defined by the Primitives Psa1 through Psa4 and th e graph function* Ω. *Then, the function* $\Psi_p: \overline{M} \rightarrow \Psi_p(\overline{M}) \subseteq Q \times S \times F \times D \times C \times Q$ *defined by*

$$\psi_p(m, m+1)=(\theta_V(m), \Omega(\theta_V(m)))$$

$$\psi_p(m+(i+1), m+(i+2))=(\rho_{5+1,---,6k}(\psi_p(m+i, m+(i+1))), \phi_{i+1}(\gamma'(\theta_E(m+(i+1)), \rho_{3k+1,---,4k}(\psi_p(m+i, m+(i+1))))), \Omega(\rho_{5k+1,---,6k}(\psi_p(m+i, m+(i+1))), \phi_{i+1}(\gamma'(\theta_E(m+(i+1)), \rho_{3k+1,---,4k}(\psi_p(m+i, m+(i+1)))))))$$

$\forall i=0, 1, ---, h-2$, *is an effective adjacency function for the graph* $G_M(Psa)$ *and belongs to the class of general recursive functions.*

b. *Let* $G_M(Es_p\, a)$ *denote the macrosystem graph of the principal environment type defined by the respective primitives and the graph function* Γ. *Then, the function* $\Psi_{E_p}: \overline{M}\to\Psi_{E_p}(\overline{M})\subseteq E\times Q\times S\times B\times F\times D\times C\times Q\times E$ *defined by*

$$\psi_{Ep}(m, m+1)=(\theta_W(m), \Gamma(\theta_W(m)))$$

$$\begin{aligned}\psi_{Ep}(m+(i+1), m+(i+2))=&(\rho_{8k+1,---,9k,7k+1,---,8k}(\psi_{Ep}(m+i, m+(i+1)))),\\ &\phi_{i+1}(\gamma'(\theta_E(m+(i+1)), \rho_{5k+1,---,6k}(\psi_{Ep}(m+i, m+(i+1))))),\\ &\Gamma(\rho_{8k+1,---,9k,7k+1,---,8k}(\psi_{Ep}(m+i, m+(i+1))),\\ &\phi_{i+1}(\gamma'(\theta_E(m+(i+1)), \rho_{5k+1,---,6k}(\psi_{Ep}(m+i, m+(i+1))))))\end{aligned}$$

$\forall i=0, 1, ---, h-2$, *is an effective adjacency function for the graph* $G_M(Es_p\ a)$ *and belongs to the class of general recursive functions.*

c. *Let* $G_M(Ts_p\ a)$ *denote the macrosystem graph of the principal time type defined by Primitives* Ts_p *a1 through* Ts_p *a4 and the graph function* Λ. *Then, the function* $\Psi_{T_p}: \overline{M}\to\Psi_{T_p}(\overline{M})\subseteq T\times E\times Q\times S\times B\times F\times D\times C\times Q\times E\times T$ *defined by*

$$\psi_{Tp}(m, m+1)=(\theta_Y(m), \Lambda(\theta_Y(m)))$$

$$\begin{aligned}\psi_{Tp}(m+(i+1), m+(i+2))=&(\rho_{10k+1,---,11k,9k+1,---,10k,8k+1,---,9k}(\psi_{Tp}(m+i, m+(i+1))),\\ &\phi_{i+1}(\gamma'(\theta_E(m+(i+1)), \rho_{6k+1,---,7k}(\psi_{Tp}(m+i, m+(i+1))))),\\ &\Lambda(\rho_{10k+1,---,11k,9k+1,---,10k,8k+1,---,9k}(\psi_{Tp}(m+i, m+(i+1))),\\ &\phi_{i+1}(\gamma'(\theta_E(m+(i+1)), \rho_{6k+1,---,7k}(\psi_{Tp}(m+i, m+(i+1)))))))\end{aligned}$$

$\forall i=0, 1, ---, h-2$, *is an effective adjacency function for the graph* $G_M(Ts_p\ a)$ and *belongs to the class of general recursive functions.*

d. *Let* $G_M(Es_a\ a)$ *denote the macrosystem graph of the alternate environment type defined by the Primitives* Es_a *a1 through* Es_a *a4 and the graph function* Λ'. *Then, the function* $\Psi_{E_a}: \overline{M}\to\Psi_{E_a}(\overline{M})\subseteq E\times T\times Q\times S\times B\times F\times D\times C\times Q\times T\times E$ *defined by*

$$\psi_{E_a}(m,m+1)=(\theta_{Y'}(m), \Lambda'(\theta_{Y'}(m)))$$

$$\begin{aligned}\psi_{E_a}(m+(i+1), m+(i+2))=&(\rho_{10k+1,---,11k,9k+1,---,10k,8k+1,---,9k}(\psi_{E_a}(m+i, m+(i+1))),\\ &\phi_{i+1}(\gamma'(\theta_E(m+(i+1)), \rho_{6k+1,---,7k}(\psi_{E_a}(m+i, m+(i+1))))),\\ &\Lambda'(\rho_{10k+1,---,11k,9k+1,---,10k,8k+1,---,9k}(\psi_{E_a}(m+i, m+(i+1))),\\ &\phi_{i+1}(\gamma'(\theta_E(m+(i+1)), \rho_{6k+1,---,7k}(\psi_{E_a}(m+i, m+(i+1)))))))\end{aligned}$$

$\forall i=0, 1, ---, h-2$, *is an effective adjacency function for the graph* $G_M(Es_a\ a)$ *and belongs to the class of general recursive functions.*

e. *Let* $G_M(Ts_a a)$ *denote the macrosystem graph of the alternate time type defined by the respective primitives and the graph function* Γ'. *Then, the function* $\Psi_{T_a}: \overline{M}\to\Psi_{Ta}(\overline{M})\subseteq T\times Q\times S\times F\times D\times C\times Q\times T$ *defined by*

$$\psi_{T_a}(m,m+1)=(\theta_N(m),\Gamma'(\theta_N(m)))$$

$$\begin{aligned}\psi_{T_a}(m+(i+1),m+(i+2))=(&\rho_{7k+1,---,8k,6k+1,---,7k}(\psi_{T_a}(m+i,m+(i+1))),\\ &\phi_{i+1}(\gamma'(\theta_E(m+(i+1)),\rho_{4k+1,---,5k}(\psi_{T_a}(m+i,m+(i+1))))),\\ &\Gamma'(\rho_{7k+1,---,8k,6k+1,---,7k}(\psi_{T^a}(m+i,m+(i+1))),\\ &\phi_{i+1}(\gamma'(\theta_E(m+(i+1)),\rho_{4k+1,---,5k}(\psi_{T_a}(m+i,m+(i+1)))))))\end{aligned}$$

$\forall i=0,1,---,h-2$, is an effective adjacency function for the graph $G_M(Ts_a a)$ and belongs to the class of general recursive functions.

The effective adjacency functions for the macrosystem graphs determine at each pair of consecutive system moments an appropriate *nk* tuple (*n*-tuples, each entry of which is a *k*-tuple) for the respective type of graph. Note that $n=6$ for the processor type, $n=9$ for the principal environment type, $n=8$ for the alternate time graph, and $n=11$ for the principal time and alternate environment types. The first *k*-tuple entry in the *n*-tuple names the initial point, the last *k*-tuple entry in the *n*-tuple names the adjacent point, and the remaining entries, collectively, describe the operating conditions associated with the line. As before, the images of the effective adjacency functions interpreted in terms of the physical situation describe an element operation of the system of a kind corresponding to the type of the macrosystem graph.

Effective Operation. Just as in the case of the microsystem graphs, the effective adjacency functions established for the macrosystem graphs are useful for determining the effective operation of the system. The effective operation of each kind is a proper sequence of the element operations of that kind as previously defined. However, in terms of the macrosystem graphs, the effective operation of a given kind is reflected by a unique path through the proper type of graph. This path is an alternating sequence of points and lines. The initial point represents the respective initial system status, and all remaining points in order represent successive system statuses. All lines taken in order represent successive operating conditions enabling the system to pass from the previous status to the next one.

The purpose here is to define a set of effective operation functions applicable to the macrosystem graphs and to determine the effective operation of each kind for the system through the graphs. Finally, it remains to show that the effective operation of each kind is the same for the two classes of graphs in the model.

Suppose $A=\{A_j|j\in J=\{1,2,---,k\}\}$ is a general system of interactive automata with an associated moment base $M=\{m+i|i\in I\}$. Consider the macrosystem graph of the alternate environment type $G_M(Es_a\,a)$ as defined by the primitives $Es_a\,a1$ through $Es_a\,a4$ and the graph function Λ'. From the primitives, the set of points is $E\subseteq\Pi_j E_j$ and the set of lines is $X'\subseteq T\times Q\times S\times B\times F\times D\times C\times Q\times T$. Let the domain set for the function Ξ_{E_a} be the subset of the natural numbers $N_{2h+1}=\{1, 2, ---, 2h+1\}$. Furthermore, let the range set for Ξ_{E_a} be the set $E\cup X'$. Let ρ_j denote the jth projection function, and it follows from Theorem B6 that $\rho_{1,---,k}(\Psi_{E_a}(m+(i+1), m+(i+2)))\in E$, $\rho_{10k+1,---,11k}(\Psi_{E_a}(m+(i+1), m+(i+2)))\in E$, and $\rho_{k+1,\ ---,\ 10k}(\Psi_{E_a}(m+(i+1), m+(i+2)))\in X'$. Thus, define $\Xi_{E_a}(1)=\rho_{1,\ ---,\ k}(\Psi_{E_a}(m, m+1))$, and this represents the initial point for the desired path. Next, define $\Xi_{E_a}(2)=\rho_{k+1,---,10k}(\Psi_{E_a}(m, m+1))$, and this represents the initial line in the desired path. The images of the even domain elements are used to determine lines in the path and likewise images of the odd elements determine points. So, for $n\geq 1$, define $\Xi_{E_a}(2n)=\rho_{k+1,---,10k}(\Psi_{E_a}(m+(n-1), m+n))$ and $\Xi_{E_a}(2n+1)=\rho_{10k+1,---,11k}(\Psi_{E_a}(m+(n-1), m+n))$. The function Ξ_{E_a} as defined determines a unique path in the graph $G_M(Es_a\, a)$ and, furthermore, the function Ξ_{E_a} is recursive. The recursive nature follows directly from that for the effective adjacency function Ξ_{E_a}.

Theorem B7. *Let $A=\{A_j|j\in J=\{1, 2, ---, k\}\}$ denote a general system of interactive automata over a system moment base $M=\{m+i|i\in I=\{0, 1, ---, h\}\}$. Consider the set $N_{2h+1}=\{1, 2, ---, 2h+1\}$ and let ρ_j denote the jth projection function. Let $G_M(Es_a\, a)$ be the macrosystem graph of the alternate environment type. From the respective primitives, the set of system environments E denotes the points in $G_M(Es_a\, a)$ and the set $X'\subseteq T\times Q\times S\times B\times F\times D\times C\times Q\times T$ denotes the lines. Let Ψ_{E_a} be the effective adjacency function for $G_M(Es_a\, a)$ as established in Theorem* B6. *Then, the function Ξ_{E_a}: $N_{2h+1}\rightarrow\Xi_{E_a}(N_{2h+1})\subseteq E\cup X'$ defined by*

$$\Xi_{E_a}(1)=\rho_{1,---,k}(\psi_{E_a}(m, m+1))$$

and for $n\geq 1$

$$\Xi_{E_a}(2n)=\rho_{k+1,---,10k}(\psi_{E_a}(m+(n-1), m+n))$$

$$\Xi_{E_a}(2n+1)=\rho_{10k+1,---,11k}(\psi_{E_a}(m+(n-1), m+n))$$

is an effective operation function for A whose ordered images describe a unique effective operation path in $G_M(Es_a\ a)$ with respect to the space-time operation for A. Furthermore, the function Ξ_{E_a} belongs to the class of general recursive functions.

The effective operation is defined as an alternating sequence of sets of status conditions and sets of operating conditions. Since this corresponds to the set of ordered images under the effective operation functions, the following is an obvious result of the preceding theorem:

Corollary B7. *The effective operation for A of the space-time kind corresponds to the effective operation path in the macrosystem graph of the alternate environment type and is given by the following sequence:*

$$\Xi_{E_a}(1), \Xi_{E_a}(2), ---, \Xi_{E_a}(2h+1)$$

In a like manner, effective operation functions are recursively defined for other types of macrosystem graphs. These functions are summarized in the following theorem:

Theorem B8. *Let $A=\{A_j | j\in J=\{1, 2, ---, k\}\}$ be a general system of interactive automata with an associated moment base $M=\{m+i | i\in I=\{0, 1, ---, h\}\}$. Consider the set $N_{2h+1}=\{1, 2, ---, 2h+1\}$ and suppose ρ_j denotes the jth projection function.*

a. *Let $G_M(Psa)$ be the macrosystem graph of the processor type. From the respective primitives, the set of system states Q denotes the points in $G_M(Psa)$, and the set $U\subseteq S\times F\times D\times C$ denotes the set of lines. Let Ψ_p be the effective adjacency function for $G_M(Psa)$ as detailed in Theorem* B6. *Then, the function Ξ_p: $N_{2h+1}\rightarrow\Xi_p(N_{2h+1})\subseteq Q\cup U$ defined by*

$$\Xi_p(1)=\rho_{1,---,k}(\psi_p(m, m+1))$$

and for $n\geq 1$

$$\Xi_p(2n)=\rho_{k+1,---,5k}(\psi_p(m+(n-1), m+n))$$

$$\Xi_p(2n+1)=\rho_{5k+1,---,6k}(\psi_p(m+(n-1), m+n))$$

is an effective operation function for A whose ordered images describe a unique effective operation path in $G_M(Psa)$ with respect to the internal operation for A. Furthermore, the function Ξ_p belongs to the class of general recursive functions.

b. *Let $G_M(Es_p\ a)$ be the macrosystem graph of the principal environment type. From the respective primitives, the set of system environments E denotes the points in $G_M(Es_p\ a)$, and the set $X\subseteq Q\times S\times B\times F\times D\times C\times Q$ denotes the set of lines. Let Ψ_{E_p} be the effective adjacency function for $G_M(Es_p\ a)$ as established in Theorem* B6. *Then, the function Ξ_{E_p}: $N_{2h+1}\rightarrow \Xi_{E_p}(N_{2h+1})\subseteq E\cup X$ defined by*

$$\Xi_{Ep}(1)=\rho_{1,---,k}(\psi_{Ep}(m, m+1))$$

and for $n\geq 1$

$$\Xi_{Ep}(2n)=\rho_{k+1,---,8k}(\psi_{Ep}(m+(n-1), m+n))$$

$$\Xi_{Ep}(2n+1)=\rho_{8k+1,---,9k}(\psi_{Ep}(m+(n-1), m+n))$$

is an effective operation function for A whose ordered images describe a unique effective operation path in $G_M(Es_p\ a)$ with respect to the space operation for A. Furthermore, the function Ξ_{E_p} belongs to the class of general recursive functions.

c. *Let $G_M(Ts_p\ a)$ be a macrosystem graph of the principal time type. From the respective primitives, the set of system times T denotes the points in*

$G_M(Ts_p\ a)$, and the set $Z \subseteq E \times Q \times S \times B \times F \times D \times C \times Q \times E$ denotes the set of lines. Let Ψ_{T_p} be the effective adjacency function for $G_M(Ts_p\ a)$ as established in Theorem B6. *Then, the function Ξ_{T_p}: $N_{2h+1} \rightarrow \Xi_{T_p}(N_{2h+1}) \subseteq T \cup Z$ defined by*

$$\Xi_{Tp}(1) = \rho_{1,\cdots,k}(\psi_{Tp}(m, m+1))$$

and for $n \geq 1$

$$\Xi_{Tp}(2n) = \rho_{k+1,\cdots,10k}(\psi_{Tp}(m+(n-1), m+n))$$

$$\Xi_{Tp}(2n+1) = \rho_{10k+1,\cdots,11k}(\psi_{Tp}(m+(n-1), m+n))$$

is an effective operation function for A whose ordered images describe a unique effective operation path in $G_M(Ts_p\ a)$ with respect to the time-space operation for A. Furthermore, the function Ξ_{T_p} belongs to the class of general recursive functions.

d. *Let $G_M(Ts_a\ a)$ be a macrosystem graph of the alternate time type. From the respective primitives, the set of system times T denotes the points in $G_M(Ts_a\ a)$, and the set $Z' \subseteq Q \times S \times F \times D \times C \times Q$ denotes the set of lines. Let Ψ_{T_a} be the effective adjacency function for $G_M(Ts_a\ a)$ as established in Theorem* B6. *Then, the function Ξ_{T_a}: $N_{2h+1} \rightarrow \Xi_{T_a}(N_{2h+1}) \subseteq T \cup Z'$ defined by*

$$\Xi_{T_a}(1) = \rho_{1,\cdots,k}(\psi_{T_a}(m, m+1))$$

and for $n \geq 1$

$$\Xi_{T_a}(2n) = \rho_{k+1,\cdots,7k}(\psi_{T_a}(m+(n-1), m+n))$$

$$\Xi_{T_a}(2n+1) = \rho_{7k+1,\cdots,8k}(\psi_{T^a}(m+(n-1), m+n))$$

is an effective operation function for A whose ordered images describe a unique effective operation path in $G_M(Ts_a\ a)$ with respect to the time operation for A. Furthermore, the function Ξ_{T_a} belongs to the class of general recursive functions.

Corollary B8. *The effective operations for A of the internal, space, time, and time-space kind correspond to the effective operation paths in the macrosystem graph of the processor, principal environment, alternate time, and principal time types and are given by the following respective sequences:*

a. $\Xi_p(1), \Xi_p(2), \cdots, \Xi_p(2h+1)$
b. $\Xi_{E_p}(1), \Xi_{E_p}(2), \cdots, \Xi_{E_p}(2h+1)$
c. $\Xi_{T_a}(1), \Xi_{T_a}(2), \cdots, \Xi_{T_a}(2h+1)$
d. $\Xi_{T_p}(1), \Xi_{T_p}(2), \cdots, \Xi_{T_p}(2h+1)$

As a final remark, the effective operation of various kinds for a general system of interactive automata is the same, independent of whether they are determined by the unique set of k effective operation paths in the microsystem graphs or the unique path in the macrosystem. Clearly, in comparison, the odd terms of the sequences of Corollaries B4a and B4b and those of Corollaries B7 and B8 contain the same elements. In the former, the system status conditions are represented by sets, each consisting of k elements, whereas in the latter sequences, the system status conditions are represented by k-tuples of the same elements. In comparing the even terms, the sequences of the former consist of a set of k n-tuples representing the operating conditions, whereas those sequences for the latter have odd terms in the form of n k-tuples. Furthermore, the nk-tuples actually are n k-tuples where the k-tuple parts define system operating conditions. By applying proper permutations to the even terms of either sequence, the even terms for the other sequence are similarly arranged.

TABLE B1. Performance descriptors for component automaton A_1

A_1=man				
$M_1=\{m_1, m_1+1, \ldots, m_1+8\}$				$E_1 \subseteq R^3$
$Q_1=\{q_{10}, q_{11}, q_{12}, q_{13}, q_{14}\}$				$S_1=\{s_{10}, s_{11}, s_{12}, s_{13}, s_{14}, s_{15}\}$
q_{10}=readiness state				s_{10}=desire for knowledge
q_{11}=observing state				s_{11}=no information
q_{12}=requesting state				s_{12}=no light on
q_{13}=information collecting state				s_{13}=light on
q_{14}=knowledge increment state				s_{14}=information
				s_{15}=continue on
Initial conditions				
$E_1(m_1)=e_{10}$	$T_1(m_1)=t_{10}$		$Q_1(m_1)=q_{10}$	$S_1(m_1)=s_{10}$
Initial environment function				Interpreting functions
$\phi_{10}(e_{10})=s_{10}$ (e_{10}=man's mind)				$\alpha_{11}(b_{13})=s_{13}$
$\phi_{10}(e_{11})=s_{11}$ (e_{11}=cathode ray display)				$\alpha_{11}(b_{15})=s_{15}$
$\phi_{10}(e_{12})=s_{12}$ (e_{12}=request keyboard)				$\alpha_{12}(b_{21})=s_{11}$

V_1	$\sigma_1(V_1)$	$\delta_1(V_1)$	$\tau_1(V_1)$	$\omega_1(V_1)$	
					$\alpha_{12}(b_{22})=s_{12}$
$q_{10}\,s_{10}$	f_{15}	d_{11}	c_{13}	q_{11}	$\alpha_{12}(b_{24})=s_{14}$
$q_{11}\,s_{11}$	f_{10}	d_{11}	c_{12}	q_{12}	$\alpha_{12}(b_{25})=s_{15}$
$q_{12}\,s_{12}$	f_{13}	d_{10}	c_{11}	q_{12}	
$q_{12}\,s_{13}$	f_{10}	$d_{1,-1}$	c_{13}	q_{13}	
$q_{13}\,s_{11}$	f_{10}	d_{10}	c_{12}	q_{13}	
$q_{13}\,s_{14}$	f_{10}	$d_{1,-1}$	c_{12}	q_{14}	
$q_{14}\,s_{15}$	f_{10}	d_{11}	c_{12}	q_{11}	

B.4 EXAMPLE

Apply the effective operation methods to the example system of Section 5.2, where automaton A_1 represents a man and A_2 a computer. $A=\{A_1, A_2\}$ is a heterogeneous system of automata representing a human-computer system and is analyzed over the moment base $M=\{m+i \mid i \in I=\{0, 1, ---, 8\}\}$. It is assumed that the environment space for the human-computer system is three-dimensional, and there are no media effects upon the responses of either component. The interface corresponds to shared environments for the human and computer, and A is properly viewed as a system of Interactive automata.

Table B1 presents information A_1 (the man) in the form of his minimal set of performance descriptors. Table B2 presents the corresponding information for A_2 (the computer). The microsystem graph of the alternate environment type for the example is shown in Figure 5.2, and the effective operation of the

TABLE B2. Performance descriptors for component automaton A_2

A_2=computer	
$M_2=\{m_2, m_2+1, \ldots, m_2+8\}$	$E_2 \subseteq R^3$
$Q_2=\{q_{20}, q_{21}, q_{22}, q_{23}, q_{24}, q_{25}\}$	$S_2=\{s_{20}, s_{21}, s_{22}, s_{23}, s_{24}, s_{25}\}$
q_{20}=readiness state	s_{20}=executive stimulus
q_{21}=observing state	s_{21}=no information
q_{22}=locating information state	s_{22}=no light on
q_{23}=displaying information state	s_{23}=light on
q_{24}=destroying information state	s_{24}=information
q_{25}=idlcness state	s_{25}=continue on

Initial conditions			
$E_2(m_2)=e_{20}$	$T_2(m_2)=t_{21}$	$Q_2(m_2)=q_{20}$	$S_2(m_2)=s_{20}$

Initial environment function					Interpreting functions
$\phi_{20}(e_{20})=s_{20}$ (e_{20}=core memory)					$\alpha_{22}(b_{21})=s_{21}$
$\phi_{20}(e_{21})=s_{21}$ (e_{21}=cathode ray display)					$\alpha_{22}(b_{22})=s_{22}$
$\phi_{20}(e_{22})=s_{22}$ (e_{22}=request keyboard)					$\alpha_{22}(b_{24})=s_{24}$
V_2	$\sigma_2(V_2)$	$\delta_2(V_2)$	$\tau_2(V_2)$	$\omega_2(V_2)$	$\alpha_{22}(b_{25})=s_{25}$
$q_{20}\,s_{20}$	f_{25}	d_{22}	c_{21}	q_{21}	$\alpha_{21}(b_{13})=s_{23}$
$q_{21}\,s_{22}$	f_{20}	d_{20}	c_{22}	q_{25}	$\alpha_{21}(b_{15})=s_{25}$
$q_{21}\,s_{23}$	f_{22}	$d_{2,-2}$	c_{21}	q_{22}	
$q_{22}\,s_{25}$	f_{20}	d_{21}	c_{22}	q_{23}	
$q_{23}\,s_{21}$	f_{24}	d_{20}	c_{22}	q_{24}	
$q_{24}\,s_{24}$	f_{21}	d_{21}	c_{22}	q_{21}	
$q_{25}\,s_{22}$	f_{20}	d_{20}	c_{23}	q_{21}	

TABLE B3. Effective operation of the space-time kind derived from $G_M(Es_a i)$ of Figure 5.2 by Corollary B4a

$\{e_{10}e_{20}\}$	$\{(t_{10}q_{10}s_{10}b_{15}f_{15}d_{11}c_{13}q_{11}t_{13}), (t_{21}q_{20}s_{20}b_{25}f_{25}d_{22}c_{21}q_{21}t_{22})\}$
$\{e_{11}e_{22}\}$	$\{(t_{13}q_{11}s_{11}b_{10}f_{10}d_{11}c_{12}q_{12}t_{15}), (t_{22}q_{21}s_{22}b_{20}f_{20}d_{20}c_{22}q_{25}t_{24})\}$
$\{e_{12}e_{22}\}$	$\{(t_{15}q_{12}s_{12}b_{13}f_{13}d_{10}c_{11}q_{12}t_{16}), (t_{24}q_{25}s_{22}b_{20}f_{20}d_{20}c_{23}q_{21}t_{27})\}$
$\{e_{12}e_{22}\}$	$\{(t_{16}q_{12}s_{13}b_{10}f_{10}d_{1-1}c_{13}q_{13}t_{19}), (t_{27}q_{21}s_{23}b_{22}f_{22}d_{2-2}c_{21}q_{22}t_{28})\}$
$\{e_{11}e_{20}\}$	$\{(t_{19}q_{13}s_{11}b_{10}f_{10}d_{10}c_{12}q_{13}t_{1,11}), (t_{28}q_{22}s_{25}b_{20}f_{20}d_{21}c_{22}q_{23}t_{2,10})\}$
$\{e_{11}e_{21}\}$	$\{(t_{1,11}q_{13}s_{11}b_{10}f_{10}d_{10}c_{12}q_{13}t_{1,13}), (t_{2,10}q_{23}s_{21}b_{24}f_{24}d_{20}c_{22}q_{24}t_{2,12})\}$
$\{e_{11}e_{21}\}$	$\{(t_{1,13}q_{13}s_{14}b_{10}f_{10}d_{1-1}c_{12}q_{14}t_{1,15}), (t_{2,12}q_{24}s_{24}b_{21}f_{21}d_{21}c_{22}q_{21}t_{2,14})\}$
$\{e_{10}e_{22}\}$	$\{(t_{1,15}q_{14}s_{15}b_{10}f_{10}d_{11}c_{12}q_{11}t_{1,17}), (t_{2,14}q_{21}s_{22}b_{20}f_{20}d_{20}c_{22}q_{25}t_{2,16})\}$

TABLE B4. Effective operation of the space-time kind derived from G_M $(Es_a a)$ of Figure 5.3 by Corollary B7

$(e_{10}e_{20})$	$(t_{10}t_{21}q_{10}q_{20}s_{10}s_{20}t_{15}b_{25}f_{15}f_{25}d_{11}d_{22}c_{13}c_{21}q_{11}q_{21}t_{13}t_{22})$
$(e_{11}e_{22})$	$(t_{13}t_{22}q_{11}q_{21}s_{11}s_{22}b_{10}b_{20}f_{10}f_{20}d_{11}d_{20}c_{12}c_{22}q_{12}q_{25}t_{15}t_{24})$
$(e_{12}e_{22})$	$(t_{15}t_{24}q_{12}q_{25}s_{12}s_{22}b_{13}b_{20}f_{13}f_{20}d_{10}d_{20}c_{11}c_{23}q_{12}q_{21}t_{16}t_{27})$
$(e_{12}e_{22})$	$(t_{16}t_{27}q_{12}q_{21}s_{13}s_{23}b_{10}b_{22}f_{10}f_{22}d_{1-1}d_{2-2}c_{13}c_{21}q_{13}q_{22}t_{19}t_{28})$
$(e_{11}e_{20})$	$(t_{19}t_{28}q_{13}q_{22}s_{11}s_{25}b_{10}b_{20}f_{10}f_{20}d_{10}d_{21}c_{12}c_{22}q_{13}q_{23}t_{1,11}t_{2,10})$
$(e_{11}e_{21})$	$(t_{1,11}t_{2,10}q_{13}q_{23}s_{11}s_{21}b_{10}b_{24}f_{10}f_{24}d_{10}d_{20}c_{12}c_{22}q_{13}q_{24}t_{1,13}t_{2,12})$
$(e_{11}e_{21})$	$(t_{1,13}t_{2,12}q_{13}q_{24}s_{14}s_{24}b_{10}b_{21}f_{10}f_{21}d_{1-1}d_{21}c_{12}c_{22}q_{14}q_{21}t_{1,15}t_{2,14})$
$(e_{10}e_{22})$	$(t_{1,15}t_{2,14}q_{14}q_{21}s_{15}s_{22}b_{10}b_{20}f_{10}f_{20}d_{11}d_{20}c_{12}c_{22}q_{11}q_{25}t_{1,17}t_{2,16})$

space-time kind, as derived from this graph using Corollary B4a, is detailed in Table B3. Similarly, the macrosystem graph of the alternate environment type is shown in Figure 5.3, and the effective operation, as derived from this graph using Corollary B7, is detailed in Table B4.

EXERCISES

B.1 Detail algorithms from Theorems B2–B4 and corollaries for calculating effective operations of the microsystem model for the a) processor graph, b) alternate environment graph, and c) principal time graph. (The exercise parallels Exercise A5.)

B.2 Write programs from the algorithms established in Exercise B1 in the language of your choice for the a) processor graph, b) alternate environment graph, and c) principal time graph. You may use the example of a two-component system given in this appendix for checking your program.

REFERENCES

Berge, C. (1962), *The Theory of Graphs and Its Applications*, London: Methuen & Co., LTD.

Bulkeley, W.M. (1998), "Peering Ahead," *Technology, The Wall Street Journal*, November 16.

Busacker, R.G. and Saaty, T.L. (1965), *Finite Graphs and Networks, An Introduction with Applications*, New York: McGraw-Hill.

Bush, V. (1929), *Operational Circuit Analysis*, New York: Wiley.

Chapuis, A. and Droz, E. (1958), *Automata—A Historical and Technological Study*, Translation by Alec Reid, Editions du Griffon, Neuchatel, Switzerland.

Chen, E. (1999), "Digital Signing," *IEEE Spectrum*, August, p. 42.

Chestnut, H. and Mayer, R.W. (1959), *Servomechanisms and Regulating Systems Design*, New York: Wiley.

Chrystal, D., Guenther, T. and Koenig, E.C. (1962a), "Applying Control Concepts to an Organization," *Total Systems*, Detroit, MI: American Data Processing, Inc., pp. 110–129.

Chrystal, D., Guenther, T. and Koenig, E.C. (1962b), "Automatic Manufacturing Control," *Computer Applications*, Detroit, MI: American Data Processing, Inc., pp. 94–98.

Comerford, R. (2000), "The Internet, Technology 2000," *IEEE Spectrum*, January, p. 40.

Cray, S. (1978), "News in Perspective," *Datamation*, April, p. 187.

Frederick, T.J. and Koenig, E.C. (1971a), "Analysis for an Effective Operation of a General Automaton: Recursive Methods Applied to a Graph Model," *Journal of Cybernetics*, Vol. 1, Nos. 1, 2, pp. 53–70, 49–66.

Frederick, T.J. and Koenig, E.C. (1971b), "General System of Interactive Automata: Analysis for an Effective Operation," *Cybernetica*, Namur, Belgium, Vol. 14, No. 4, pp. 297–334.

Geller, D.P. (1975), "Realization with Feedback Encoding. I: Analogues of the Classical Theory," *SIAM J. Comput.*, Vol. 4, No. 1.

Gentzen, G. (1934), "Untersuchungenüber das Logische Schliessen," *Mathematische Zeitschrift*, Vol. 39, pp. 176–210, 405–431.

George, F.H. (1980), "Philosophical Foundations of Cybernetics", Vol. 1 of *Cybernetics and Systems Series*, J. Rose (Ed.).

Knowledge Structures for Communications in Human-Computer Systems: General Automata-Based, by Eldo C. Koenig

Ginsburg, S. (1959), A Synthesis Technique for Minimal State Sequential Machines, *IRE Trans., Electron. Computers*, Vol. EC-8, no. 1, pp. 13–24.

Ginsburg, S. (1962), *An Introduction to Mathematical Machine Theory*, Old Tappan, NJ: Addison-Wesley.

Glushkov, V. (1966), *Introduction to Cybernetics*, St. Louis, MO: Academic Press.

Gould, R. (1959), "Application of Graph Theory to the Synthesis of Contact Networks," *Proceedings of an International Symposium on the Theory of Switching*, pp. 244–292, Cambridge, MA: Harvard University Press.

Harary, F., Norman, R. and Cartwright, D. (1965), *Structural Models: An Introduction to the Theory of Directed Graphs*, New York: Wiley.

Hartmanis, J. and Stearns, R.E. (1966), *Algebraic Structure Theory of Sequential Machines*, Dallas TX: Prentice-Hall.

Hellemans, A. (1999), "Internet Security Code Is Cracked," *Science*, 3 September, Vol. 285, p. 1472.

Immergut, D.J. (1996), "Bridging the Web's Language Gap," *The Wall Street Journal*, November 12.

Kaiser, J. (1999), "Irish Lass Invents Crypto Code," *Science*, 29 January, Vol. 283, p. 599.

Kleene, S.C. (1967), *Mathematical Logic*, New York: Wiley.

Koenig, E.C. and Frederick, T.J. (1969), "Formal Analysis for a General System of Interactive Automata," *Proceedings 3rd Hawaii International Conference*, System Science.

Koenig, E.C. and Frederick, T.J. (1970a), "Formal Analysis for a General Automaton," *Progress in Cybernetics*, Vol. 3, J. Rose (Ed.), New York: Gordon and Breach, pp. 613–640.

Koenig, E.C. and Frederick, T.J. (1970b), "A General System of Interactive Automata: Analysis for Defining the Graph Model," (abridged), *Proceedings 6th International Congress Cybernetics*, Namur, Belgium.

Koenig, E.C. and Frederick, T.J. (1971), "A General System of Interactive Automata: Analysis for Defining the Graph Model," *Cybernetica*, Namur, Belgium, Vol. 14, No. 2, pp. 89–131.

Koenig, E.C. and Frederick, T.J. (1972), "Select Properties of the Graph Model of a General Automaton," *Kybernetes*, Vol. 1, No. 1, pp. 22–33.

Koenig, E.C. (1972a), "A General Systems Theory for Studies in Cybernetics and Automation," (abridged), *Proceedings International Congress Cybernetics and Systems*, Oxford, England.

Koenig, E.C. (1972b), "Intelligence Systems: Introduction to the Modeling for the Inherited Ability for Knowledge Association," *Proceedings Brazilian Congress for General Systems and Cybernetics*. Porto Alegre, Brazil.

Koenig, E.C. and Schultz, J.V. (1972–73), "Analysis of a General Logical Discourse for Man-Machine Interaction," *Kybernetes*, Vol. 1, No. 4, pp. 231–241; Vol. 2, No. 1, pp. 27–36.

Koenig, E.C. (1973), "A Language Information Theory: Obligatory Relations," (abridged), *Proceedings 7th International Congress Cybernetics*, Namur, Belgium.

Koenig, E.C. (1973–74), "A General Systems Theory for Studies in Cybernetics and Automation," *Kybernetes*, Vol. 2, No. 4, pp. 217–223; Vol. 3, No. 1, pp. 11–16.

Koenig, E.C. (1974), "A Language Information Theory: Obligatory Relations," *Cybernetica* (Belgium), Vol. 17, No. 1, pp. 57–79.

Koenig, E.C. (1975a), "Analysis for a General Automaton with Distinguishable Receptors and Effectors," *IEEE Transactions*, Vol. SMC 5, No. 1.

Koenig, E.C. (1975b), "Cybernetic Considerations for Automata as Reflected on a Graph Model," *Proceedings 3rd International Congress Cybernetics and Systems*, Bucharest, Hungary.

Koenig, E.C. (1978), "Intelligent Systems: Knowledge Association and Related Deductive Processes," *Kybernetes*, Vol. 7, No. 2, pp. 99–106.

Koenig, E.C., Mason, J. and Jakubowski, E. (1978), "Extracting and Storing the Meaning of Sentences by Computer Based on Principles of General Automata," *Cybernetica*, Namur, Belgium, Vol. 21, No. 1, pp. 45–60.

Koenig, E.C. (1979a), "Establishing Valid Arguments by Computer and Storing their Meanings: A Premiss of Form A of Aristotelian Logic," *Kybernetes*, Vol. 8, No. 4, pp. 299–303.

Koenig, E.C. (1979b), "Knowledge Structures for Sentences Describing Systems of Interactive Automata Involving Words of "give," "receive" and "sell," "buy," *Cybernetica*, Namur, Belgium, Vol. 22,1, pp. 5–75.

Koenig, E.C. (1980), "Knowledgc Structures Based on General Automata Compared with Other Types," *Proceedings of the International Congress on Applied Systems Research and Cybernetics*, Acapulco, Mexico.

Koenig, E.C. (1982), "Associating Current Knowledge with that of Past Experience Based on Knowledge About Automata," *Kybernetes*, Vol. 11, No. 3.

Koenig, E.C. (1983), "Generating Inferences from Knowledge Structures Based on General Automata," *Cybernetica*, Namur, Belgium, Vol. 26, No. 2, pp. 79–97.

Koenig, E.C. (1984), "Some Principles for Robotics Based on General Automata," *Proceedings of the 6th International Congress of Cybernetics and Systems*, Paris, France, pp. 653–659.

Koenig, E.C. (1986a), "A Reclassification of Words for the Age of Computers and Robots Based on General Automata," *Proceedings of the 11th International Congress on Cybernetics*, Namur, Belgium.

Koenig, E.C. (1986b), "Some Principles for Robotics Based on General Automata," *Robotica*, Vol. 4, Part 1.

Koenig, E.C. (1987), "A Model Knowledge Structure for Robots," *Proceedings of the Annual Conference*, 1987 *IEEE, Systems Man and Cybernetics*, 20–23 Oct., Alexandria, VA.

Koenig, E.C. (1989), "Analysis for Correct Reasoning by Robots: Modus Ponens, Modus Tollens," *Proceedings of the* 1989 *IEEE International Phoenix Conference on Computers and Communications*, pp. 584–589.

Koenig, E.C. (1990), "Analysis for Correct Reasoning by Robots: Hypothetical Syllogism with Modus Ponens, Modus Tollens," *Proceedings of the 8th International Congress of Cybernetics and Systems*, Hunter College, City University of New York, New York.

Koenig, E.C. (1994a), "Analysis for Correct Reasoning in Interactive Man-Robot Systems: Disjunctive Syllogism with Modus Ponens and Modus Tollens," *Artificial Intelligence in Industrial Decision Making, Control and Automation*, S. Tzafestas and H. Verbruggen (Eds.), The Netherlands: Kluwer Academic Publishers.

Koenig, E.C. (1994b), "Parallel Processing Considerations for Interactive Man-Robot Systems," *Systems Analysis—Modeling—Simulation*, Vol. 16, No. 2.

Koenig, E.C. (1997a), "Analysis for Natural Language Communication in Interactive Man-Machine Systems: Knowledge Structures and the Generation of Sentences," *Knowledge Based Systems: Advanced Concepts, Techniques and Applications*, S. G. Tzafestas (Ed.), River Edge, NJ: World Scientific Publishing.

Koenig, E.C. (1997b), "Formal Analysis and Modeling for Communications in Interactive Systems," *Information Sciences: Informatics and Computer Sciences*, Vol. 96, No. 3 and 4, pp. 153–167.

Koenig, E.F. (1976), "Theory of Graphs: Applications to Automaton Systems," *Kybernetes*.

Kohavi, Z. and Winograd, J. (1973), "Establishing Bounds Concerning Finite Automata," *Journal of Computer and System Sciences*, Vol. 7, no. 3, June, pp. 288–299.

Krohn, K.B. and Rhodes, J.L. (1965), "Algebraic Theory of Machines," *Proceedings of the Symposium on Mathematical Theory of Automata*, New York, Apr. 25–26.

Miller, R.E. (1965), *Switching Theory*, Vol. 2, New York: Wiley.

Minsky, M.L. (1967), *Computation: Finite and Infinite Machines*, Dallas, TX: Prentice-Hall.

Moore, E.F. (1956), "Gedanken-Experiments on Sequential Machines," *Automata Studies*, Princeton, NJ: Princeton University Press, pp. 129–153.

Moore, E.F. (1964), *Sequential Machines: Selected Papers*, Old Tappan, NJ: Addison-Wesley.

Myhill, J. (1957), Finite Automata and the Representation of Events, *WADC Tech. Rept.* 57–624, pp. 112–137.

Peter, R. (1967), *Recursive Functions*, Saint Louis, MO: Academic Press.

Rogers, Jr., H. (1967), *Theory of Recursive Functions and Effective Computability*, New York: McGraw-Hill.

Rosenberg, J. (1962), *German, How to Speak & Write It*, New York: Dover Publications.

Sabliere, J. (1966), *De L'Automata A L'Automatisation*, Editeur, Paris: Gauthier-Villars.

Shannon, C.E. (1938), "A Symbolic Analysis of Relay and Switching Circuits," *Trans. AIEE*, Vol. 57, pp. 713–723.

Shannon, C.E. and McCarthy, J. (Eds.) (1956), *Automata Studies* (*Annals of Mathematical Studies*, 34), Princeton, NJ.

Sikorski, R. and Peters, R. (1999), "Digital Security," *Science*, January 15, February 19, Vol. 283, p. 48, p. 1133.

Simmons, R.F. (1973), "Semantic Networks: Their Computation and Use for Understanding English Sentences," *Computer Models of Thought and Language*, R. C. Schank and K. M. Colby (Eds.), San Francisco, CA: Freeman.

Turing, A.M. (1936), "On Computable Numbers, with an Application to the Entscheidungsproblem," *Proceedings London Mathematical Society*, Ser. 2–42, pp. 230–265.

Von Neumann, J. (1961), "The General and Logical Theory of Automata," *John von Neumann Collected Works*, Vol. V, St. Louis, MO: Pergamon Press.

Wiener, N. (1948), *Cybernetics*, New York: Wiley.

Wiener, N. (1954), *The Human Use of Human Beings*, Scranton, PA: Avon Books.

Wuethrich, Bernice (2000), "Learning the World's Languages—Before They Vanish," *Science*, 19 May, Vol. 288, pp. 1156–1159.

INDEX

Knowledge Structures for Communications in Human-Computer Systems: General Automata-Based, by Eldo C. Koenig

Printed in the United States
By Bookmasters